T0156133

Lecture Notes in Mathematics

Volume 2279

Editors-in-Chief

Jean-Michel Morel, CMLA, ENS, Cachan, France
Bernard Teissier, IMJ-PRG, Paris, France

Series Editors

Karin Baur, University of Leeds, Leeds, UK
Michel Brion, UGA, Grenoble, France
Camillo De Lellis, IAS, Princeton, NJ, USA
Alessio Figalli, ETH Zurich, Zurich, Switzerland
Annette Huber, Albert Ludwig University, Freiburg, Germany
Davar Khoshnevisan, The University of Utah, Salt Lake City, UT, USA
Ioannis Kontoyiannis, University of Cambridge, Cambridge, UK
Angela Kunoth, University of Cologne, Cologne, Germany
Ariane Mézard, IMJ-PRG, Paris, France
Mark Podolskij, University of Luxembourg, Esch-sur-Alzette, Luxembourg
Sylvia Serfaty, NYU Courant, New York, NY, USA
Gabriele Vezzosi, UniFI, Florence, Italy
Anna Wienhard, Ruprecht Karl University, Heidelberg, Germany

This series reports on new developments in all areas of mathematics and their applications - quickly, informally and at a high level. Mathematical texts analysing new developments in modelling and numerical simulation are welcome. The type of material considered for publication includes:

1. Research monographs
2. Lectures on a new field or presentations of a new angle in a classical field
3. Summer schools and intensive courses on topics of current research.

Texts which are out of print but still in demand may also be considered if they fall within these categories. The timeliness of a manuscript is sometimes more important than its form, which may be preliminary or tentative.

More information about this series at http://www.springer.com/series/304

Donald Yau

Involutive Category Theory

Springer

Donald Yau
Department of Mathematics
Ohio State University at Newark
Newark, OH, USA

ISSN 0075-8434 ISSN 1617-9692 (electronic)
Lecture Notes in Mathematics
ISBN 978-3-030-61202-3 ISBN 978-3-030-61203-0 (eBook)
https://doi.org/10.1007/978-3-030-61203-0

Mathematics Subject Classification: 16W10, 18-01, 18D10, 18D50, 46C50, 18C15, 46L05

© The Editor(s) (if applicable) and The Author(s), under exclusive license to Springer Nature Switzerland AG 2020
This work is subject to copyright. All rights are solely and exclusively licensed by the Publisher, whether the whole or part of the material is concerned, specifically the rights of translation, reprinting, reuse of illustrations, recitation, broadcasting, reproduction on microfilms or in any other physical way, and transmission or information storage and retrieval, electronic adaptation, computer software, or by similar or dissimilar methodology now known or hereafter developed.
The use of general descriptive names, registered names, trademarks, service marks, etc. in this publication does not imply, even in the absence of a specific statement, that such names are exempt from the relevant protective laws and regulations and therefore free for general use.
The publisher, the authors, and the editors are safe to assume that the advice and information in this book are believed to be true and accurate at the date of publication. Neither the publisher nor the authors or the editors give a warranty, expressed or implied, with respect to the material contained herein or for any errors or omissions that may have been made. The publisher remains neutral with regard to jurisdictional claims in published maps and institutional affiliations.

This Springer imprint is published by the registered company Springer Nature Switzerland AG
The registered company address is: Gewerbestrasse 11, 6330 Cham, Switzerland

To Eun Soo and Jacqueline

Preface

For an object X in a category, such as a set or a vector space, an involution is a degree 2 self-map, i.e., a map $j : X \longrightarrow X$ such that $j \circ j = \mathrm{Id}_X$. An involutive category is a categorification of an object with an involution. More precisely, an involutive category is a category C equipped with

- a functor $I : \mathsf{C} \longrightarrow \mathsf{C}$, called the involution functor, and
- a natural isomorphism $\iota : \mathrm{Id}_{\mathsf{C}} \xrightarrow{\cong} I^2$, called the unit of I,

such that the coherence condition:

$$\iota I = I\iota : I \longrightarrow I^3,$$

called the triangle identity, is satisfied. Involutive categories are important because they are the natural setting to study objects with involutive structures. A basic example is the involutive category $\mathsf{Vect}(\mathbb{C})$ of complex vector spaces, with the involution functor I given by taking conjugate vector spaces. Star-algebras, which satisfy

$$(ab)^* = b^*a^*,$$

and their differential graded analogues both arise naturally as involutive objects in some involutive categories. These $*$-algebras are important in applications to physics, in particular as algebras of observables in quantum theory.

This book is about involutive categories and involutive operads, with applications to the GNS construction and algebraic quantum field theory. While there are many books on category theory, this book is the first one that focuses on involutive categories. Some of the materials—including parts of Chapters 2, 4, 7, and 8—are based on the work of Jacobs [Jac12] and Benini, Schenkel, and Woike [BSW19b]. Other parts—including Chapters 3, 5, and 6 on coherence and strictification of involutive categories and involutive (symmetric) monoidal categories—are new materials that have not appeared anywhere else before.

This book is written at the first- to second-year graduate level. With detailed explanation and exercises at the end of every chapter, this book is suitable for independent study and graduate courses. It is also a useful reference for mathematicians and physicists whose work involves involutive objects.

A brief description of each chapter follows.

Chapter 1. Category Theory

To make this book as self-contained as possible, in this chapter we recall some basic category theory, including monoidal categories and monads. Since we will discuss colimits related to involutive categories, in this chapter we provide a simple and constructive proof that the category Cat of small categories has all small colimits. This is a well-known and often used fact, but a direct proof is not easily accessible in the literature.

Chapter 2. Involutive Categories

Involutive analogues of basic categorical concepts, including functors, natural transformations, and adjunctions, are defined in this chapter. In involutive categories, there is a natural concept of an involutive object, which consists of an object X and a morphism $j : X \longrightarrow IX$, such that the diagram:

$$\begin{array}{ccc} X & \xrightarrow{\;j\;} & IX \\ & \searrow_{\iota_X} & \downarrow{\scriptstyle Ij} \\ & & I^2X \end{array}$$

is commutative. For a given involutive category (C, I, ι), the category $\mathsf{Inv}(\mathsf{C})$ of involutive objects inherits from C an involutive structure and other categorical properties. One main result of this chapter is Theorem 2.3.8. It states that for an adjunction between involutive categories, if one of the functors is an involutive functor, then so is the other one, and the adjunction is an involutive adjunction. As a consequence, if an involutive functor is an equivalence of categories in the usual sense, then it is an involutive equivalence.

Chapter 3. Coherence of Involutive Categories

With each type of categories comes a set of coherence questions. In this chapter, we prove several coherence properties of involutive categories. An involutive category is strict if the unit is the identity natural transformation $\mathrm{Id} : \mathrm{Id}_{\mathsf{C}} \longrightarrow I^2 = \mathrm{Id}_{\mathsf{C}}$. We construct the free strict involutive category, the cofree strict involutive category, and the free involutive category of a category. Then, we observe that there is a unique strict involutive equivalence from the free involutive category to the free strict involutive category of a category. Next, we show that involutive categories are algebras over a monad. We also show that the category of involutive categories and strict involutive functors is locally finitely presentable and that its forgetful functor to Cat admits a right adjoint.

Chapter 4. Involutive Monoidal Categories

This chapter is about the involutive analogues of monoidal categories. For an involutive monoidal category C, its category $\mathsf{Inv}(\mathsf{C})$ of involutive objects inherits from C the structure of an involutive monoidal category. Monoids in $\mathsf{Inv}(\mathsf{C})$ are the same as involutive objects in $\mathsf{Mon}(\mathsf{C})$, which is the category of monoids in C with an inherited involutive structure. If C is an involutive symmetric monoidal category, then $\mathsf{Mon}(\mathsf{C})$ has another involutive structure that incorporates the symmetry isomorphism in C. The corresponding involutive objects are called *reversing involutive monoids*. Complex unital $*$-algebras and their differential graded analogues are both examples of reversing involutive monoids. This chapter ends with a section on involutive monads, which yield more general objects with involutive structures.

Chapter 5. Coherence of Involutive Monoidal Categories

In this chapter, we prove several coherence properties of involutive monoidal categories. First, we construct the free involutive monoidal category and the free involutive strict monoidal category of a category and show that they are equivalent via a strict involutive strict monoidal equivalence. Then, we prove that all formal diagrams in an involutive monoidal category are commutative. Next, we prove a strictification result that says that every involutive monoidal category is equivalent to an involutive strict monoidal category via an involutive adjoint equivalence. This chapter ends with the constructions of the free involutive monoidal category and the free involutive strict monoidal category of an involutive category.

Chapter 6. Coherence of Involutive Symmetric Monoidal Categories

This chapter is the symmetric analogue of the previous chapter. We prove a number of coherence results regarding involutive symmetric monoidal categories. In particular, we prove a strictification result that says that every involutive symmetric monoidal category is equivalent to an involutive strict symmetric monoidal category via an involutive adjoint equivalence. This strictification result allows us to safely ignore parentheses in multiple monoidal products in an involutive symmetric monoidal category. This is particularly important when we discuss involutive operads in Chapter 8.

Chapter 7. Categorical Gelfand-Naimark-Segal Construction

The GNS construction is an important result in functional analysis that has consequences in quantum mechanics. This chapter contains a categorical analogue of the GNS construction as an isomorphism between two categories. This result is due to Jacobs [Jac12], and we provide detailed proofs of some technical Lemmas that are not included in that paper.

Chapter 8. Involutive Operads

An operad is a generalization of a category that encodes algebraic structures with multiple inputs and one output. Without assuming any prior knowledge of operads, in this chapter we first recall the definitions of an operad and of an algebra over an operad and provide a list of examples. Involutive operads are involutive monoids in an involutive monoidal category of colored symmetric sequences. We observe that every involutive operad yields an involutive monad, whose involutive algebras are defined as those of the involutive operad. For example, reversing involutive monoids are involutive algebras over the involutive associative operad, and similarly for the diagram version. Next, we prove that every involutive symmetric monoidal functor induces an involutive functor between the respective involutive categories of involutive operads. As examples, we observe that the involutive associative operad and the algebraic quantum field theory involutive operad of [BSW19b] both arise from this change-of-category construction.

Newark, OH, USA Donald Yau

Contents

Chapter 1
Category Theory

In this chapter, we fix some notations and briefly recall some basic category theory, including monoidal categories and monads. Everything in this chapter is well-documented in existing books, so there are no proofs, with the following two exceptions.

(1) Lemma 1.2.26 is a useful fact in a symmetric monoidal category that we will use in a later chapter. Its proof is included to illustrate some diagram-chasing techniques that we will later use.
(2) Section 1.4 provides a simple and constructive proof that the category Cat of small categories has all small colimits. This is a well-known and often used fact, but a straightforward proof is not easily accessible in the literature.

1.1 Basic Concepts of Category Theory

In this section, we briefly recall some basic concepts of category theory, including functors, natural transformations, adjunctions, equivalences, and (co)limits. These topics are covered in many introductory books on category theory, such as [Awo10, Gra18, Lei14, Rie16, Rom17, Sim11], where we refer the reader for more detailed discussion.

A *category* C consists of:

- a class $\mathsf{Ob}(\mathsf{C})$ of *objects* in C;
- a set $\mathsf{C}(X, Y)$, also denoted by $\mathsf{C}(X; Y)$, of *morphisms* with *domain* $X = \mathsf{dom}(f)$ and *codomain* $Y = \mathsf{cod}(f)$ for any objects $X, Y \in \mathsf{Ob}(\mathsf{C})$;
- an assignment called *composition*

$$\mathsf{C}(Y, Z) \times \mathsf{C}(X, Y) \xrightarrow{\circ} \mathsf{C}(X, Z) \;, \qquad \circ(g, f) = g \circ f$$

© The Author(s), under exclusive license to Springer Nature Switzerland AG 2020

D. Yau, *Involutive Category Theory*, Lecture Notes in Mathematics 2279,
https://doi.org/10.1007/978-3-030-61203-0_1

for objects X, Y, Z in C;
- an *identity morphism* $\mathrm{Id}_X \in \mathsf{C}(X, X)$ for each object X in C.

These data are required to satisfy the following two conditions.

Associativity: For morphisms f, g, and h, the equality

$$h \circ (g \circ f) = (h \circ g) \circ f$$

holds, provided the compositions are defined.

Unity: For each morphism $f \in \mathsf{C}(X, Y)$, the equalities

$$\mathrm{Id}_Y \circ f = f = f \circ \mathrm{Id}_X$$

hold.

In a category C, the collection of morphisms is denoted by $\mathsf{Mor}(\mathsf{C})$. For an object $X \in \mathsf{Ob}(\mathsf{C})$ and a morphism $f \in \mathsf{Mor}(\mathsf{C})$, we often write $X \in \mathsf{C}$ and $f \in \mathsf{C}$. We also denote a morphism $f \in \mathsf{C}(X, Y)$ by

$$f : X \longrightarrow Y, \qquad X \xrightarrow{f} Y, \quad \text{and} \quad X \xrightarrow{-f\rightarrow} Y.$$

Morphisms $f : X \longrightarrow Y$ and $g : Y \longrightarrow Z$ are called *composable*, and $g \circ f \in \mathsf{C}(X, Z)$ is also denoted by gf, called their *composite*. A morphism $f : X \longrightarrow Y$ is:

- an *epimorphism* if for any pair of morphisms $g, h : Y \longrightarrow Z$ with $gf = hf$, it follows that $g = h$.
- a *monomorphism* if for any pair of morphisms $g, h : W \longrightarrow X$ with $fg = fh$, it follows that $g = h$.

The identity morphism Id_X of an object X is also denoted simply by Id. When we draw a diagram in a category C, the identity morphism Id_X of an object X is sometimes denoted by X. A morphism $f : X \longrightarrow Y$ in a category C is called an *isomorphism* if there exists a morphism $g : Y \longrightarrow X$ such that $gf = \mathrm{Id}_X$ and $fg = \mathrm{Id}_Y$. A category is *discrete* if it contains no non-identity morphisms. A *groupoid* is a category in which every morphism is an isomorphism.

For a category C, its *opposite category* is denoted by C^{op}. It has the same objects as C and morphism sets $\mathsf{C}^{\mathsf{op}}(X, Y) = \mathsf{C}(Y, X)$, with identity morphisms and composition inherited from C. A *small category* is a category whose class of objects forms a set.

For categories C and D, a *functor* $F : \mathsf{C} \longrightarrow \mathsf{D}$ consists of:

- an assignment

$$\mathsf{Ob}(\mathsf{C}) \longrightarrow \mathsf{Ob}(\mathsf{D}), \qquad X \longmapsto F(X)$$

on objects;

- an assignment

$$\mathsf{C}(X, Y) \longrightarrow \mathsf{D}\big(F(X), F(Y)\big), \qquad f \longmapsto F(f)$$

on morphisms.

These data are required to satisfy the following two conditions.

Composition: The equality

$$F(gf) = F(g)F(f)$$

of morphisms holds, provided the compositions are defined.

Identities: For each object $X \in \mathsf{C}$, the equality

$$F(\mathrm{Id}_X) = \mathrm{Id}_{F(X)}$$

in $\mathsf{D}\big(F(X), F(X)\big)$ holds.

We often write $F(X)$ and $F(f)$ as FX and Ff, respectively, to simplify the notations. Functors are composed by composing the assignments on objects and on morphisms. We write Cat for the category whose objects are small categories and whose morphisms are functors.

Suppose $F, G : \mathsf{C} \longrightarrow \mathsf{D}$ are functors between the same two categories. A *natural transformation* $\theta : F \longrightarrow G$ consists of a morphism $\theta_X : FX \longrightarrow GX$ in D for each object $X \in \mathsf{C}$ such that the diagram

$$\begin{array}{ccc} FX & \xrightarrow{\theta_X} & GX \\ {\scriptstyle Ff}\downarrow & & \downarrow{\scriptstyle Gf} \\ FY & \xrightarrow{\theta_Y} & GY \end{array}$$

in D is commutative for each morphism $f : X \longrightarrow Y$ in C. In other words, the equality

$$Gf \circ \theta_X = \theta_Y \circ Ff$$

holds in $\mathsf{D}(FX, GY)$. Each morphism θ_X is called a *component* of θ. The *identity natural transformation* $\mathrm{Id}_F : F \longrightarrow F$ of a functor F has each component an identity morphism. A *natural isomorphism* is a natural transformation in which every component is an isomorphism.

Suppose $\phi : G \longrightarrow H$ is a natural transformation for a functor $H : \mathsf{C} \longrightarrow \mathsf{D}$. With $\theta : F \longrightarrow G$ as above, the *vertical composition*

$$\phi\theta : F \longrightarrow H$$

is the natural transformation with components

$$(\phi\theta)_X = \phi_X \circ \theta_X : FX \longrightarrow HX \quad \text{for} \quad X \in \mathsf{C}.$$

Suppose $\theta' : F' \longrightarrow G'$ is a natural transformation for functors $F', G' : \mathsf{D} \longrightarrow \mathsf{E}$. The *horizontal composition*

$$\theta' * \theta : F'F \longrightarrow G'G$$

is the natural transformation whose component $(\theta' * \theta)_X$ for an object $X \in \mathsf{C}$ is defined as either composite in the commutative diagram

$$\begin{array}{ccc} F'FX & \xrightarrow{\theta'_{FX}} & G'FX \\ {\scriptstyle F'\theta_X}\downarrow & & \downarrow{\scriptstyle G'\theta_X} \\ F'GX & \xrightarrow{\theta'_{GX}} & G'GX \end{array}$$

in E.

For a category C and a small category D, a D-*diagram in* C is a functor $\mathsf{D} \longrightarrow \mathsf{C}$. The *diagram category* C^{D} has D-diagrams in C as objects, natural transformations between such functors as morphisms, and vertical compositions of natural transformations as composition.

An *adjunction* is a tuple (L, R, ϕ) consisting of:

- A pair of functors in opposite directions

$$\mathsf{C} \underset{R}{\overset{L}{\rightleftarrows}} \mathsf{D}\,.$$

- A bijection

$$\mathsf{D}(LX, Y) \xrightarrow[\cong]{\phi_{X,Y}} \mathsf{C}(X, RY)$$

  that is natural in the objects $X \in \mathsf{C}$ and $Y \in \mathsf{D}$.

Such an adjunction is also called an *adjoint pair*, with L the *left adjoint* and R the *right adjoint*. We sometimes denote such an adjunction by $L \dashv R$ or (L, R). We always display the left adjoint on top, pointing to the right. If an adjunction is displayed vertically, then the left adjoint is written on the left-hand side.

In an adjunction $L \dashv R$ as above, the natural bijection ϕ yields natural transformations

$$\mathrm{Id}_{\mathsf{C}} \xrightarrow{\ \varepsilon\ } RL \quad \text{and} \quad LR \xrightarrow{\ \eta\ } \mathrm{Id}_{\mathsf{D}}\ ,$$

called the *unit* and the *counit*. The natural transformations

$$R \xrightarrow{\ \varepsilon R\ } RLR \xrightarrow{\ R\eta\ } R \quad \text{and} \quad L \xrightarrow{\ L\varepsilon\ } LRL \xrightarrow{\ \eta L\ } L$$

are equal to Id_R and Id_L, respectively. These identities are known as the *triangle identities*.

Characterizations of adjunctions are given in [Bor94a] (Chapter 3) and [Mac98] (IV.1). We will use the following characterization of an adjunction from [Mac98] (IV.1 Theorem 2(i)) several times in this book. An adjunction (L, R, ϕ) is completely determined by

- the functors $L : \mathsf{C} \longrightarrow \mathsf{D}$ and $R : \mathsf{D} \longrightarrow \mathsf{C}$, and
- the natural transformation $\varepsilon : \mathrm{Id}_{\mathsf{C}} \longrightarrow RL$

such that for each morphism $f : X \longrightarrow RY$ in C with $X \in \mathsf{C}$ and $Y \in \mathsf{D}$, there exists a unique morphism $f' : LX \longrightarrow Y$ in D such that the diagram

$$\begin{array}{ccc} X & \xrightarrow{\ \varepsilon_X\ } & RLX \\ \Vert & & \downarrow{\scriptstyle Rf'} \\ X & \xrightarrow{\ f\ } & RY \end{array}$$

in C is commutative.

A functor $F : \mathsf{C} \longrightarrow \mathsf{D}$ is called an *equivalence* if there exist

- a functor $G : \mathsf{D} \longrightarrow \mathsf{C}$ and
- natural isomorphisms $\mathrm{Id}_{\mathsf{C}} \xrightarrow{\ \cong\ } GF$ and $\mathrm{Id}_{\mathsf{D}} \xrightarrow{\ \cong\ } FG$.

In this case, G is both a left adjoint and a right adjoint of F, and (F, G) is called an *adjoint equivalence* if F is left adjoint to G with the given natural isomorphisms $\mathrm{Id}_{\mathsf{C}} \longrightarrow GF$ and $FG \longrightarrow \mathrm{Id}_{\mathsf{D}}$ as the unit and the counit, respectively. A useful fact is that a functor F is an equivalence if and only if it is both:

- *fully faithful*, which means that each function $\mathsf{C}(X, Z) \longrightarrow \mathsf{D}(FX, FZ)$ on morphism sets is a bijection;
- *essentially surjective*, which means that for each object $Y \in \mathsf{D}$, there exists an isomorphism $FX \xrightarrow{\ \cong\ } Y$ for some object $X \in \mathsf{C}$.

Suppose $F : \mathsf{D} \longrightarrow \mathsf{C}$ is a functor. A *colimit of* F, if it exists, is a pair $(\operatorname{colim} F, \delta)$ consisting of

- an object $\operatorname{colim} F \in \mathsf{C}$ and
- a morphism $\delta_d : Fd \longrightarrow \operatorname{colim} F$ in C for each object $d \in \mathsf{D}$

that satisfies the following two conditions.

(1) For each morphism $f : d \longrightarrow d'$ in D, the diagram

$$\begin{array}{ccc} Fd & \xrightarrow{\delta_d} & \mathsf{colim}\, F \\ {\scriptstyle Ff}\downarrow & & \| \\ Fd' & \xrightarrow{\delta_{d'}} & \mathsf{colim}\, F \end{array}$$

in C is commutative.

(2) The pair $(\mathsf{colim}\, F, \delta)$ is *universal* among such pairs. In other words, if (X, δ') is another such pair that satisfies property (1), then there exists a unique morphism $h : \mathsf{colim}\, F \longrightarrow X$ in C such that the diagram

$$\begin{array}{ccc} Fd & \xrightarrow{\delta_d} & \mathsf{colim}\, F \\ \| & & \downarrow{\scriptstyle h} \\ Fd & \xrightarrow{\delta'_d} & X \end{array}$$

is commutative for each object $d \in \mathsf{D}$.

A *limit of* F, if it exists, is defined dually by turning all the morphisms backward. A *small (co)limit* is a (co)limit of a functor whose domain category is a small category. A category C is *(co)complete* if it has all small (co)limits. For a functor $F : \mathsf{D} \longrightarrow \mathsf{C}$, its colimit, if it exists, is also denoted by $\mathsf{colim}_{x\in\mathsf{D}}\, Fx$ and $\mathsf{colim}_{\mathsf{D}}\, F$, and similarly for limits.

A left adjoint $F : \mathsf{C} \longrightarrow \mathsf{D}$ preserves all the colimits that exist in C. In other words, if $H : \mathsf{E} \longrightarrow \mathsf{C}$ has a colimit, then $FH : \mathsf{E} \longrightarrow \mathsf{D}$ also has a colimit, and the natural morphism

$$\underset{e\in\mathsf{E}}{\mathsf{colim}}\, FHe \longrightarrow F\Big(\underset{e\in E}{\mathsf{colim}}\, He\Big)$$

is an isomorphism. Similarly, a right adjoint $G : \mathsf{D} \longrightarrow \mathsf{C}$ preserves all the limits that exist in D.

EXAMPLE 1.1.1. Here are some special types of colimits in a category C.

- An *initial object* $\varnothing^{\mathsf{C}}$ in C is a colimit of the functor $\varnothing \longrightarrow \mathsf{C}$, where $\varnothing$ is the empty category with no objects and no morphisms. It is characterized by the universal property that for each object X in C, there is a unique morphism $\varnothing^{\mathsf{C}} \longrightarrow X$ in C.
- A *coproduct* is a colimit of a functor whose domain category is a discrete category. We use the symbols $\coprod$ and $\amalg$ to denote coproducts.
- A *pushout* is a colimit of a functor whose domain category has the form

$$\bullet \longleftarrow \bullet \longrightarrow \bullet$$

with three objects and two non-identity morphisms.

- A *coequalizer* is a colimit of a functor whose domain category has the form

$$\bullet \rightrightarrows \bullet$$

with two objects and two non-identity morphisms.

Terminal objects, products, pullbacks, and equalizers are the corresponding limit concepts. ◇

1.2 Monoidal Categories

One may think of a monoidal category as a categorical generalization of a monoid, in which there is a way to multiply together objects and morphisms. From a more sophisticated viewpoint, a monoidal category is exactly a bicategory [Ben67] with one object. In this section, we recall the definitions of a (symmetric) monoidal category, a (commutative) monoid, a (symmetric) monoidal functor, and a monoidal natural transformation. We also recall Mac Lane's Coherence Theorem for monoidal categories and discuss some examples. For monoidal categories and their coherence, our references are [Bor94b, JS93, Kel64, Mac63, Mac98].

DEFINITION 1.2.1. A *monoidal category* is a tuple

$$(\mathsf{C}, \otimes, \mathbb{1}, \alpha, \lambda, \rho)$$

consisting of:

- a category C;
- a functor $\otimes : \mathsf{C} \times \mathsf{C} \longrightarrow \mathsf{C}$ called the *monoidal product*;
- an object $\mathbb{1} \in \mathsf{C}$ called the *monoidal unit*;
- a natural isomorphism

$$(1.2.2) \qquad (X \otimes Y) \otimes Z \xrightarrow[\cong]{\alpha_{X,Y,Z}} X \otimes (Y \otimes Z)$$

  for all objects $X, Y, Z \in \mathsf{C}$ called the *associativity isomorphism*;
- natural isomorphisms

$$(1.2.3) \qquad \mathbb{1} \otimes X \xrightarrow[\cong]{\lambda_X} X \quad \text{and} \quad X \otimes \mathbb{1} \xrightarrow[\cong]{\rho_X} X$$

  for all objects $X \in \mathsf{C}$ called the *left unit isomorphism* and the *right unit isomorphism*, respectively.

These data are required to satisfy the following axioms.

Unity Axioms: The diagram

$$\begin{array}{ccc} (X\otimes \mathbb{1})\otimes Y & \xrightarrow{\alpha_{X,\mathbb{1},Y}} & X\otimes(\mathbb{1}\otimes Y)\\ {\scriptstyle \rho_X\otimes Y}\downarrow & & \downarrow{\scriptstyle X\otimes\lambda_Y}\\ X\otimes Y & = & X\otimes Y\end{array} \tag{1.2.4}$$

is commutative for all objects $X, Y \in \mathsf{C}$. Moreover, the equality

$$\lambda_{\mathbb{1}} = \rho_{\mathbb{1}} : \mathbb{1}\otimes\mathbb{1} \xrightarrow{\cong} \mathbb{1}$$

holds.

Pentagon Axiom: The pentagon

$$\begin{array}{ccccc} & & (W\otimes X)\otimes(Y\otimes Z) & & \\ & {\scriptstyle \alpha_{W\otimes X,Y,Z}}\nearrow & & \searrow{\scriptstyle \alpha_{W,X,Y\otimes Z}} & \\ ((W\otimes X)\otimes Y)\otimes Z & & & & W\otimes(X\otimes(Y\otimes Z))\\ {\scriptstyle \alpha_{W,X,Y}\otimes Z}\downarrow & & & & \uparrow{\scriptstyle W\otimes\alpha_{X,Y,Z}}\\ (W\otimes(X\otimes Y))\otimes Z & & \xrightarrow{\alpha_{W,X\otimes Y,Z}} & & W\otimes((X\otimes Y)\otimes Z)\end{array}$$

is commutative for all objects $W, X, Y, Z \in \mathsf{C}$.

A *strict monoidal category* is a monoidal category in which α, λ, and ρ are all identity morphisms.

REMARK 1.2.5. Consider Definition 1.2.1.

(1) We usually suppress the associativity isomorphism α and the unit isomorphisms λ and ρ from the notation, and abbreviate a monoidal category to $(\mathsf{C}, \otimes, \mathbb{1})$ or even just C. On the other hand, to emphasize the ambient monoidal category C, we decorate the monoidal structure accordingly as $\otimes^{\mathsf{C}}$, $\mathbb{1}^{\mathsf{C}}$, α^{C}, λ^{C}, and ρ^{C}.
(2) The axiom $\lambda_{\mathbb{1}} = \rho_{\mathbb{1}}$ is a consequence of the unity axiom (1.2.4) and the pentagon axiom, so it is redundant [Kel64].
(3) The following unity diagrams are commutative.

$$\begin{array}{ccc} (\mathbb{1}\otimes X)\otimes Y & \xrightarrow{\alpha_{\mathbb{1},X,Y}} & \mathbb{1}\otimes(X\otimes Y)\\ {\scriptstyle \lambda_X\otimes Y}\downarrow & & \downarrow{\scriptstyle \lambda_{X\otimes Y}}\\ X\otimes Y & = & X\otimes Y\end{array} \qquad \begin{array}{ccc} (X\otimes Y)\otimes\mathbb{1} & \xrightarrow{\alpha_{X,Y,\mathbb{1}}} & X\otimes(Y\otimes\mathbb{1})\\ {\scriptstyle \rho_{X\otimes Y}}\downarrow & & \downarrow{\scriptstyle X\otimes\rho_Y}\\ X\otimes Y & = & X\otimes Y\end{array} \tag{1.2.6}$$

See [Mac98] VII.1 Eq. (9) and Exercise 1. ◇

CONVENTION 1.2.7. In a monoidal category, an *empty tensor product*, written as $X^{\otimes 0}$ or $X^{\otimes\varnothing}$, means the monoidal unit $\mathbb{1}$. ◇

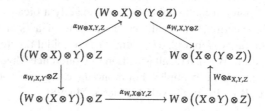

DEFINITION 1.2.8. A *monoid* in a monoidal category C is a triple $(X, \mu, 1)$ with:

- X an object in C;
- $\mu : X \otimes X \longrightarrow X$ a morphism, called the *multiplication*;
- $1 : \mathbb{1} \longrightarrow X$ a morphism, called the *unit*.

These data are required to make the following associativity and unity diagrams commutative.

$$\begin{array}{ccccc} (X\otimes X)\otimes X & \xrightarrow{\alpha} & X\otimes(X\otimes X) & \xrightarrow{X\otimes\mu} & X\otimes X \\ {\scriptstyle \mu\otimes X}\downarrow & & & & \downarrow{\scriptstyle \mu} \\ X\otimes X & & \xrightarrow{\mu} & & X \end{array} \qquad \begin{array}{ccccc} \mathbb{1}\otimes X & \xrightarrow{1\otimes X} & X\otimes X & \xleftarrow{X\otimes 1} & X\otimes\mathbb{1} \\ {\scriptstyle \lambda}\downarrow{\scriptstyle \cong} & & \downarrow{\scriptstyle \mu} & & {\scriptstyle \cong}\downarrow{\scriptstyle \rho} \\ X & = & X & = & X \end{array}$$

A morphism of monoids

$$f : (X, \mu^X, 1^X) \longrightarrow (Y, \mu^Y, 1^Y)$$

is a morphism $f : X \longrightarrow Y$ in C that preserves the multiplications and the units in the sense that the diagrams

$$\begin{array}{ccc} X\otimes X & \xrightarrow{f\otimes f} & Y\otimes Y \\ {\scriptstyle \mu^X}\downarrow & & \downarrow{\scriptstyle \mu^Y} \\ X & \xrightarrow{f} & Y \end{array} \qquad \begin{array}{ccc} \mathbb{1} & \xrightarrow{1^X} & X \\ \| & & \downarrow{\scriptstyle f} \\ \mathbb{1} & \xrightarrow{1^Y} & Y \end{array}$$

are commutative. The category of monoids in a monoidal category C is denoted by Mon(C).

DEFINITION 1.2.9. For monoidal categories C and D, a *monoidal functor*

$$(F, F_2, F_0) : \mathsf{C} \longrightarrow \mathsf{D}$$

consists of:

- a functor $F : \mathsf{C} \longrightarrow \mathsf{D}$;
- a natural transformation

$$F(X) \otimes F(Y) \xrightarrow{F_2} F(X \otimes Y) \in \mathsf{D}, \tag{1.2.10}$$

  where X and Y are objects in C;
- a morphism

$$\mathbb{1}^{\mathsf{D}} \xrightarrow{F_0} F\mathbb{1}^{\mathsf{C}} \in \mathsf{D}\,. \tag{1.2.11}$$

These data are required to satisfy the following associativity and unity axioms.

Associativity: The diagram

(1.2.12)
$$\begin{array}{ccc}
(F(X)\otimes F(Y))\otimes F(Z) & \xrightarrow{\alpha^{\mathsf{D}}} & F(X)\otimes (F(Y)\otimes F(Z)) \\
\downarrow{\scriptstyle F_2\otimes F(Z)} & & \downarrow{\scriptstyle F(X)\otimes F_2} \\
F(X\otimes Y)\otimes F(Z) & & F(X)\otimes F(Y\otimes Z) \\
\downarrow{\scriptstyle F_2} & & \downarrow{\scriptstyle F_2} \\
F((X\otimes Y)\otimes Z) & \xrightarrow{F(\alpha^{\mathsf{C}})} & F(X\otimes (Y\otimes Z))
\end{array}$$

is commutative for all objects $X, Y, Z \in \mathsf{C}$.

Left Unity: The diagram

(1.2.13)
$$\begin{array}{ccc}
\mathbb{1}^{\mathsf{D}}\otimes F(X) & \xrightarrow{\lambda^{\mathsf{D}}_{F(X)}} & F(X) \\
\downarrow{\scriptstyle F_0\otimes F(X)} & & \uparrow{\scriptstyle F(\lambda^{\mathsf{C}}_X)} \\
F(\mathbb{1}^{\mathsf{C}})\otimes F(X) & \xrightarrow{F_2} & F(\mathbb{1}^{\mathsf{C}}\otimes X)
\end{array}$$

is commutative for all objects $X \in \mathsf{C}$.

Right Unity: The diagram

(1.2.14)
$$\begin{array}{ccc}
F(X)\otimes \mathbb{1}^{\mathsf{D}} & \xrightarrow{\rho^{\mathsf{D}}_{F(X)}} & F(X) \\
\downarrow{\scriptstyle F(X)\otimes F_0} & & \uparrow{\scriptstyle F(\rho^{\mathsf{C}}_X)} \\
F(X)\otimes F(\mathbb{1}^{\mathsf{C}}) & \xrightarrow{F_2} & F(X\otimes \mathbb{1}^{\mathsf{C}})
\end{array}$$

is commutative for all objects $X \in \mathsf{C}$.

A *strong monoidal functor* is a monoidal functor in which the morphisms F_0 and F_2 are all isomorphisms. A *strict monoidal functor* is a monoidal functor in which the morphisms F_0 and F_2 are all identity morphisms.

REMARK 1.2.15. We often abbreviate a monoidal functor (F, F_2, F_0) to F. A monoidal functor is sometimes called a *lax monoidal functor* in the literature to emphasize the fact that the morphisms F_2 and F_0 are not necessarily invertible. A strong monoidal functor is also known as a *tensor functor* [JS93]. ◇

PROPOSITION 1.2.16. *Each monoidal functor* $(F, F_2, F_0) : \mathsf{C} \longrightarrow \mathsf{D}$ *induces a functor*

$$\mathsf{Mon}(\mathsf{C}) \xrightarrow{F} \mathsf{Mon}(\mathsf{D})$$

that sends a monoid $(X, \mu, 1)$ *in* C *to the monoid* $\big(FX, \mu^{FX}, 1^{FX}\big)$ *in* D *with unit*

$$\mathbb{1}^{\mathsf{D}} \xrightarrow{F_0} F\mathbb{1}^{\mathsf{C}} \xrightarrow{F1} FX \qquad (\text{composite } 1^{FX})$$

and multiplication

$$FX \otimes FX \xrightarrow{F_2} F(X \otimes X) \xrightarrow{F\mu} FX \,. \qquad (\text{composite } \mu^{FX})$$

DEFINITION 1.2.17. For monoidal functors $F, G : \mathsf{C} \longrightarrow \mathsf{D}$, a *monoidal natural transformation* $\theta : F \longrightarrow G$ is a natural transformation between the underlying functors such that the diagrams

(1.2.18)
$$\begin{array}{ccc} F(X) \otimes F(Y) & \xrightarrow{\theta_X \otimes \theta_Y} & G(X) \otimes G(Y) \\ {\scriptstyle F_2}\downarrow & & \downarrow{\scriptstyle G_2} \\ F(X \otimes Y) & \xrightarrow{\theta_{X \otimes Y}} & G(X \otimes Y) \end{array}$$

and

(1.2.19)
$$\begin{array}{ccc} \mathbb{1}^{\mathsf{D}} & \xrightarrow{F_0} & F\mathbb{1}^{\mathsf{C}} \\ \Vert & & \downarrow{\scriptstyle \theta_{\mathbb{1}^{\mathsf{C}}}} \\ \mathbb{1}^{\mathsf{D}} & \xrightarrow{G_0} & G\mathbb{1}^{\mathsf{C}} \end{array}$$

are commutative for all objects $X, Y \in \mathsf{C}$.

The following coherence/strictification result for monoidal categories is due to Mac Lane; see [Mac63], [Mac98] (XI.3 Theorem 1), and [JS93].

THEOREM 1.2.20 (Coherence for Monoidal Categories). *For each monoidal category* C, *there exist a strict monoidal category* C_{st} *and an adjoint equivalence*

$$\mathsf{C} \underset{R}{\overset{L}{\rightleftarrows}} \mathsf{C}_{\mathsf{st}}$$

with (i) both L *and* R *strong monoidal functors and (ii)* $RL = \mathrm{Id}_{\mathsf{C}}$.

In other words, every monoidal category can be strictified via an adjoint equivalence consisting of strong monoidal functors.

- A second version of the Coherence Theorem [Mac98] (VII.2 Theorem 1) describes explicitly the free monoidal category generated by one object.

- A third version of the Coherence Theorem [Mac98] (VII.2 Corollary) states that every diagram in a monoidal category involving monoidal products of the associativity isomorphism, the unit isomorphisms, their inverses, and identity morphisms, and their composites, commutes.
- A fourth version of the Coherence Theorem states that, for each category C, the unique strict monoidal functor from the free monoidal category generated by C to the free strict monoidal category generated by C is an equivalence of categories [JS93] (Theorem 1.2).

Generalizations of these coherence results to other types of monoidal categories, such as braided, ribbon, and coboundary monoidal categories, are discussed in [Yau∞].

Next we consider symmetric monoidal categories.

DEFINITION 1.2.21. A *symmetric monoidal category* is a pair (C, ξ) in which:

- $C = (C, \otimes, \mathbb{1}, \alpha, \lambda, \rho)$ is a monoidal category as in Definition 1.2.1.
- ξ is a natural isomorphism

$$(1.2.22) \qquad\qquad X \otimes Y \xrightarrow[\cong]{\xi_{X,Y}} Y \otimes X$$

for objects $X, Y \in C$, called the *symmetry isomorphism*.

These data are required to satisfy the following three axioms.

Symmetry Axiom: The diagram

$$(1.2.23)$$

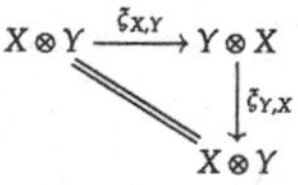

$$X \otimes Y \xrightarrow{\xi_{X,Y}} Y \otimes X$$
$$\searrow \qquad \downarrow{\xi_{Y,X}}$$
$$X \otimes Y$$

is commutative for all objects $X, Y \in C$.

Unit Axiom: The diagram

$$(1.2.24)$$

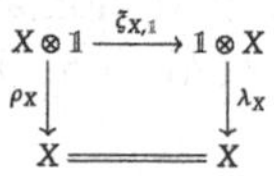

$$\begin{array}{ccc} X \otimes \mathbb{1} & \xrightarrow{\xi_{X,\mathbb{1}}} & \mathbb{1} \otimes X \\ {\scriptstyle \rho_X} \downarrow & & \downarrow {\scriptstyle \lambda_X} \\ X & =\!=\!= & X \end{array}$$

is commutative for all objects $X \in C$.

Hexagon Axiom: The diagram

$$(1.2.25)$$

$$X \otimes (Z \otimes Y) \xrightarrow{X \otimes \xi_{Z,Y}} X \otimes (Y \otimes Z)$$
$$\nearrow{\scriptstyle \alpha} \qquad\qquad\qquad \searrow{\scriptstyle \alpha^{-1}}$$
$$(X \otimes Z) \otimes Y \qquad\qquad\qquad (X \otimes Y) \otimes Z$$
$$\searrow{\scriptstyle \xi_{X \otimes Z, Y}} \qquad\qquad \nearrow{\scriptstyle \xi_{Y,X} \otimes Z}$$
$$Y \otimes (X \otimes Z) \xrightarrow{\alpha^{-1}} (Y \otimes X) \otimes Z$$

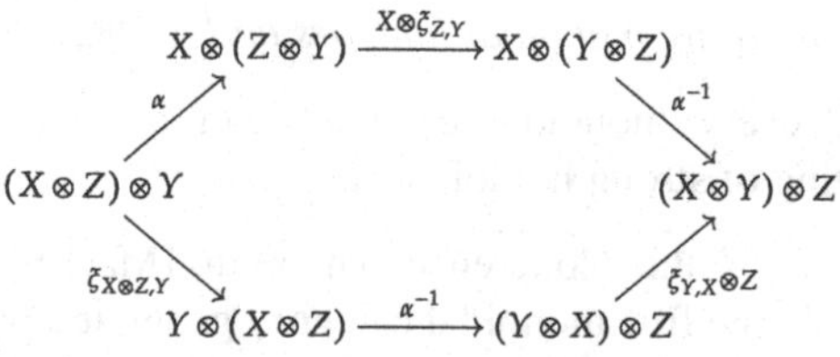

is commutative for all objects $X, Y, Z \in \mathsf{C}$.

The following consequence of the Symmetry Axiom and the Hexagon Axiom will be used in the proof of Proposition 4.7.1.

LEMMA 1.2.26. *In a symmetric monoidal category* C *with associativity isomorphism* α *and symmetry isomorphism* ξ, *the diagram*

$$\begin{array}{ccccc} (A\otimes B)\otimes C & \xrightarrow{\alpha} & A\otimes(B\otimes C) & \xrightarrow{A\otimes\xi} & A\otimes(C\otimes B) \\ \downarrow{\scriptstyle \xi\otimes C} & & & & \downarrow{\scriptstyle \xi} \\ (B\otimes A)\otimes C & \xrightarrow{\xi} & C\otimes(B\otimes A) & \xleftarrow{\alpha} & (C\otimes B)\otimes A \end{array}$$

is commutative for all objects $A, B, C \in \mathsf{C}$.

PROOF. To save space in the following diagram, we use concatenation as an abbreviation for the monoidal product, so $(AB)C$ means $(A \otimes B) \otimes C$.

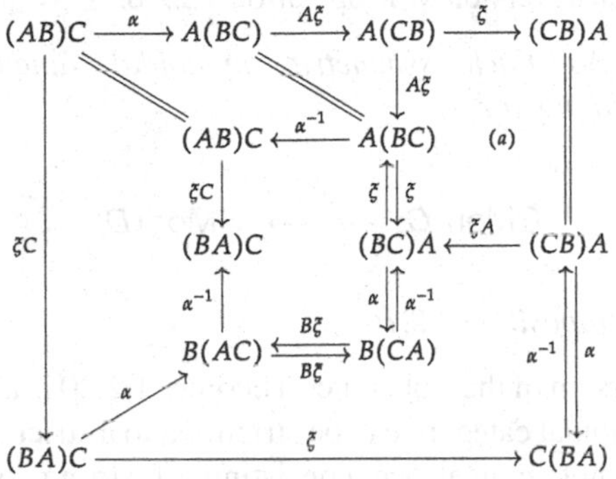

The sub-diagram (a) is commutative by the naturality of ξ. The middle rectangle and the bottom-right sub-diagram are commutative by the Hexagon Axiom. The other sub-diagrams are commutative by definition or that $\xi \circ \xi = \mathrm{Id}$. □

DEFINITION 1.2.27. A *commutative monoid* in a symmetric monoidal category (C, ξ) is a monoid $(X, \mu, 1)$ in C such that the multiplication μ is commutative in the sense that the diagram

$$\begin{array}{ccc} X\otimes X & \xrightarrow{\xi_{X,X}} & X\otimes X \\ \downarrow{\scriptstyle \mu} & & \downarrow{\scriptstyle \mu} \\ X & = & X \end{array}$$

is commutative. A morphism of commutative monoids is a morphism of the underlying monoids. The category of commutative monoids in C is denoted by $\mathsf{CMon}(\mathsf{C})$.

DEFINITION 1.2.28. For symmetric monoidal categories C and D, a *symmetric monoidal functor* $(F, F_2, F_0) : \mathsf{C} \longrightarrow \mathsf{D}$ is a monoidal functor between the underlying monoidal categories that is compatible with the symmetry isomorphisms, in the sense that the diagram

$$\begin{array}{ccc} F(X) \otimes F(Y) & \xrightarrow[\cong]{\xi_{FX,FY}} & F(Y) \otimes F(X) \\ {\scriptstyle F_2}\downarrow & & \downarrow{\scriptstyle F_2} \\ F(X \otimes Y) & \xrightarrow[\cong]{F\xi_{X,Y}} & F(Y \otimes X) \end{array} \tag{1.2.29}$$

is commutative for all objects $X, Y \in \mathsf{C}$. A symmetric monoidal functor is said to be *strong* (resp., *strict*) if the underlying monoidal functor is so.

Next is the symmetric version of Proposition 1.2.16.

PROPOSITION 1.2.30. *Each symmetric monoidal functor* $(F, F_2, F_0) : C \longrightarrow D$ *induces a functor*

$$\mathsf{CMon}(C) \xrightarrow{F} \mathsf{CMon}(D)$$

that is defined as in Proposition 1.2.16.

The symmetric version of the Coherence Theorem 1.2.20 is also true. It states that every symmetric monoidal category can be strictified to a strict symmetric monoidal category via an adjoint equivalence consisting of strong symmetric monoidal functors.

- Another version of the Coherence Theorem for symmetric monoidal categories states that every diagram in a symmetric monoidal category involving monoidal products of the associativity isomorphism, the unit isomorphisms, the symmetry isomorphism, their inverses, and identity morphisms, as well as their composites, commutes. This is a special case of [JS93] (Corollary 2.6), which is stated more generally for braided monoidal categories.
- Yet another version of the Coherence Theorem for symmetric monoidal categories states that, for each category C, the unique strict symmetric monoidal functor from the free symmetric monoidal category generated by C to the free strict symmetric monoidal category generated by C is an equivalence of categories. This is the symmetric special case of [JS93] (Theorem 2.5), which is again stated for braided monoidal categories.

EXAMPLE 1.2.31. Here are some examples of symmetric monoidal categories.

- $(\mathsf{Set}, \times, *)$: category of sets and functions. A monoid in Set is a monoid in the usual sense, also called a unital monoid if one wants to emphasize the unit.
- $(\mathsf{Poset}, \times, *)$: category of partially ordered sets and ordered-preserving functions.
- $(\mathsf{Rel}, \times, *)$: category of sets with relations as morphisms. For two sets X and Y, a relation from X to Y is a subset of the product $X \times Y$.
- $(\mathsf{Cat}, \times, \mathbf{1})$: category of small categories and functors. Here $\mathbf{1}$ is a category with one object and only the identity morphism.
- $(\mathsf{Chain}(\mathbb{K}), \otimes, \mathbb{K})$: category of chain complexes over a field $\mathbb{K}$ and chain maps [Wei97]. A monoid in $\mathsf{Chain}(\mathbb{K})$ is a unital differential graded $\mathbb{K}$-algebra.
- $(\mathsf{Vect}(\mathbb{K}), \otimes, \mathbb{K})$: category of $\mathbb{K}$-vector spaces and $\mathbb{K}$-linear maps. A monoid in $\mathsf{Vect}(\mathbb{K})$ is a unital $\mathbb{K}$-algebra.
- $(\mathsf{Hilb}, \widehat{\otimes}, \mathbb{C})$: category of complex Hilbert spaces and bounded linear maps, where $\widehat{\otimes}$ is the completed tensor product of Hilbert spaces [Wei80].

Note that the last three examples are *not* strict symmetric monoidal categories because the associativity isomorphisms are not equalities. ◇

EXAMPLE 1.2.32. For a monoidal category $(\mathsf{C}, \otimes, \mathbb{1}, \alpha, \lambda, \rho)$, its *opposite monoidal category* is the opposite category C^{op} whose monoidal product is the same as that in C on objects. It is given by $f^{\mathsf{op}} \otimes g^{\mathsf{op}} = (f \otimes g)^{\mathsf{op}}$ for morphisms f, g in C. Here for a morphism $f \in \mathsf{C}(X, Y)$, we write $f^{\mathsf{op}} \in \mathsf{C}^{\mathsf{op}}(Y, X)$ for the corresponding morphism. The monoidal unit in C^{op} is the monoidal unit $\mathbb{1}$ in C. The associativity isomorphism and the left/right unit isomorphisms in C^{op} are given by

$$(X \otimes Y) \otimes Z \xrightarrow{(\alpha^{-1}_{X,Y,Z})^{\mathsf{op}}} X \otimes (Y \otimes Z) \quad \text{and} \quad \mathbb{1} \otimes X \xrightarrow{(\lambda^{-1}_X)^{\mathsf{op}}} X \xleftarrow{(\rho^{-1}_X)^{\mathsf{op}}} X \otimes \mathbb{1} .$$

If we regard the monoidal category C as a bicategory with one object, then the opposite monoidal category C^{op} is called the *conjugate* of the one-object bicategory C in [Ben67] (3.1).

If (C, ξ) is a symmetric monoidal category, then C^{op} equipped with the symmetry isomorphism

$$X \otimes Y \xrightarrow{(\xi^{-1}_{X,Y})^{\mathsf{op}}} Y \otimes X$$

for objects $X, Y \in \mathsf{C}$ is a symmetric monoidal category. ◇

In Example 8.1.22 below, we will see an important example of a monoidal category that is not symmetric.

DEFINITION 1.2.33. A symmetric monoidal category C is *closed* if, for each object X, the functor

$$- \otimes X : \mathsf{C} \longrightarrow \mathsf{C}$$

admits a right adjoint, denoted by $[X, -]$ and called the *internal hom*.

EXAMPLE 1.2.34. All the symmetric monoidal categories in Example 1.2.31 are closed, in which the right adjoint of $- \otimes X$ is the most straightforward hom-construction. For the category $\mathsf{Chain}(\mathbb{K})$ of chain complexes, one has to be careful about the sign convention, the details of which can be found in [Hov99] (Proposition 4.2.13). ◇

EXAMPLE 1.2.35. Here is an example of a strict monoidal category that is not symmetric. Suppose C is a small category. Denote by $\mathsf{End}(\mathsf{C})$ the category whose

- objects are functors $\mathsf{C} \longrightarrow \mathsf{C}$;
- morphisms are natural transformations between such functors.

The category $\mathsf{End}(\mathsf{C})$ has a monoidal structure whose

- monoidal product is composition of functors for $\mathsf{Ob}(\mathsf{End}(\mathsf{C}))$, and is horizontal composition of natural transformations for $\mathsf{Mor}(\mathsf{End}(\mathsf{C}))$;
- monoidal unit is the identity functor Id_{C}.

This is a strict monoidal category, since composition of functors is strictly associative and unital, but it is not symmetric in general. ◇

1.3 Monads

For further discussion of monads, the reader may consult [BW85, Bor94b, Mac98, Sch72].

DEFINITION 1.3.1. A *monad* in a category C is a triple (T, μ, ϵ) in which

- $T : \mathsf{C} \longrightarrow \mathsf{C}$ is a functor and
- $\mu : T^2 \longrightarrow T$, called the *multiplication*, and $\epsilon : \mathrm{Id}_{\mathsf{C}} \longrightarrow T$, called the *unit*, are natural transformations

such that the associativity and unity diagrams

$$\begin{array}{ccc} T^3 & \xrightarrow{T\mu} & T^2 \\ {\scriptstyle \mu T}\downarrow & & \downarrow{\scriptstyle \mu} \\ T^2 & \xrightarrow{\mu} & T \end{array} \qquad \begin{array}{ccccc} \mathrm{Id}_{\mathsf{C}} \circ T & \xrightarrow{\epsilon T} & T^2 & \xleftarrow{T\epsilon} & T \circ \mathrm{Id}_{\mathsf{C}} \\ \| & & \downarrow{\scriptstyle \mu} & & \| \\ T & = & T & = & T \end{array}$$

are commutative. When there is no danger of confusion, we abbreviate such a monad to T.

PROPOSITION 1.3.2. *Suppose* C *is a small category. Then monads in* C *are exactly the monoids in the strict monoidal category* $(\mathsf{End}(\mathsf{C}), \circ, \mathrm{Id}_{\mathsf{C}})$ *in Example 1.2.35.*

PROOF. The associativity and unity diagrams in Definition 1.3.1 for a monad in C are equal to the associativity and unity diagrams in Definition 1.2.8 for a monoid in the strict monoidal category $\mathsf{End}(\mathsf{C})$. □

DEFINITION 1.3.3. Suppose (T, μ, ϵ) is a monad in a category C.

(1) A *T-algebra* is a pair (X, θ) consisting of
   - an object X in C and
   - a morphism $\theta : TX \longrightarrow X$, called the *structure morphism*,

   such that the associativity and unity diagrams

(1.3.4)
$$\begin{array}{ccc} T^2X & \xrightarrow{T\theta} & TX \\ {\scriptstyle \mu_X}\downarrow & & \downarrow{\scriptstyle \theta} \\ TX & \xrightarrow{\theta} & X \end{array} \qquad \begin{array}{ccc} X & \xrightarrow{\epsilon_X} & TX \\ & \searrow\!\!\!= & \downarrow{\scriptstyle \theta} \\ & & X \end{array}$$

   are commutative.

(2) A *morphism of T-algebras*
$$f : (X, \theta^X) \longrightarrow (Y, \theta^Y)$$
   is a morphism $f : X \longrightarrow Y$ in C such that the diagram
$$\begin{array}{ccc} TX & \xrightarrow{Tf} & TY \\ {\scriptstyle \theta^X}\downarrow & & \downarrow{\scriptstyle \theta^Y} \\ X & \xrightarrow{f} & Y \end{array}$$
   is commutative.

(3) The category of T-algebras is denoted by $\mathsf{Alg}(T)$.

DEFINITION 1.3.5. For a monad (T, μ, ϵ) in a category C, the *Eilenberg-Moore adjunction* is the adjunction
$$\mathsf{C} \underset{U}{\overset{T}{\rightleftarrows}} \mathsf{Alg}(T)$$
in which:

- The right adjoint U is the forgetful functor $U(X, \theta) = X$.
- The left adjoint sends an object $X \in \mathsf{C}$ to the free T-algebra
$$\big(TX, \mu_X : T^2X \longrightarrow TX\big).$$

The following result about the existence of limits in the category of algebras is [Bor94b] Proposition 4.3.1.

PROPOSITION 1.3.6. *Suppose T is a monad in a category C. If C admits some type of limits, then the category $\mathsf{Alg}(T)$ admits the same type of limits, which are furthermore preserved by the forgetful functor U.*

Existence of colimits in the category of algebras is more subtle. The following result is [Bor94b] Proposition 4.3.4.

PROPOSITION 1.3.7. *Suppose T is a monad in a cocomplete category C. Then the category $\mathsf{Alg}(T)$ is cocomplete if and only if it has all coequalizers.*

To state another result that guarantees the existence of colimits in the category of algebras, we first need a few definitions, all of which can be found in [Bor94a, Bor94b].

DEFINITION 1.3.8. Suppose C is a category.

(1) The *kernel pair* of a morphism $f : X \longrightarrow Y$ in C is the pullback $P \rightrightarrows X$, if it exists, of f with itself.
(2) A *regular epimorphism* is an epimorphism that can be written as the coequalizer of some pair of morphisms.
(3) C is called a *regular category* if the following conditions are satisfied:
    (i) Every morphism has a kernel pair.
    (ii) Every kernel pair has a coequalizer.
    (iii) The pullback of a regular epimorphism along any morphism exists and is again a regular epimorphism.
(4) A *section* of a morphism $f : X \longrightarrow Y$ is a morphism $s : Y \longrightarrow X$ such that $fs = \mathrm{Id}_Y$.

The following result about the existence of (co)limits in the category of algebras is [Bor94b] Theorem 4.3.5.

THEOREM 1.3.9. *Suppose C is a complete and cocomplete regular category in which every regular epimorphism has a section. Then for each monad T in C, the category $\mathsf{Alg}(T)$ is complete, cocomplete, and regular.*

EXAMPLE 1.3.10. Categories that satisfy the hypotheses of Theorem 1.3.9 include Set, $\mathsf{Vect}(\mathbb{K})$, and $\mathsf{Chain}(\mathbb{K})$. On the other hand, the category Cat of small categories is not regular [Bor94b] (2.4.6), although it is complete and cocomplete. See Theorem 1.4.9 below. ◇

1.4 Colimits in Cat

The purpose of this section is to provide an explicit description of colimits in the category Cat of small categories and functors. Since a colimit can also be expressed as a coequalizer involving coproducts, when they exist, we will concentrate on coequalizers in Cat. To describe these coequalizers explicitly, first we need some definitions. For a morphism f in a category, recall that $\mathsf{dom}(f)$ and $\mathsf{cod}(f)$ denote its domain and codomain, respectively.

DEFINITION 1.4.1. Suppose $F, G : \mathsf{C} \rightrightarrows \mathsf{D}$ are two functors between small categories.

(1) Define $\overset{\text{ob}}{\sim}$ as the equivalence relation on $\mathsf{Ob}(\mathsf{D})$ generated by

$$F(X) \sim G(X) \quad \text{for} \quad X \in \mathsf{Ob}(\mathsf{C}).$$

For an object $Y \in \mathsf{D}$, its $\overset{\text{ob}}{\sim}$-equivalence class is denoted by $[Y]$.

(2) Define

$$\overline{\mathsf{Mor}}(\mathsf{D}) = \coprod_{n \geq 1} \mathsf{Mor}(\mathsf{D})^{\times n}$$

as the set of non-empty finite sequences of morphisms in D that are not necessarily composable.

(3) For $\underline{f} = (f_1, \ldots, f_m) \in \overline{\mathsf{Mor}}(\mathsf{D})$ with each f_i a morphism in D, define its *domain* and *codomain* as

$$\mathsf{dom}(\underline{f}) = \mathsf{dom}(f_1) \quad \text{and} \quad \mathsf{cod}(\underline{f}) = \mathsf{cod}(f_m).$$

We say that $\underline{f}$ is $\overset{\text{ob}}{\sim}$*-composable* if

$$\mathsf{cod}(f_i) \overset{\text{ob}}{\sim} \mathsf{dom}(f_{i+1}) \quad \text{for} \quad i = 1, \ldots, m-1.$$

(4) Define

$$\widetilde{\mathsf{Mor}}(\mathsf{D}) \subseteq \overline{\mathsf{Mor}}(\mathsf{D})$$

as the subset of $\overset{\text{ob}}{\sim}$-composable sequences.

(5) Define $\simeq$ as the equivalence relation on $\widetilde{\mathsf{Mor}}(\mathsf{D})$ generated by $\sim_1$ and $\sim_2$ as follows. Suppose $\underline{f} = (f_1, \ldots, f_m)$ and $\underline{g} = (g_1, \ldots, g_n)$ are in $\widetilde{\mathsf{Mor}}(\mathsf{D})$.

- We define

$$\underline{f} \sim_1 \underline{g}$$

if $m = n + 1$, and there exists $1 \le i \le m - 1$ such that

(i) $(f_1, \ldots, f_{i-1}) = (g_1, \ldots, g_{i-1})$,
(ii) $(f_{i+2}, \ldots, f_m) = (g_{i+1}, \ldots, g_n)$,
(iii) $\mathsf{cod}(f_i) = \mathsf{dom}(f_{i+1})$, and
(iv) $f_{i+1} f_i = g_i \in \mathsf{Mor}(\mathsf{D})$.

- We define

$$\underline{f} \sim_2 \underline{g}$$

if $m = n$, and there exists $1 \le i \le m$ such that

(i) $f_j = g_j$ for $i \neq j \in \{1, \ldots, m\}$, and
(ii) $f_i = F(h)$ and $g_i = G(h)$ for some $h \in \mathsf{Mor}(\mathsf{C})$.

(6) For $\underline{f} \in \widetilde{\mathsf{Mor}}(\mathsf{D})$, its $\simeq$-equivalence class is denoted by $[\underline{f}]$.

EXPLANATION 1.4.2. Let us explain Definition 1.4.1 with more details.

(1) For two objects $X, Y \in \mathsf{D}$, $X \overset{\mathsf{ob}}{\sim} Y$ if and only if either one of the following two statements holds.

(i) $X = Y$.
(ii) There exists a sequence of objects

$$X = W_0, W_1, \ldots, W_n = Y$$

in D with $n > 0$ such that, for each $0 \le i \le n - 1$, there exists an object $V_i \in \mathsf{C}$ with

$$(W_i, W_{i+1}) = \big(F(V_i), G(V_i)\big) \quad \text{or} \quad \big(G(V_i), F(V_i)\big).$$

(2) $\underline{f} \sim_1 \underline{g}$ means that

- $\underline{f}$ contains a consecutive pair of composable morphisms, and
- $\underline{g}$ is obtained from $\underline{f}$ by replacing this consecutive pair with their composite in D.

The condition

$$(f_1, \ldots, f_{i-1}) = (g_1, \ldots, g_{i-1})$$

is empty if $i = 1$, and the condition

$$(f_{i+2}, \ldots, f_m) = (g_{i+1}, \ldots, g_n)$$

is empty if $i = m - 1$.

(3) $\underline{f} \sim_2 \underline{g}$ means that

- $\underline{f}$ contains an entry of the form $F(h)$ for some $h \in \mathsf{Mor}(\mathsf{C})$, and
- $\underline{g}$ is obtained from $\underline{f}$ by replacing this entry with $G(h)$.

(4) $\underline{f} \simeq \underline{g}$ if and only if either one of the following two statements holds.

(i) $\underline{f} = \underline{g}$.
(ii) There exists a sequence

$$\underline{f} = \psi_0, \psi_1, \ldots, \psi_r = \underline{g}$$

with each $\psi_i \in \widetilde{\mathsf{Mor}}(\mathsf{D})$ and $r > 0$ such that, for each $0 \le j \le r-1$, either

$$\psi_j \sim_{\epsilon_j} \psi_{j+1} \quad \text{or} \quad \psi_{j+1} \sim_{\epsilon_j} \psi_j \quad \text{with} \quad \epsilon_j \in \{1, 2\}.$$

(5) If $\underline{f} \simeq \underline{g}$, then

$$\mathsf{dom}(\underline{f}) \overset{\mathsf{ob}}{\sim} \mathsf{dom}(\underline{g}) \quad \text{and} \quad \mathsf{cod}(\underline{f}) \overset{\mathsf{ob}}{\sim} \mathsf{cod}(\underline{g}) \tag{1.4.3}$$

because $\sim_1$ and $\sim_2$ do not change the $\overset{\mathsf{ob}}{\sim}$-equivalence classes of the domain and the codomain. ◇

The concepts in Definition 1.4.1 are used below.

DEFINITION 1.4.4. Suppose $F, G : \mathsf{C} \rightrightarrows \mathsf{D}$ are two functors between small categories. Define a small category E as follows.

Objects: $\mathsf{Ob}(\mathsf{E}) = \mathsf{Ob}(\mathsf{D})/\overset{\mathsf{ob}}{\sim}$, i.e., the set of $\overset{\mathsf{ob}}{\sim}$-equivalence classes of objects in D.
Morphisms: $\mathsf{Mor}(\mathsf{E}) = \widetilde{\mathsf{Mor}}(\mathsf{D})/\simeq$, i.e., the set of $\simeq$-equivalence classes in $\widetilde{\mathsf{Mor}}(\mathsf{D})$.
(Co)domain: For $[\underline{f}] \in \mathsf{Mor}(\mathsf{E})$, its domain and codomain are defined as

$$\mathsf{dom}([\underline{f}]) = [\mathsf{dom}(\underline{f})] \quad \text{and} \quad \mathsf{cod}([\underline{f}]) = [\mathsf{cod}(\underline{f})].$$

Identity: For $[X] \in \mathsf{Ob}(\mathsf{E})$, its identity morphism $\mathrm{Id}_{[X]}$ is $[\mathrm{Id}_X]$.
Composition: The composite of two composable morphisms

$$[X] \xrightarrow{[\underline{f}]} [Y] \xrightarrow{[\underline{g}]} [Z]$$

in E is defined as

$$[\underline{g}] \circ [\underline{f}] = [(\underline{f}, \underline{g})],$$

in which $(\underline{f}, \underline{g})$ is the concatenation of $\underline{f}$ and $\underline{g}$

LEMMA 1.4.5. *E in Definition 1.4.4 is a small category.*

PROOF. For a morphism $[\underline{f}] \in \mathsf{Mor}(\mathsf{E})$, its domain and codomain are well-defined by (1.4.3). For an object $[X] \in \mathsf{Ob}(\mathsf{E})$, its identity morphism is well-defined by $\sim_2$. Indeed, for each object $V \in \mathsf{C}$, we have that

$$\mathrm{Id}_{F(V)} = F(\mathrm{Id}_V) \sim_2 G(\mathrm{Id}_V) = \mathrm{Id}_{G(V)}.$$

So $[\mathrm{Id}_X]$ is independent of the choice of the representative $X \in [X]$ by $\sim_2$.

For the composition, the concatenation $(\underline{f}, \underline{g})$ is $\overset{\mathsf{ob}}{\sim}$-composable because both $\mathsf{cod}(\underline{f})$ and $\mathsf{dom}(\underline{g})$ are in the same equivalence class $[Y]$. That the $\simeq$-equivalence class $\big[(\underline{f}, \underline{g})\big]$ is independent of the choices of the representatives $\underline{f} \in [\underline{f}]$ and $\underline{g} \in [\underline{g}]$ follows from the definitions of $\sim_1$ and $\sim_2$. Composition of morphisms in E is associative because concatenation of sequences is associative.

The unit equalities

$$[\underline{f}] \circ \mathrm{Id}_{\mathsf{dom}([\underline{f}])} = [\underline{f}] = \mathrm{Id}_{\mathsf{cod}([\underline{f}])} \circ [\underline{f}]$$

hold by $\sim_1$. In fact, once a representative $\underline{f} = (f_1, \ldots, f_m) \in [\underline{f}]$ is chosen, we may choose the representative

$$\mathrm{Id}_{\mathsf{dom}(\underline{f})} = \mathrm{Id}_{\mathsf{dom}(f_1)} \in \mathrm{Id}_{\mathsf{dom}([\underline{f}])}.$$

So a representative of the composite $[\underline{f}] \circ \mathrm{Id}_{\mathsf{dom}([\underline{f}])}$ is

$$\big(\mathrm{Id}_{\mathsf{dom}(f_1)}, f_1, \ldots, f_m\big) \sim_1 \underline{f}.$$

A similar argument works for the other unit equality. □

LEMMA 1.4.6. *In the context of Definition 1.4.4, the assignments*

$$\begin{cases} \mathsf{Ob}(\mathsf{D}) \ni X \longmapsto [X] \in \mathsf{Ob}(\mathsf{E}), \\ \mathsf{Mor}(\mathsf{D}) \ni f \longmapsto [f] \in \mathsf{Mor}(\mathsf{E}) \end{cases}$$

define a functor $\pi : \mathsf{D} \longrightarrow \mathsf{E}$.

PROOF. That π preserves domains, codomains, and identity morphisms follows from the definition. That π preserves compositions of morphisms follows from $\sim_1$.

□

THEOREM 1.4.7. *In the context of Definition 1.4.4 and Lemma 1.4.6, the diagram*

$$\mathsf{C} \overset{F}{\underset{G}{\rightrightarrows}} \mathsf{D} \xrightarrow{\ \pi\ } \mathsf{E}$$

is a coequalizer in Cat.

PROOF. The equality $\pi F = \pi G$ holds on objects by the definition of $\overset{ob}{\sim}$. This equality holds on morphisms by the definition of $\sim_2$.

To check that (E, π) has the required universal property of a coequalizer, suppose

$$H : D \quad\longrightarrow\quad B \in \mathsf{Cat} \quad \text{such that} \quad HF = HG.$$

We must show that there exists a unique functor $\overline{H} : E \longrightarrow B$ such that the triangle in the diagram

$$
\begin{array}{ccc}
C \overset{F}{\underset{G}{\rightrightarrows}} D & \overset{\pi}{\longrightarrow} & E \\
& H \downarrow \quad {}^{\nearrow} \overline{H} & \\
& B &
\end{array}
$$

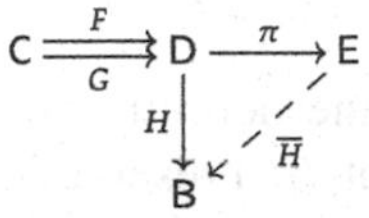

is commutative. First note that H induces a unique function $\overline{H}$ on objects that makes the diagram

(1.4.8)
$$
\begin{array}{ccc}
\mathsf{Ob}(D) & \overset{\text{quotient}}{\longrightarrow} & \mathsf{Ob}(E) = \mathsf{Ob}(D)/\overset{ob}{\sim} \\
H \downarrow & {}^{\swarrow} \overline{H} & \\
\mathsf{Ob}(B) & &
\end{array}
$$

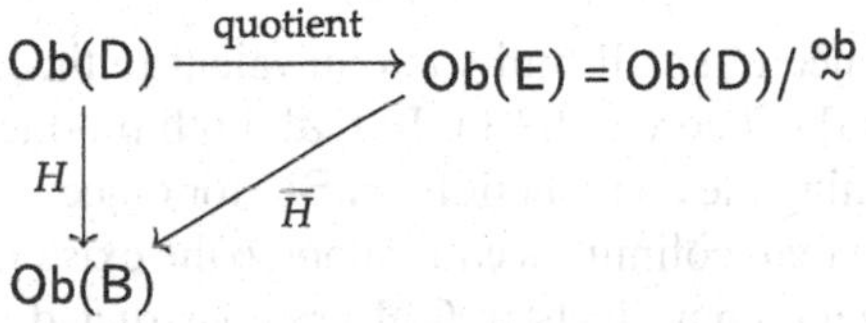

commutative. Indeed, since $HF(X) = HG(X)$ for each object $X \in C$, $X \overset{ob}{\sim} Y$ implies $H(X) = H(Y)$ by Explanation 1.4.2(1).

We define the desired functor $\overline{H} : E \longrightarrow B$ by

$$\overline{H}([X]) = H(X) \quad \text{for} \quad [X] \in \mathsf{Ob}(E),$$

$$\overline{H}([\underline{f}]) = (Hf_m) \circ \cdots \circ (Hf_1)$$

for $[\underline{f}] = [(f_1, \ldots, f_m)] \in \mathsf{Mor}(E)$. To see that $\overline{H}$ is well-defined on morphisms, first observe that for each $\overset{ob}{\sim}$-composable sequence $\underline{f}$, the morphisms $(Hf_m, \ldots, Hf_1)$ are composable in B by (1.4.8). That $\overline{H}([\underline{f}])$ is independent of the choice of the representative $\underline{f} \in [\underline{f}]$ follows from

- the fact that H preserves compositions of morphisms (for $\sim_1$) and
- the equality $HF = HG$ on morphisms in C (for $\sim_2$).

Moreover, $\overline{H}$ preserves identity morphisms because H does, and $\overline{H}$ preserves compositions of morphisms by the associativity of composition in B. By construction, the equality $\overline{H}\pi = H$ holds.

For uniqueness, suppose $H' : \mathsf{E} \longrightarrow \mathsf{B}$ is a functor such that $H'\pi = H$. We already mentioned that $\overline{H}$ is unique on objects. For a $\overset{\mathsf{ob}}{\sim}$-composable sequence $\underline{f} = (f_1, \ldots, f_m)$, there are equalities

$$\begin{aligned} H'([\underline{f}]) &= H'\big([f_m] \circ \cdots \circ [f_1]\big) \\ &= H'([f_m]) \circ \cdots \circ H'([f_1]) \\ &= H(f_m) \circ \cdots \circ H(f_1) \\ &= \overline{H}([\underline{f}]). \end{aligned}$$

The first equality follows from the definition of composition in E. The second equality follows from the fact that H' preserves compositions of morphisms. The third equality follows from the equality $H'\pi = H$ on morphisms. The last equality is the definition of $\overline{H}([\underline{f}])$. Therefore, $H' = \overline{H}$ as functors. □

THEOREM 1.4.9. *The category* Cat *is complete and cocomplete.*

PROOF. The existence of small limits is equivalent to the existence of products and equalizers [Bor94a] (Theorem 2.8.1). In Cat, both products and equalizers are constructed by performing the constructions in Set for objects and morphisms. Similarly, the existence of small colimits is equivalent to the existence of coproducts and coequalizers. Once again, coproducts in Cat are constructed by taking coproducts of object sets and of morphism sets. Coequalizers in Cat exist by Theorem 1.4.7. □

EXAMPLE 1.4.10. Suppose $\mathbf{1} = \{0\}$ is a discrete category with one object 0, and $\mathbf{2} = \big\{\, 0 \xrightarrow{f} 1 \,\big\}$ is a category with two objects and one non-identity morphism f. Consider the functors $F, G : \mathbf{1} \rightrightarrows \mathbf{2}$ defined by $F(0) = 0$ and $G(0) = 1$. For the coequalizer E of F and G, the object set

$$\mathsf{Ob}(\mathsf{E}) = \mathsf{Ob}(\mathbf{2})/\overset{\mathsf{ob}}{\sim} \; = \{*\}$$

has a single object. Every non-empty finite sequence of morphisms in $\mathbf{2}$ is $\overset{\mathsf{ob}}{\sim}$-composable, so

$$\widetilde{\mathsf{Mor}}(\mathbf{2}) = \overline{\mathsf{Mor}}(\mathbf{2}).$$

The quotient by the equivalence relation $\simeq$ is

$$\mathsf{Mor}(\mathsf{E}) = \widetilde{\mathsf{Mor}}(\mathbf{2})/\!\simeq \; = \big\{[f^n] : n \geq 0\big\},$$

in which

$$f^n = \begin{cases} (f, \ldots, f) & \text{with } n \text{ copies of } f \text{ if } n > 0, \\ \mathrm{Id}_0 & \text{if } n = 0. \end{cases}$$

The identity morphism of the object $*$ is $[\mathrm{Id}_0]$. Composition in the coequalizer E is given by

$$[f^n] \circ [f^m] = [f^{n+m}].$$

The functor $\pi : \mathbf{2} \longrightarrow \mathsf{E}$ sends both objects $0, 1$ to $* \in \mathsf{E}$, and sends the morphism $f \in \mathbf{2}$ to the morphism $[f] \in \mathsf{E}$. ◇

1.5 Exercises and Notes

(1) Prove Propositions 1.2.16 and 1.2.30.
(2) Prove that a category has small (co)limits if and only if it has (co)products and (co)equalizers.
(3) Prove Proposition 1.3.7.
(4) Suppose M is a complete and cocomplete symmetric monoidal category, in which the monoidal product commutes with small colimits on each side. Along the lines of Theorem 1.4.7, give an explicit construction of coequalizers in the category of small M-enriched categories and M-enriched functors.
(5) Give an explicit construction of a pushout in Cat. Do the same for M-enriched categories.

1.5.1 Notes

The explicit description of coequalizers in Cat in Theorem 1.4.7 is based on (i) the generalized congruences in [BBP99] and (ii) the quotient construction in [Sch72] (6.4.4 and 8.2.4). Another proof of the existence of small (co)limits in Cat is given in [Rie16] (4.5.16), using the fact that Cat is a reflective subcategory of the category of simplicial sets, which has all small (co)limits. Although the existence of small colimits in Cat is well-known, we included the detailed explanation for two reasons. First, an explicit and straightforward description of coequalizers in Cat is not easily accessible. Second, we will use Theorem 1.4.9 below, so a good understanding of coequalizers in Cat will be useful.

Chapter 2
Involutive Categories

In this chapter, we present the definitions and basic properties of involutive categories, involutive functors, involutive natural transformations, involutive adjunctions, and involutive objects in involutive categories. An important observation is Theorem 2.3.8, which says that an adjoint of an involutive functor must also be an involutive functor such that the adjunction is an involutive adjunction. As a consequence, an involutive functor that is also an equivalence of categories is automatically an involutive equivalence.

2.1 Definition and Examples

DEFINITION 2.1.1. An *involutive structure* on a category C is a pair (I, ι) in which:

- $I : \mathsf{C} \longrightarrow \mathsf{C}$ is a functor, called the *involution functor*.
- $\iota : \mathrm{Id}_{\mathsf{C}} \xrightarrow{\cong} I^2$ is a natural isomorphism, called the *unit* of I, that satisfies the equality

(2.1.2)
$$I \xrightarrow{\iota_I = I\iota} I^3 ,$$

called the *triangle identity*.

If we need to emphasize the ambient category C, we will decorate I and ι as I_{C} and ι^{C}, respectively. We also define the following.

(1) An involutive structure is *strict* if the unit ι is the identity natural transformation $\mathrm{Id}_{\mathrm{Id}_{\mathsf{C}}} : \mathrm{Id}_{\mathsf{C}} \longrightarrow I^2 = \mathrm{Id}_{\mathsf{C}}$.
(2) An *involutive category* is a category equipped with an involutive structure.

© The Author(s), under exclusive license to Springer Nature Switzerland AG 2020
D. Yau, *Involutive Category Theory*, Lecture Notes in Mathematics 2279,
https://doi.org/10.1007/978-3-030-61203-0_2

(3) A *strict involutive category* is a category equipped with a strict involutive structure.

REMARK 2.1.3. A concept similar to an involutive category is a dagger category. A *dagger category* is a category C equipped with a functor $(-)^\dagger : \mathsf{C}^{op} \longrightarrow \mathsf{C}$ that is the identity on objects and that satisfies $f^{\dagger\dagger} = f$ for morphisms f. Roughly speaking, in a dagger category there is a natural way to turn each morphism around. Dagger categories are useful in the study of Hilbert spaces and quantum systems [AC04, AC08, Bae06, Sel07, Sel10]. In Example 2.5.7 we will observe that small dagger categories are exactly the involutive objects in a certain involutive category. ◇

The following observation says that each involution functor is self-adjoint.

PROPOSITION 2.1.4. *Suppose* (C, I, ι) *is an involutive category. Then:*

(1) There is an adjunction $I \dashv I$ *with unit* ι *and counit* ι^{-1}.
(2) I *preserves all the limits and colimits that exist in* C.

PROOF. To show that there is an adjunction $I \dashv I$, we must show that there exists a bijection

$$\mathsf{C}(IX, Y) \cong \mathsf{C}(X, IY)$$

for objects X, Y in C. For morphisms $f : IX \longrightarrow Y$ and $g : X \longrightarrow IY$, we define the composites:

$$f^\sharp : X \xrightarrow{\iota_X} I^2X \xrightarrow{If} IY \qquad g^\flat : IX \xrightarrow{Ig} I^2Y \xrightarrow{\iota_Y^{-1}} Y \tag{2.1.5}$$

The assignments

$$f \longmapsto f^\sharp \quad \text{and} \quad g \longmapsto g^\flat$$

give the desired bijection. Indeed, the commutative diagram

$$\begin{array}{ccc} IX & \xrightarrow{f} & Y \\ {\scriptstyle I\iota_X = \iota_{IX}} \downarrow & & \downarrow {\scriptstyle \iota_Y} \quad \uparrow {\scriptstyle \iota_Y^{-1}} \\ I^3X & \xrightarrow{I^2 f} & I^2Y \end{array}$$

shows that $(f^\sharp)^\flat = f$, in which $I\iota_X = \iota_{IX}$ by the triangle identity (2.1.2). The equality $(g^\flat)^\sharp = g$ is proved similarly.

The second assertion follows from the first assertion, since left adjoints preserve colimits and right adjoints preserve limits. □

COROLLARY 2.1.6. *Suppose X is an object in an involutive category (C, I, ι). Under the adjunction bijections*

$$\mathsf{C}(I^2X, X) \xleftarrow[\cong]{(-)^\flat} \mathsf{C}(IX, IX) \xrightarrow[\cong]{(-)^\sharp} \mathsf{C}(X, I^2X)\ ,$$

there are equalities

$$\mathrm{Id}^\sharp_{IX} = \iota_X \quad \textit{and} \quad \mathrm{Id}^\flat_{IX} = \iota_X^{-1}.$$

PROPOSITION 2.1.7. *Suppose (C, I, ι) is an involutive category, and D is a small category. Then the diagram category C^D inherits an involutive structure (I', ι') defined by*

$$I'X = I \circ X \quad \textit{for} \quad X \in \mathsf{C}^\mathsf{D},$$
$$(\iota'_X)(d) = \iota_{X(d)} : X(d) \longrightarrow \big[(I')^2X\big](d) \quad \textit{for} \quad d \in \mathsf{D}.$$

PROOF. The functoriality of I' and the naturality of ι' follow from their definitions. The triangle identity (2.1.2) for (I', ι') follows from that for (I, ι). □

The rest of this section contains examples of involutive categories.

EXAMPLE 2.1.8 (Trivial Involution). Every category C can be equipped with the *trivial involutive structure* $\big(\mathrm{Id}_\mathsf{C}, \mathrm{Id}_{\mathrm{Id}_\mathsf{C}}\big)$ with the identity functor as the involution functor and the identity natural transformation of Id_C as the unit. ◇

EXAMPLE 2.1.9 (Commutative Group Action). Suppose G is a commutative group, regarded as a groupoid with one object and with the elements in G as the morphisms. A G-diagram in a category C is exactly an object in C equipped with a G-action. There is a strict involutive structure $\overline{(-)}$ on the diagram category C^G that takes a G-diagram $X : G \longrightarrow \mathsf{C}$ to the G-diagram

$$\overline{X} : G \xrightarrow{(-)^{-1}} G \xrightarrow{X} \mathsf{C}\,,$$

in which g^{-1} is the multiplicative inverse of an element $g \in G$. In other words, $\overline{X}$ is the same object X with the inverted G-action. When G is regarded as a one-object groupoid, that the inversion map $(-)^{-1} : G \longrightarrow G$ defines an isomorphism of groupoids is where we used the assumption that G is commutative. For example:

(1) When G is the trivial group {id}, this involutive structure on $\mathsf{C}^{\{\mathrm{id}\}} = \mathsf{C}$ is the trivial involutive structure in Example 2.1.8.
(2) When G is the circle group S^1 with C a suitable category of topological spaces or spectra, this involutive structure was used in [Bar08] to study $O(2)$-equivariant spaces/spectra. ◇

EXAMPLE 2.1.10 (Conjugate Vector Spaces). The category $\mathsf{Vect}(\mathbb{C})$ of complex vector spaces has a strict involutive structure $\overline{(-)}$ that sends each complex vector space V to its conjugate $\overline{V}$, which has the same underlying additive group as V. For $z \in \mathbb{C}$ and $v \in V$, the $\mathbb{C}$-linear structure in $\overline{V}$ is given by

$$z \cdot_{\overline{V}} v = \overline{z} \cdot_V v$$

with $\overline{z}$ the complex conjugate of z, and $\cdot_V$ and $\cdot_{\overline{V}}$ the scalar multiplications in V and $\overline{V}$, respectively. For a $\mathbb{C}$-linear map f, $\overline{f}$ has the same underlying set map as f. Moreover:

(1) The category $\mathsf{Chain}(\mathbb{C})$ of chain complexes of complex vector spaces has a strict involutive structure that restricts to $\overline{(-)}$ in each degree.
(2) The strict involutive structure $\overline{(-)}$ extends to the category Hilb of complex Hilbert spaces. For a Hilbert space H, the inner product on the conjugate space $\overline{H}$ is the complex conjugate of the inner product on H. ◇

EXAMPLE 2.1.11 (Opposite Categories). The category Cat of small categories has a strict involutive structure $(-)^{\mathsf{op}}$ given by taking the opposite category. For a functor $F : \mathsf{C} \longrightarrow \mathsf{D}$, the functor $F^{\mathsf{op}} : \mathsf{C}^{\mathsf{op}} \longrightarrow \mathsf{D}^{\mathsf{op}}$ is the same as F on objects, and is given by $F^{\mathsf{op}}(f^{\mathsf{op}}) = (F(f))^{\mathsf{op}}$ for morphisms $f \in \mathsf{C}(X, Y)$. Furthermore, restricted to the subcategory Poset of partially ordered sets, this gives Poset a strict involutive structure by reversing the partial orderings. ◇

EXAMPLE 2.1.12 (Bicategories). A bicategory B is *small* if it has a set of objects. It is *locally small* if for each pair of objects X and Y, the category $\mathsf{B}(X, Y)$ is a small category. Consider the category Bicat of bicategories that are both small and locally small, and morphisms between them [Ben67, Lac10]. Then Bicat has three strict involutive structures:

(1) There is an involution functor I^{op} that sends a bicategory B to the bicategory B^{op} obtained by reversing only the 1-cells. In [Ben67] B^{op} is called the *transpose*. This generalizes Example 2.1.11 if we consider a category C as a bicategory in which each morphism set $\mathsf{C}(X, Y)$ is regarded as a discrete category, i.e., a category with only identity morphisms.
(2) There is another involution functor I^{co} given by $I^{\mathsf{co}}\mathsf{B} = \mathsf{B}^{\mathsf{co}}$, which is the bicategory obtained from B by reversing only the 2-cells. In [Ben67] B^{co} is called the *conjugate*.
(3) Finally, there is an involution functor I^{coop} given by $I^{\mathsf{coop}}\mathsf{B} = \mathsf{B}^{\mathsf{coop}}$, which is the bicategory obtained from B by reversing both the 1-cells and the 2-cells. ◇

2.2 Involutive Functors

In this section, we define the involutive versions of functors and the category of small involutive categories.

DEFINITION 2.2.1. For two involutive categories C and D, an *involutive functor*

$$(F, \nu) : (\mathsf{C}, I_{\mathsf{C}}, \iota^{\mathsf{C}}) \longrightarrow (\mathsf{D}, I_{\mathsf{D}}, \iota^{\mathsf{D}})$$

is a pair in which:

- $F : \mathsf{C} \longrightarrow \mathsf{D}$ is a functor.
- $\nu : FI_{\mathsf{C}} \longrightarrow I_{\mathsf{D}}F$ is a natural transformation, called the *distributive law*, such that the diagram

(2.2.2)
$$\begin{array}{ccc} F & \xrightarrow{\iota^{\mathsf{D}}_F} & I^2_{\mathsf{D}}F \\ {\scriptstyle F\iota^{\mathsf{C}}}\downarrow & & \uparrow{\scriptstyle I_{\mathsf{D}}\nu} \\ FI^2_{\mathsf{C}} & \xrightarrow{\nu_{I_{\mathsf{C}}}} & I_{\mathsf{D}}FI_{\mathsf{C}} \end{array}$$

is commutative.

EXAMPLE 2.2.3. Suppose (C, I, ι) is an involutive category.

(1) Then the involution functor defines an involutive functor

$$(I, \mathrm{Id}_{I^2}) \; : (\mathsf{C}, I, \iota) \longrightarrow (\mathsf{C}, I, \iota),$$

since the axiom (2.2.2) is just the triangle identity (2.1.2) in this case.

(2) Similarly, the identity functor defines an involutive functor

$$(\mathrm{Id}_{\mathsf{C}}, \mathrm{Id}_I) : (\mathsf{C}, I, \iota) \longrightarrow (\mathsf{C}, I, \iota).$$

The identity functor on an involutive category is always regarded as an involutive functor this way. ◇

EXAMPLE 2.2.4. Suppose C is a category, and $f : G \longrightarrow H$ is a homomorphism of commutative groups. In the context of Example 2.1.9, pre-composition with the group homomorphism f yields a functor

$$f^* : \mathsf{C}^H \longrightarrow \mathsf{C}^G$$

from H-objects in C to G-objects in C. When C^H and C^G are equipped with the strict involutive structures in Example 2.1.9, the functor f^* is an involutive functor whose distributive law is the identity natural transformation. ◇

The first observation about involutive functors is that the distributive law must have invertible components.

PROPOSITION 2.2.5. *If $(F, v) : C \longrightarrow D$ is an involutive functor, then the distributive law v is a natural isomorphism.*

PROOF. We first define a natural transformation $\pi : I_D F \longrightarrow F I_C$ as the composite

$$
\begin{array}{ccc}
I_D F & \xrightarrow{\ \pi\ } & F I_C \\
{\scriptstyle I_D F \iota^C}\big\downarrow & & \big\uparrow{\scriptstyle (\iota^D_{F I_C})^{-1}} \\
I_D F I_C^2 & \xrightarrow[I_D v_{I_C}]{} & I_D^2 F I_C.
\end{array}
$$

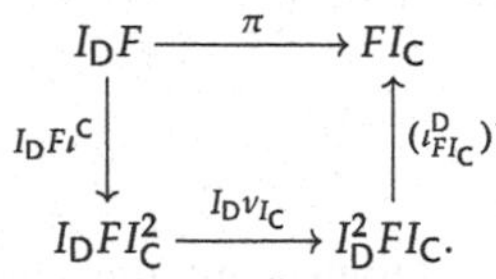

The equality

$$
\pi \circ v = \mathrm{Id}_{F I_C}
$$

is proved by the following commutative diagram.

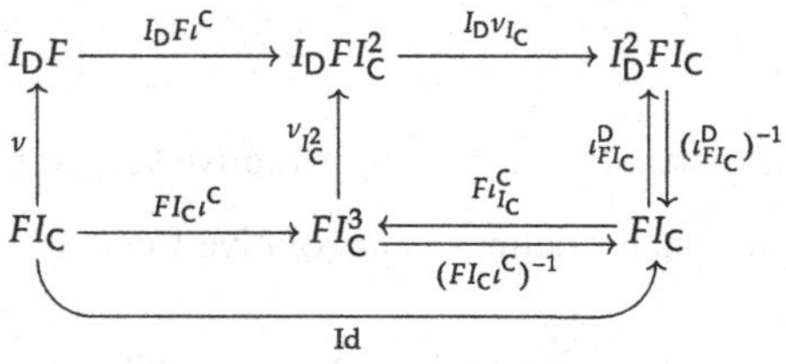

In the above diagram, the equality

$$
F \iota^C_{I_C} = F I_C \iota^C
$$

holds by the triangle identity (2.1.2). The left and the right squares are commutative by the naturality of v and the axiom (2.2.2), respectively. A similar diagram proves the equality $v \circ \pi = \mathrm{Id}_{I_D F}$. $\square$

The next observation says that involutive functors can be composed.

PROPOSITION 2.2.6. *Suppose*

$$
(C, I_C, \iota^C) \xrightarrow{\ (F, v^F)\ } (D, I_D, \iota^D) \xrightarrow{\ (G, v^G)\ } (E, I_E, \iota^E)
$$

are two involutive functors. Then the composite functor $GF : C \longrightarrow E$ becomes an involutive functor when equipped with the distributive law v^{GF} given by the composite

(2.2.7)
$$\nu^{GF} : GFI_C \xrightarrow{G\nu^F} GI_DF \xrightarrow{\nu^G_F} I_EGF\,.$$

PROOF. The axiom (2.2.2) for ν^{GF} is proved by the following commutative diagram.

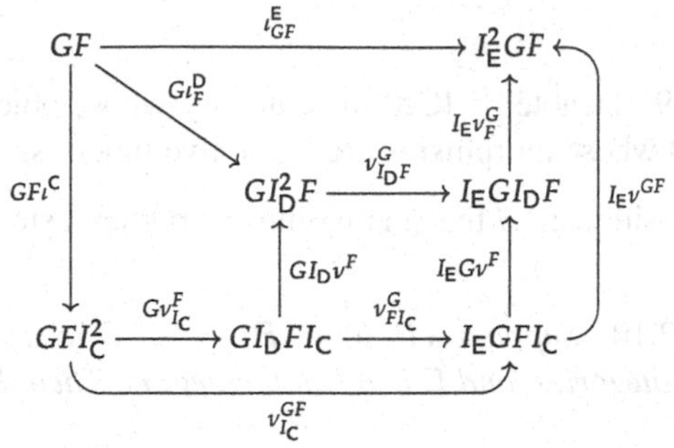

In the above diagram, the two trapezoids are commutative by the axiom (2.2.2) for ν^F and ν^G. The lower-right square is commutative by the naturality of ν^G. □

The composite of two involutive functors is always regarded as an involutive functor equipped with the distributive law (2.2.7). To define the category of small involutive categories, we first need to make sure that the distributive law construction ν^{GF} is associative with respect to composition of involutive functors.

PROPOSITION 2.2.8. *Suppose*

$$(B, I_B, \iota^B) \xrightarrow{(H,\nu^H)} (C, I_C, \iota^C) \xrightarrow{(F,\nu^F)} (D, I_D, \iota^D) \xrightarrow{(G,\nu^G)} (E, I_E, \iota^E)$$

are three involutive functors. Then the equality

$$\nu^{(GF)H} = \nu^{G(FH)}$$

holds.

PROOF. By the definition (2.2.7), the desired equality is the commutativity of the following diagram.

$$\begin{array}{ccc} GFHI_B & \xrightarrow{GF\nu^H} & GFI_CH \\ {\scriptstyle G\nu^{FH}}\downarrow & & \downarrow{\scriptstyle \nu^{GF}_H} \\ GI_DFH & \xrightarrow{\nu^G_{FH}} & I_EGFH \end{array}$$

Expanding ν_H^{GF} and ν^{FH} using the definition (2.2.7), it follows that both $\nu_H^{GF} \circ GF\nu^H$ and $\nu_{FH}^G \circ G\nu^{FH}$ are equal to the composite

$$GFHI_{\mathsf{B}} \xrightarrow{GF\nu^H} GFI_{\mathsf{C}}H \xrightarrow{G\nu_H^F} GI_{\mathsf{D}}FH \xrightarrow{\nu_{FH}^G} I_{\mathsf{E}}GFH\ .$$

□

DEFINITION 2.2.9. Denote by ICat the category whose objects are small involutive categories and whose morphisms are involutive functors.

Recall from Proposition 2.1.7 that an involutive structure extends to each diagram category.

PROPOSITION 2.2.10. *Suppose* $(F, \nu) : \mathsf{B} \longrightarrow \mathsf{C}$ *is an involutive functor between involutive categories, and* D *is a small category. Then the induced functor*

$$\mathsf{B}^{\mathsf{D}} \xrightarrow{F^{\mathsf{D}} = F\circ(-)} \mathsf{C}^{\mathsf{D}}$$

inherits from F *the structure of an involutive functor.*

PROOF. For $X \in \mathsf{B}^{\mathsf{D}}$, the X-component of the distributive law

$$\nu^{F^{\mathsf{D}}} : F^{\mathsf{D}} I'_{\mathsf{B}} \longrightarrow I'_{\mathsf{C}} F^{\mathsf{D}}$$

is the natural transformation whose d-component is the morphism

$$\nu_{X(d)} : FI_{\mathsf{B}}X(d) \longrightarrow I_{\mathsf{C}}FX(d) \in \mathsf{C}$$

for each object $d \in \mathsf{D}$. The naturality and the involutive functor axiom (2.2.2) of $\nu^{F^{\mathsf{D}}}$ follow from those of ν. □

2.3 Involutive Natural Transformations and Adjunctions

Next we define the involutive versions of a natural transformation and of an adjunction.

DEFINITION 2.3.1. Suppose

$$(\mathsf{C}, I_{\mathsf{C}}, \iota^{\mathsf{C}}) \underset{(G,\nu^G)}{\overset{(F,\nu^F)}{\rightrightarrows}} (\mathsf{D}, I_{\mathsf{D}}, \iota^{\mathsf{D}})$$

are two involutive functors. An *involutive natural transformation*

$$\theta : (F, \nu^F) \longrightarrow (G, \nu^G)$$

is a natural transformation $\theta : F \longrightarrow G$ of the underlying functors such that the diagram

(2.3.2)
$$\begin{array}{ccc} FI_{\mathsf{C}} & \xrightarrow{\theta_{I_{\mathsf{C}}}} & GI_{\mathsf{C}} \\ {\scriptstyle \nu^F}\downarrow & & \downarrow{\scriptstyle \nu^G} \\ I_{\mathsf{D}}F & \xrightarrow{I_{\mathsf{D}}\theta} & I_{\mathsf{D}}G \end{array}$$

is commutative. An *involutive natural isomorphism* is an involutive natural transformation with an underlying natural isomorphism.

Involutive natural transformations are closed under vertical and horizontal compositions of natural transformations; see Exercise 3 in Section 2.7.

EXAMPLE 2.3.3. If (C, I, ι) is an involutive category, then the unit $\iota : \mathrm{Id}_{\mathsf{C}} \xrightarrow{\cong} I^2$ is an involutive natural isomorphism, since the diagram (2.3.2) is the triangle identity (2.1.2) in this case. ◇

EXAMPLE 2.3.4. Suppose (C, I, ι) is a small involutive category. Denote by $\mathsf{IEnd}(\mathsf{C})$ the category whose

- objects are involutive functors $(F, \nu) : \mathsf{C} \longrightarrow \mathsf{C}$;
- morphisms are involutive natural transformations between such involutive functors.

As in Example 1.2.35, the category $\mathsf{IEnd}(\mathsf{C})$ has a strict monoidal structure with

- monoidal product given by composition of involutive functors on C for $\mathsf{Ob}\big(\mathsf{IEnd}(\mathsf{C})\big)$, and by horizontal composition of involutive natural transformations for $\mathsf{Mor}\big(\mathsf{IEnd}(\mathsf{C})\big)$;
- monoidal unit the identity functor Id_{C}. ◇

DEFINITION 2.3.5. Suppose

$$(\mathsf{C}, I_{\mathsf{C}}, \iota^{\mathsf{C}}) \underset{(G,\nu^G)}{\overset{(F,\nu^F)}{\rightleftarrows}} (\mathsf{D}, I_{\mathsf{D}}, \iota^{\mathsf{D}})$$

are two involutive functors. An *involutive adjunction*

$$(F, \nu^F) \dashv (G, \nu^G)$$

is an adjunction $F \dashv G$ of the underlying functors such that the unit $\varepsilon : \mathrm{Id}_{\mathsf{C}} \longrightarrow GF$ and the counit $\eta : FG \longrightarrow \mathrm{Id}_{\mathsf{D}}$ are both involutive natural transformations. In this case, we call (G, ν^G) an *involutive right adjoint* of (F, ν^F), and call (F, ν^F) an *involutive left adjoint* of (G, ν^G).

We now provide some characterizations of involutive adjunctions.

LEMMA 2.3.6. *Suppose* $F : \mathsf{C} \rightleftarrows \mathsf{D} : G$ *is an adjunction with* $(\mathsf{C}, I_{\mathsf{C}}, \iota^{\mathsf{C}})$ *and* $(\mathsf{D}, I_{\mathsf{D}}, \iota^{\mathsf{D}})$ *involutive categories, and* (F, ν^F) *and* (G, ν^G) *involutive functors. Then the following statements are equivalent:*

(1) $(F, \nu^F) \dashv (G, \nu^G)$ *is an involutive adjunction.*
(2) The diagrams

(2.3.7)
$$\begin{array}{ccc} I_{\mathsf{C}} & \xrightarrow{I_{\mathsf{C}}\varepsilon} & I_{\mathsf{C}}GF \\ {\scriptstyle \varepsilon_{I_{\mathsf{C}}}}\downarrow & & \uparrow{\scriptstyle \nu^G_F} \\ GFI_{\mathsf{C}} & \xrightarrow{G\nu^F} & GI_{\mathsf{D}}F \end{array} \qquad \begin{array}{ccc} FGI_{\mathsf{D}} & \xrightarrow{\eta_{I_{\mathsf{D}}}} & I_{\mathsf{D}} \\ {\scriptstyle F\nu^G}\downarrow & & \uparrow{\scriptstyle I_{\mathsf{D}}\eta} \\ FI_{\mathsf{C}}G & \xrightarrow{\nu^F_G} & I_{\mathsf{D}}FG \end{array}$$

are commutative.

PROOF. Using (2.2.6) to expand the distributive laws ν^{FG} and ν^{GF} in terms of ν^F and ν^G, the two diagrams in (2.3.7) are equal to the involutive natural transformation condition (2.3.2) for the unit $\varepsilon : \mathrm{Id}_{\mathsf{C}} \longrightarrow GF$ and the counit $\eta : FG \longrightarrow \mathrm{Id}_{\mathsf{D}}$ of the adjunction. □

The next observation says that an adjoint of an involutive functor is automatically an involutive functor such that the adjunction is an involutive adjunction.

THEOREM 2.3.8. *Suppose* $(\mathsf{C}, I_{\mathsf{C}}, \iota^{\mathsf{C}})$ *and* $(\mathsf{D}, I_{\mathsf{D}}, \iota^{\mathsf{D}})$ *are involutive categories, and* $F : \mathsf{C} \rightleftarrows \mathsf{D} : G$ *is an adjunction of the underlying categories. Then the following statements are equivalent:*

(1) (F, ν^F) *is an involutive functor for some natural transformation* $FI_{\mathsf{C}} \xrightarrow{\nu^F} I_{\mathsf{D}}F$.

(2) (G, ν^G) *is an involutive functor for some natural transformation* $GI_{\mathsf{D}} \xrightarrow{\nu^G} I_{\mathsf{C}}G$.

Furthermore, if either (1) or (2) holds, then $(F, \nu^F) \dashv (G, \nu^G)$ *is an involutive adjunction.*

PROOF. To prove (1) $\Longrightarrow$ (2), suppose (F, ν^F) is an involutive functor. Using the adjunctions $F \dashv G$ and $I_{\mathsf{C}} \dashv I_{\mathsf{C}}$ from Proposition 2.1.4, we define ν^G as the adjoint of the composite

(2.3.9)
$$FI_{\mathsf{C}}GI_{\mathsf{D}} \xrightarrow{\nu^F_{GI_{\mathsf{D}}}} I_{\mathsf{D}}FGI_{\mathsf{D}} \xrightarrow{I_{\mathsf{D}}\eta_{I_{\mathsf{D}}}} I_{\mathsf{D}}^2 \xrightarrow{(\iota^{\mathsf{D}})^{-1}} \mathrm{Id}_{\mathsf{D}}$$

in which $\eta : FG \longrightarrow \mathrm{Id}_D$ is the counit of the adjunction.

To check the axiom (2.2.2) for (G, ν^G), suppose $X \in C$ and $Y \in D$ are objects. Consider the following diagram.

(2.3.10)

$$
\begin{array}{ccc}
C(X, G(Y)) & \xrightarrow{(\iota^C_{G(Y)})_*} & C(X, I^2_C G(Y)) \\
\downarrow & & \downarrow \\
D(F(X), Y) & \xleftarrow{(F\iota^C_X)^*} & D(FI^2_C(X), Y) \\
\uparrow{(\iota^D_{F(X)})^*} & & \uparrow{(\nu^F_{I_C(X)})^*} \\
D(I^2_D F(X), Y) & \xrightarrow{(I_D \nu^F_X)^*} & D(I_D FI_C(X), Y) \\
\uparrow & & \downarrow \\
C(X, GI^2_D(Y)) & \xrightarrow{(\nu^G_{I_D(Y)})_*} & C(X, I_C GI_D(Y))
\end{array}
$$

with the outer maps $(G\iota^D_Y)_*$ on the left and $(I_C \nu^G_Y)_*$ on the right.

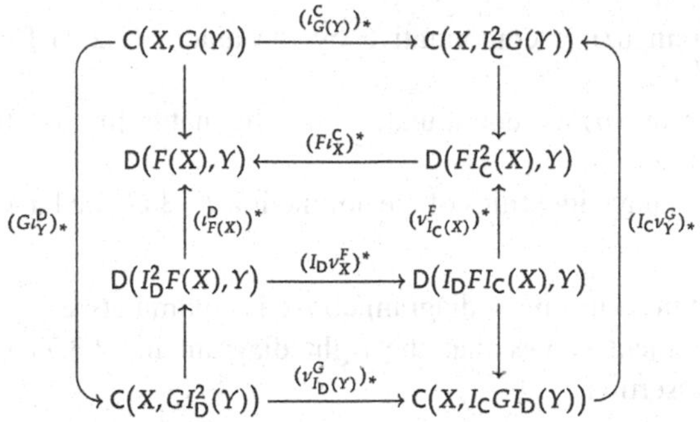

In the above diagram:

- The four unnamed maps are bijections defined by the adjunctions $F \dashv G$, $I_C \dashv I_C$, and $I_D \dashv I_D$.
- The top and the left rectangles are commutative by adjunction.
- The bottom and the right rectangles are commutative by the definition of ν^G in terms of ν^F.
- The middle square is commutative by the axiom (2.2.2) for the involutive functor (F, ν^F).

Since ι^C, ι^D, and ν^F are natural isomorphisms by Proposition 2.2.5, it follows that the outer-most diagram is also commutative. Since $X \in C$ and $Y \in D$ are arbitrary, this proves that (G, ν^G) satisfies (2.2.2), so (1) $\Longrightarrow$ (2).

An argument similar to the previous paragraph shows (2) $\Longrightarrow$ (1).

For the last assertion, suppose (1) holds, which also means (2) holds. To show that (F, ν^F) and (G, ν^G) form an involutive adjunction, by Lemma 2.3.6 it suffices to check the commutativity of the two diagrams in (2.3.7). To prove that the left diagram in (2.3.7) is commutative, we consider its adjoint diagram, which is the outer-most diagram below.

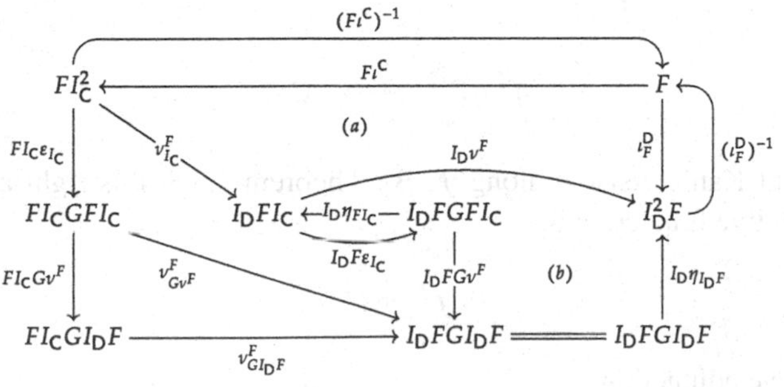

Here we used the expression (2.3.9) for the adjoint of ν_F^G, and $\varepsilon : \mathrm{Id}_{\mathsf{C}} \longrightarrow GF$ is the unit of the adjunction.

- The lower-left triangle and the sub-diagram immediately above it are commutative by definition.
- The sub-diagram (a) is commutative by the axiom (2.2.2) for the involutive functor (F, ν^F).
- The sub-diagram (b) is commutative by the naturality of the counit $\eta : FG \longrightarrow \mathrm{Id}_{\mathsf{D}}$.
- By one of the triangle identities of the adjunction $F \dashv G$, we have that $\eta_F \circ F\varepsilon = \mathrm{Id}_F$.

We conclude that the outer-most diagram above is commutative.

A similar argument shows that the right diagram in (2.3.7) is commutative, proving the last assertion. □

EXAMPLE 2.3.11. As in Example 2.2.4, suppose C is a category, and $f : G \longrightarrow H$ is a homomorphism of commutative groups. The category C^G is regarded as an involutive category under the inversion in G, and similarly for C^H. Pre-composition with the group homomorphism f yields an involutive functor

$$f^* : \mathsf{C}^H \longrightarrow \mathsf{C}^G$$

whose distributive law is the identity natural transformation. Assuming that C has all small colimits, the functor f^* admits a left adjoint

$$f_! : \mathsf{C}^G \longrightarrow \mathsf{C}^H$$

given by left Kan extension along f [Bor94a] (Theorem 3.7.2). By Theorem 2.3.8, this left adjoint $f_!$ is also an involutive functor, and

$$f_! \dashv f^*$$

is an involutive adjunction.

Similarly, assuming that C has all small limits, the functor f^* admits a right adjoint

$$f^! : \mathsf{C}^G \longrightarrow \mathsf{C}^H$$

given by right Kan extension along f. By Theorem 2.3.8, this right adjoint $f^!$ is also an involutive functor, and

$$f^* \dashv f^!$$

is an involutive adjunction. ◇

2.4 Involutive Equivalences

Recall that a functor $F : \mathsf{C} \longrightarrow \mathsf{D}$ is called an adjoint *equivalence of categories* if it admits either a left adjoint or a right adjoint such that both the unit and the counit of the adjunction are natural isomorphisms. In this case, the adjunction is called an *adjoint equivalence*. A basic fact in category theory says that a functor F is an equivalence of categories if and only if it is fully faithful and essentially surjective.

- Fully faithful means that for each pair of possibly-equal objects X and Y in C, the map

$$F : \mathsf{C}(X, Y) \longrightarrow \mathsf{D}(FX, FY)$$

  is a bijection.
- Essentially surjective means that for each object $Z \in \mathsf{D}$, there exist an object $W \in \mathsf{C}$ and an isomorphism $F(W) \cong Z$.

We now define the involutive version of an equivalence, and provide some characterizations.

DEFINITION 2.4.1. An involutive functor

$$(F, \nu^F) : (\mathsf{C}, I_{\mathsf{C}}, \iota^{\mathsf{C}}) \longrightarrow (\mathsf{D}, I_{\mathsf{D}}, \iota^{\mathsf{D}})$$

is called an *involutive equivalence* if it admits an involutive right adjoint (G, ν^G) such that both the unit and the counit of the involutive adjunction $F \dashv G$ are natural isomorphisms. In this case,

$$(F, \nu^F) \dashv (G, \nu^G)$$

is called an *involutive adjoint equivalence*.

COROLLARY 2.4.2. *Suppose*

$$(F, \nu^F) : (\mathsf{C}, I_{\mathsf{C}}, \iota^{\mathsf{C}}) \longrightarrow (\mathsf{D}, I_{\mathsf{D}}, \iota^{\mathsf{D}})$$

is an involutive functor. Then the following statements are equivalent:

(1) $F : \mathsf{C} \longrightarrow \mathsf{D}$ *is an equivalence of categories.*
(2) (F, ν^F) *is an involutive equivalence.*
(3) (F, ν^F) *admits an involutive left adjoint such that both the unit and the counit of the involutive adjunction are natural isomorphisms.*

PROOF. (2) $\Longrightarrow$ (1) follows immediately from the definition. For (1) $\Longrightarrow$ (2), suppose F is an equivalence of categories. By the characterizations of equivalences of categories in [Bor94a] (Proposition 3.4.3), we may assume that F admits a right adjoint G such that both the unit and the counit of the adjunction are natural isomorphisms. Since F is an involutive functor, by Theorem 2.3.8 the right adjoint G is also an involutive functor such that the adjunction is an involutive adjunction with invertible unit and counit. So (F, ν^F) is an involutive equivalence. The equivalence between (1) and (3) is proved similarly, since an equivalence of categories can also be characterized as a functor that admits a left adjoint with invertible unit and counit.

□

2.5 Involutive Objects

DEFINITION 2.5.1. Suppose $(\mathsf{C}, I_{\mathsf{C}}, \iota^{\mathsf{C}})$ is an involutive category.

(1) An *involutive object* in C is a pair (X, j^X) with

- $X \in \mathsf{C}$ an object and
- $j^X : X \longrightarrow I_{\mathsf{C}}(X)$ a morphism in C, called the *involutive structure map*,

such that the diagram

(2.5.2)
$$X \xrightarrow{j^X} I_{\mathsf{C}}(X) \xrightarrow{I_{\mathsf{C}} j^X} I^2_{\mathsf{C}}(X), \qquad \iota^{\mathsf{C}}_X : X \to I^2_{\mathsf{C}}(X)$$

is commutative.

(2) A *morphism* $f : (X, j^X) \longrightarrow (Y, j^Y)$ of involutive objects in C is a morphism $f : X \longrightarrow Y$ in C such that the diagram

(2.5.3)
$$\begin{array}{ccc} X & \xrightarrow{f} & Y \\ {\scriptstyle j^X}\downarrow & & \downarrow{\scriptstyle j^Y} \\ I_{\mathsf{C}}(X) & \xrightarrow{I_{\mathsf{C}} f} & I_{\mathsf{C}}(Y) \end{array}$$

is commutative.

(3) The category of involutive objects in $(\mathsf{C}, I_{\mathsf{C}}, \iota)$ and their morphisms is denoted by $\mathsf{Inv}(\mathsf{C}, I_{\mathsf{C}}, \iota^{\mathsf{C}})$ or $\mathsf{Inv}(\mathsf{C})$.

(4) Denote by $U : \mathsf{Inv}(\mathsf{C}) \longrightarrow \mathsf{C}$ the forgetful functor with $U(X, j^X) = X$ and $U(f) = f$.

The following result contains some basic properties of involutive objects. The reader is asked to provide the proof in Exercise 6 in Section 2.7.

PROPOSITION 2.5.4. *Suppose $(\mathsf{C}, I_{\mathsf{C}}, \iota^{\mathsf{C}})$ is an involutive category.*

(1) If (X, j^X) is an involutive object in C, then j^X is an isomorphism in C whose inverse is the composite

$$I_{\mathsf{C}}(X) \xrightarrow{I_{\mathsf{C}} j^X} I_{\mathsf{C}}^2(X) \xrightarrow{(\iota_X^{\mathsf{C}})^{-1}} X .$$

(2) A pair $\left(X,\ X \xrightarrow{j^X} I_{\mathsf{C}}(X)\right)$ *is an involutive object in* C *if and only if the adjoint morphism* $I_{\mathsf{C}}(X) \xrightarrow{(j^X)^\flat} X$ *makes the following diagram commutative.*

$$\begin{array}{c} I_{\mathsf{C}}^2(X) \xrightarrow{I_{\mathsf{C}}(j^X)^\flat} I_{\mathsf{C}}X \xrightarrow{(j^X)^\flat} X \\ \text{with } (\iota_X^{\mathsf{C}})^{-1} : I_{\mathsf{C}}^2(X) \to X \end{array}$$

(3) Suppose (X, j^X) *and* (Y, j^Y) *are involutive objects, and* $f : X \longrightarrow Y$ *is a morphism in* C*. Then* f *is a morphism of involutive objects if and only if the diagram*

$$\begin{array}{ccc} I_{\mathsf{C}}(X) & \xrightarrow{I_{\mathsf{C}} f} & I_{\mathsf{C}}(Y) \\ {\scriptstyle (j^X)^\flat}\downarrow & & \downarrow{\scriptstyle (j^Y)^\flat} \\ X & \xrightarrow{f} & Y \end{array}$$

is commutative.

(4) Suppose D *is another involutive category, and* $\mathsf{C} \times \mathsf{D}$ *is given the componentwise involutive structure. Then there is a canonical isomorphism of categories*

$$\mathsf{Inv}(\mathsf{C} \times \mathsf{D}) \cong \mathsf{Inv}(\mathsf{C}) \times \mathsf{Inv}(\mathsf{D}).$$

Some examples of involutive objects follow.

EXAMPLE 2.5.5 (Objects with Involutions). For any category C, an involutive object in the trivial involutive category $\left(\mathsf{C}, \mathrm{Id}_{\mathsf{C}}, \mathrm{Id}_{\mathrm{Id}_{\mathsf{C}}}\right)$ is exactly an object $X \in \mathsf{C}$ equipped with an involution $j : X \longrightarrow X$, i.e., $j^2 = \mathrm{Id}_X$. ◇

EXAMPLE 2.5.6 (Strict Involutive Categories). Recall that an involutive category (C, I, ι) is *strict* if the unit $\iota : \mathrm{Id}_{\mathsf{C}} \xrightarrow{\cong} I^2 = \mathrm{Id}_{\mathsf{C}}$ is the identity natural transformation. Also recall that Cat denotes the category of all small categories and functors between them. Then the involutive objects in the trivial involutive category $\left(\mathsf{Cat}, \mathrm{Id}, \mathrm{Id}_{\mathrm{Id}}\right)$ are exactly the small strict involutive categories. ◇

EXAMPLE 2.5.7 (Dagger Categories). Denote by Cat_i the subcategory of Cat consisting of all the objects (i.e., all small categories) and only the functors that are identity on objects. The involutive structure $\left((-)^{\mathsf{op}}, \mathrm{Id}_{\mathrm{Id}}\right)$ on Cat restricts to an involutive structure on Cat_i, since for a functor $F : \mathsf{C} \longrightarrow \mathsf{D}$, $F^{\mathsf{op}} : \mathsf{C}^{\mathsf{op}} \longrightarrow \mathsf{D}^{\mathsf{op}}$ is the same as F on objects. Then the involutive objects in the involutive category $(\mathsf{Cat}_i, (-)^{\mathrm{op}}, \mathrm{Id}_{\mathrm{Id}})$ are exactly the small dagger categories discussed in Remark 2.1.3. ◇

EXAMPLE 2.5.8 (Commutative Group Action). Consider the involutive category $(\mathsf{C}^G, \overline{(-)}, \mathrm{Id})$ in Example 2.1.9 for a commutative group G. An involutive object is a pair (X, j) with $j : X \longrightarrow \overline{X} \in \mathsf{C}^G$ a morphism, in which $\overline{X}$ is the object X with the inverted G-action, such that $\overline{j} \circ j = \mathrm{Id}_X$. For example, if C is a suitable category of topological spaces and if G is the circle group S^1, then this category $\mathsf{Inv}(\mathsf{C}^{S^1})$ of involutive objects is equivalent to the category of $O(2)$-equivariant spaces [Bar08] (Example 7.4.2). ◇

EXAMPLE 2.5.9 (Conjugate Vector Spaces). Consider the involutive category $(\mathsf{Vect}(\mathbb{C}), \overline{(-)}, \mathrm{Id})$ in Example 2.1.10. Each complex vector space V with a basis $\{v_i\}$ yields an involutive object with involutive structure map

$$V \xrightarrow{\;j\;} \overline{V}, \qquad j\Big(\sum_i z_i v_i\Big) = \sum_i \overline{z_i} v_i$$

for $z_i \in \mathbb{C}$. ◇

An involutive category and its category of involutive objects are related by free-forgetful and forgetful-cofree adjunctions as follows.

THEOREM 2.5.10. *For an involutive category (C, I, ι), consider the forgetful functor $U : \mathsf{Inv}(\mathsf{C}) \longrightarrow \mathsf{C}$.*

(1) If C has coproducts, then U admits a left adjoint

$$L : \mathsf{C} \longrightarrow \mathsf{Inv}(\mathsf{C})$$

that sends an object $X \in \mathsf{C}$ to the involutive object

$$L(X) = \Big(X \amalg IX,\ X \amalg IX \xrightarrow{\;j^{L(X)}\;} IX \amalg I^2X\Big)$$

with $j^{L(X)}$ determined by Id_{IX} and $\iota_X : X \longrightarrow I^2X$.

(2) If C has products, then U admits a right adjoint

$$R : \mathsf{C} \longrightarrow \mathsf{Inv}(\mathsf{C})$$

that sends an object $X \in \mathsf{C}$ to the involutive object

$$R(X) = \Big(X \times IX,\ X \times IX \xrightarrow{\;j^{R(X)}\;} IX \times I^2X\Big)$$

with $j^{R(X)}$ determined by Id_{IX} and $\iota_X : X \longrightarrow I^2X$.

PROOF. We will prove (1); the proof of (2) is similar.

For (1), we first check that $L(X)$ is an involutive object in C. Recall from Proposition 2.1.4 that the involution functor I preserves all the colimits that exist in C, so in particular it preserves coproducts. There is a canonical isomorphism

$$I(X \amalg IX) \cong IX \amalg I^2X,$$

which we used above to express the codomain of the morphism $j^{L(X)}$. The composite

$$X \amalg IX \xrightarrow{j^{L(X)}} IX \amalg I^2X \xrightarrow{Ij^{L(X)}} I^2X \amalg I^3X$$

is equal to

$$\iota_X \amalg \iota_{IX} \cong \iota_{X \amalg IX},$$

in which the last isomorphism follows from the naturality of ι. So $L(X)$ is an involutive object in C.

For a morphism $f : X \longrightarrow Y$ in C, $L(f) : L(X) \longrightarrow L(Y)$ is the morphism

$$X \amalg IX \xrightarrow{f \amalg I(f)} Y \amalg IY .$$

This makes $L : \mathsf{C} \longrightarrow \mathsf{Inv}(\mathsf{C})$ into a functor.

To check that L is a left adjoint of U, suppose (Z, j^Z) is an involutive object in C, and $g : X \longrightarrow Z = U(Z, j^Z)$ is a morphism in C. It suffices to show that the morphism

$$X \amalg IX \xrightarrow{(g,(j^Z)^{-1} \circ Ig)} Z \in \mathsf{C}$$

is the unique morphism $L(X) \longrightarrow (Z, j^Z)$ of involutive objects that restricts to g on the coproduct summand X. For the existence part, we must show that the diagram

$$\begin{array}{ccc} X \amalg IX & \xrightarrow{(g,(j^Z)^{-1} \circ Ig)} & Z \\ {\scriptstyle j^X}\downarrow & & \downarrow{\scriptstyle j^Z} \\ IX \amalg I^2X & \xrightarrow{(Ig, I(j^Z)^{-1} \circ I^2 g)} & IZ \end{array}$$

is commutative. Restricted to the summand IX in the upper-left corner, both composites are equal to

$$Ig = j^Z \circ (j^Z)^{-1} \circ Ig.$$

Restricted to the summand X in the upper-left corner, the two composites form the outer-most diagram below.

$$\begin{array}{ccccc} X & \xrightarrow{g} & Z & = & Z \\ \downarrow{\scriptstyle \iota_X} & & \downarrow{\scriptstyle \iota_Z} & & \downarrow{\scriptstyle j^Z} \\ I^2X & \xrightarrow{I^2 g} & I^2 Z & \xrightarrow{I(j^Z)^{-1}} & IZ. \end{array}$$

The left square is commutative by the naturality of the unit $\iota : \mathrm{Id}_{\mathsf{C}} \longrightarrow I^2$, and the right square is commutative because (Z, j^Z) is an involutive object.

For the uniqueness part, suppose $h : L(X) \longrightarrow (Z, j^Z)$ is a morphism of involutive objects whose restriction to the coproduct summand X is equal to g. In the commutative diagram

$$\begin{array}{ccc} X \amalg IX & \xrightarrow{h} & Z \\ \downarrow{\scriptstyle j^X} & & \downarrow{\scriptstyle j^Z} \\ IX \amalg I^2X & \xrightarrow{Ih} & IZ, \end{array}$$

restriction to X in the upper-left corner yields the commutative square

$$\begin{array}{ccc} X & \xrightarrow{h|_X = g} & Z \\ \downarrow{\scriptstyle \iota_X} & & \downarrow{\scriptstyle j^Z} \\ I^2X & \xrightarrow{I(h|_{IX})} & IZ. \end{array}$$

The adjoint form of the previous commutative square is the commutative square

$$\begin{array}{ccc} IX & \xrightarrow{Ig} & IZ \\ \| & & \downarrow{\scriptstyle (j^Z)^\flat} \\ IX & \xrightarrow{h|_{IX}} & Z. \end{array}$$

By the definition of $(-)^\flat$ in (2.1.5) and Proposition 2.5.4(1), there are equalities

$$(j^Z)^\flat = \iota_Z^{-1} \circ Ij^Z = (j^Z)^{-1}.$$

So we conclude that

$$h|_{IX} = (j^Z)^{-1} \circ Ig,$$

proving the desired uniqueness. This shows that L is a left adjoint of U. □

2.6 Lifting Structures to Involutive Objects

In this section, we observe that the category of involutive objects inherits many categorical structures from the original involutive category. The first result says that the category of involutive objects inherits limits and colimits from the original involutive category.

PROPOSITION 2.6.1. *If an involutive category* (C, I, ι) *has all small (co)limits, then so does its category* $\mathsf{Inv}(\mathsf{C})$ *of involutive objects. Furthermore, if* C *has all small limits and colimits, then the (co)limits in* $\mathsf{Inv}(\mathsf{C})$ *are created and preserved by the forgetful functor* $U : \mathsf{Inv}(\mathsf{C}) \longrightarrow \mathsf{C}$.

PROOF. For the first assertion, suppose C has all small colimits. The proof that $\mathsf{Inv}(\mathsf{C})$ has all small limits when C does is similar.

Suppose $G : \mathsf{D} \longrightarrow \mathsf{Inv}(\mathsf{C})$ is a functor with D a small category. For the functor $UG : \mathsf{D} \longrightarrow \mathsf{C}$, the colimit

$$X = \mathsf{colim}_{\mathsf{D}}\, UG \in \mathsf{C}$$

exists by assumption. Its involutive structure map j^X is defined by the commutative diagrams

$$\begin{array}{ccc} UG(d) & \xrightarrow{\text{natural}} & X \\ \Big\downarrow{\scriptstyle j^{G(d)}} & & \Big\downarrow{\scriptstyle j^X} \\ IUG(d) & \xrightarrow{\text{natural}} & \mathsf{colim}_{\mathsf{D}}\, IUG \cong IX \end{array}$$

in C for $d \in \mathsf{D}$. The natural isomorphism

$$\mathsf{colim}_{\mathsf{D}}\, IUG \cong I\big(\mathsf{colim}_{\mathsf{D}}\, UG\big) = IX$$

follows from the fact that the involution functor I commutes with all the colimits that exist in C, since it is a left adjoint by Proposition 2.1.4. The commutative diagram (2.5.3) that defines a morphism of involutive objects implies that j^X is a well-defined morphism in C. That j^X satisfies the axiom (2.5.2) for an involutive structure map follows from the fact that each $G(d)$ for $d \in \mathsf{D}$ is an involutive object in C, which satisfies (2.5.2). Therefore, (X, j^X) is an involutive object in C, and each natural morphism

$$G(d) \xrightarrow{\text{natural}} (X, j^X)$$

is a morphism of involutive objects by construction. A direct inspection shows that (X, j^X) and these natural morphisms form a colimit of the functor G. By construction, these colimits in $\mathsf{Inv}(\mathsf{C})$ are created by U.

For the second assertion, since U has a left adjoint and a right adjoint by Theorem 2.5.10, it also preserves limits and colimits. □

The next observation says that the category of involutive objects is itself an involutive category such that the forgetful functor is an involutive functor.

PROPOSITION 2.6.2. *Suppose (C, I, ι) is an involutive category. Then:*

(1) The category $\mathsf{Inv}(\mathsf{C})$ inherits an involutive structure $\bigl(I_{\mathsf{Inv}(\mathsf{C})}, \iota^{\mathsf{Inv}(\mathsf{C})}\bigr)$ with

$$I_{\mathsf{Inv}(\mathsf{C})}(X, j^X) = (IX, Ij^X)$$

for each involutive object (X, j^X), and

$$\iota^{\mathsf{Inv}(\mathsf{C})}_{(X,j^X)} = \iota_X.$$

(2) With the above involutive structure, the forgetful functor $U : \mathsf{Inv}(\mathsf{C}) \longrightarrow \mathsf{C}$ is an involutive functor with identity distributive law.

(3) If C has coproducts, then the adjunction

$$\mathsf{C} \underset{U}{\overset{L}{\rightleftarrows}} \mathsf{Inv}(\mathsf{C})$$

in Theorem 2.5.10(1) is an involutive adjunction.

(4) If C has products, then the adjunction

$$\mathsf{Inv}(\mathsf{C}) \underset{R}{\overset{U}{\rightleftarrows}} \mathsf{C}$$

in Theorem 2.5.10(2) is an involutive adjunction.

PROOF. For (1), $I_{\mathsf{Inv}(\mathsf{C})}$ defines a functor on $\mathsf{Inv}(\mathsf{C})$ that takes a morphism of involutive objects $f : (X, j^X) \longrightarrow (Y, j^Y)$ to

$$I_{\mathsf{Inv}(\mathsf{C})}(f) = I(f).$$

That $\iota^{\mathsf{Inv}(\mathsf{C})}$ defines a natural isomorphism $\mathrm{Id}_{\mathsf{Inv}(\mathsf{C})} \xrightarrow{\cong} I^2_{\mathsf{Inv}(\mathsf{C})}$ follows from the commutativity of the outer diagram

$$\begin{array}{ccc} X & \xrightarrow{j^X} & IX \\ {\scriptstyle \iota_X}\downarrow & & \downarrow{\scriptstyle \iota_{IX} = I\iota_X} \\ I^2X & \xrightarrow{I^2 j^X} & I^3X \end{array}$$

in which the square is commutative by the naturality of the unit ι. The triangle identity (2.1.2) for $\left(I_{\mathsf{Inv(C)}}, \iota^{\mathsf{Inv(C)}}\right)$ follows from the triangle identity for (I, ι).

Assertion (2) follows from the equality

$$IU = UI_{\mathsf{Inv(C)}} : \mathsf{Inv(C)} \longrightarrow \mathsf{C}.$$

Assertions (3) and (4) follow from assertion (2) and Theorem 2.3.8, which says that an adjoint of an involutive functor must also be an involutive functor such that the adjunction is an involutive adjunction. □

From now on, the category $\mathsf{Inv(C)}$ of involutive objects in C is regarded as an involutive category using Proposition 2.6.2(1). The next observation says that involutive functors lift to the categories of involutive objects.

PROPOSITION 2.6.3. *Each involutive functor*

$$(F, \nu^F) : (\mathsf{C}, I_{\mathsf{C}}, \iota^{\mathsf{C}}) \longrightarrow (\mathsf{D}, I_{\mathsf{D}}, \iota^{\mathsf{D}})$$

lifts to an involutive functor

$$\left(\mathsf{Inv}(\mathsf{C}), I_{\mathsf{Inv}(\mathsf{C})}, \iota^{\mathsf{Inv}(\mathsf{C})}\right) \xrightarrow{(\mathsf{Inv}(F),\, \nu^{\mathsf{Inv}(F)})} \left(\mathsf{Inv}(\mathsf{D}), I_{\mathsf{Inv}(\mathsf{D})}, \iota^{\mathsf{Inv}(\mathsf{D})}\right)$$

defined as

$$\mathsf{Inv}(F)(X, j^X) = \left(F(X),\ F(X) \xrightarrow{Fj^X} FI_{\mathsf{C}}(X) \xrightarrow{\nu^F_X} I_{\mathsf{D}}F(X)\right) \tag{2.6.4}$$

for $(X, j^X) \in \mathsf{Inv}(\mathsf{C})$. The distributive law

$$\mathsf{Inv}(F) \circ I_{\mathsf{Inv}(\mathsf{C})} \xrightarrow{\nu^{\mathsf{Inv}(F)}} I_{\mathsf{Inv}(\mathsf{D})} \circ \mathsf{Inv}(F)$$

is given by ν^F in each component.

PROOF. For an involutive object (X, j) in C, to show that $\mathsf{Inv}(F)(X, j)$ is an involutive object in D, we must check the axiom (2.5.2), which is the outer-most diagram below.

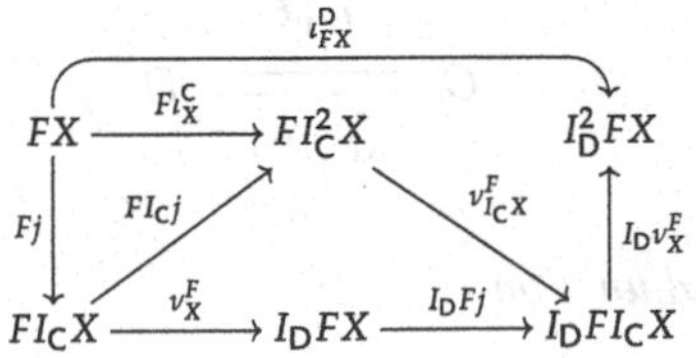

The three sub-diagrams are commutative by

- the axiom (2.5.2), which says $\iota_X^{\mathsf{C}} = (I_{\mathsf{C}} j) \circ j$,
- the naturality of ν^F, and
- the axiom (2.2.2) for the involutive functor (F, ν^F).

On a morphism $f : (X, j^X) \longrightarrow (Y, j^Y)$ of involutive objects in C, we define

$$\mathsf{Inv}(F)(f) = F(f).$$

This defines a functor

$$\mathsf{Inv}(F) : \mathsf{Inv(C)} \longrightarrow \mathsf{Inv(D)}$$

and a natural transformation $\nu^{\mathsf{Inv}(F)}$ as stated. The involutive functor axiom (2.2.2) for $\big(\mathsf{Inv}(F), \nu^{\mathsf{Inv}(F)}\big)$ follows from that for the involutive functor (F, ν^F). □

The next observation says that involutive natural transformations, involutive adjunctions, and involutive adjoint equivalences all lift to the categories of involutive objects. The reader is asked to supply the proof in Exercise 8 in Section 2.7.

PROPOSITION 2.6.5. *Suppose $(\mathsf{C}, I_{\mathsf{C}}, \iota^{\mathsf{C}})$ and $(\mathsf{D}, I_{\mathsf{D}}, \iota^{\mathsf{D}})$ are involutive categories.*

(1) Each involutive natural transformation

$$(F, \nu^F) \xrightarrow{\theta} (G, \nu^G) \quad : \mathsf{C} \rightrightarrows \mathsf{D}$$

lifts to an involutive natural transformation

$$\mathsf{Inv}(\theta) : (\mathsf{Inv}(F), \nu^{\mathsf{Inv}(F)}) \longrightarrow (\mathsf{Inv}(G), \nu^{\mathsf{Inv}(G)})$$

defined by

$$\mathsf{Inv}(\theta)(X, j^X) = \theta_X$$

for each $(X, j^X) \in \mathsf{Inv}(\mathsf{C})$.

(2) Each involutive adjunction

$$\mathsf{C} \underset{(G,\nu^G)}{\overset{(F,\nu^F)}{\rightleftarrows}} \mathsf{D}$$

lifts to an involutive adjunction

$$\left(\mathsf{Inv}(C), I_{\mathsf{Inv}(C)}, \iota^{\mathsf{Inv}(C)}\right) \underset{(\mathsf{Inv}(G), \nu^{\mathsf{Inv}(G)})}{\overset{(\mathsf{Inv}(F), \nu^{\mathsf{Inv}(F)})}{\rightleftarrows}} \left(\mathsf{Inv}(D), I_{\mathsf{Inv}(D)}, \iota^{\mathsf{Inv}(D)}\right)$$

whose unit and counit are given by those of the adjunction $F \dashv G$.

(3) In (2), if $(F, \nu^F) \dashv (G, \nu^G)$ is an involutive adjoint equivalence, then so is the lifted version.

2.7 Exercises and Notes

(1) In the proof of Proposition 2.1.4, prove that $(g^\flat)^\sharp = g$.
(2) In the proof of Proposition 2.2.5, prove the equality $\nu \circ \pi = \mathrm{Id}$.
(3) This exercise refers to the concept of a 2-category; see Section 2.3 in [JY21]. Prove that involutive natural transformations are closed under vertical and horizontal compositions. Conclude that ICat has the natural structure of a 2-category.
(4) This exercise refers to the concept of a 2-functor; see Section 4.1 in [JY21]. Using the previous exercise and Propositions 2.6.2, 2.6.3, and 2.6.5, prove that $\mathsf{Inv}(-) : \mathsf{ICat} \longrightarrow \mathsf{ICat}$ is a 2-functor.
(5) In Theorem 2.3.8:

- Prove that the top, bottom, left, and right rectangles in (2.3.10) are commutative.
- Prove (2) $\Longrightarrow$ (1).
- Assuming (1), prove that the right diagram in (2.3.7) is commutative.

(6) Prove Proposition 2.5.4.
(7) Prove assertion (2) in Theorem 2.5.10.
(8) Prove Proposition 2.6.5.

Notes. The definitions of an involutive category, an involutive functor, an involutive object, and so forth, and most of the results in this chapter are due to Jacobs [Jac12], the main exceptions being Theorem 2.3.8 and Corollary 2.4.2. In [Jac12] involutive objects are defined in adjoint form as in Proposition 2.5.4(2), and are called *self-conjugates* there. Propositions 2.1.7 and 2.2.10 about diagram involutive categories are from [BSW19b]. A strict involutive category is called a *category with involution* in [Bar08], which contains the strict versions of some of the results in this chapter.

Chapter 3
Coherence of Involutive Categories

For each type of categories with extra structures, coherence can be understood as the question of describing the free structured category of a category. The purpose of this chapter is to study the coherence of involutive categories and related topics.

In Section 3.1 we show that the forgetful functor

$$U : \mathsf{ICat}^{\mathsf{st}} \longrightarrow \mathsf{Cat},$$

from the category of small strict involutive categories and strict involutive functors to Cat, admits both a left adjoint $\mathcal{F}^{\mathsf{si}}$ and a right adjoint given by categorical coproducts and products, respectively.

In Section 3.2 we consider the category ICat_0 of small involutive categories and strict involutive functors. We construct an explicit left adjoint $\mathcal{F}^{\mathsf{icat}}$ of the forgetful functor $U : \mathsf{ICat}_0 \longrightarrow \mathsf{Cat}$. For a small category C, we call $\mathcal{F}^{\mathsf{si}}(\mathsf{C})$ and $\mathcal{F}^{\mathsf{icat}}(\mathsf{C})$ the free strict involutive category and the free involutive category generated by C, respectively. In Section 3.3 we show that there is a unique strict involutive equivalence

$$\mathcal{F}^{\mathsf{icat}}(\mathsf{C}) \xrightarrow{\ \sim\ } \mathcal{F}^{\mathsf{si}}(\mathsf{C}).$$

In other words, the free involutive category generated by C strictifies to the free strict involutive category generated by C. In Section 3.4 we show that ICat_0 is isomorphic to the category of algebras of the monad defined by the adjunction $\mathcal{F}^{\mathsf{icat}} \dashv U$.

In Section 3.5 we show that ICat_0 is locally finitely presentable. Locally finitely presentable categories have various nice properties; the reader is referred to [AR94] for more discussion. In particular, each locally finitely presentable category has all small colimits, and every object in the category may be expressed as a colimit of objects in a generating set. In Section 3.6 the local finite presentability of ICat_0 is used to show that the forgetful functor $U : \mathsf{ICat}_0 \longrightarrow \mathsf{Cat}$ admits a right adjoint.

© The Author(s), under exclusive license to Springer Nature Switzerland AG 2020

D. Yau, *Involutive Category Theory*, Lecture Notes in Mathematics 2279,
https://doi.org/10.1007/978-3-030-61203-0_3

3.1 Free Strict Involutive Categories

The purpose of this section is to describe the free and the cofree strict involutive categories generated by a category. Recall that an involutive category (C, I, ι) is called *strict* if $\iota = \mathrm{Id}_{\mathrm{Id}_\mathsf{C}}$, so $I^2 = \mathrm{Id}_\mathsf{C}$. In this case, we denote the strict involutive category by (C, I). Recall that Cat denotes the category of small categories and functors between them.

DEFINITION 3.1.1. Suppose $(\mathsf{C}, I_\mathsf{C}, \iota^\mathsf{C})$ and $(\mathsf{D}, I_\mathsf{D}, \iota^\mathsf{D})$ are involutive categories. An involutive functor

$$(F, \nu) : (\mathsf{C}, I_\mathsf{C}, \iota^\mathsf{C}) \longrightarrow (\mathsf{D}, I_\mathsf{D}, \iota^\mathsf{D})$$

is said to be *strict* if the distributive law ν is the identity natural transformation $FI_\mathsf{C} = I_\mathsf{D}F$. Furthermore:

(1) ICat_0 denotes the subcategory of ICat consisting of all the objects (i.e., all small involutive categories) and strict involutive functors.
(2) $U : \mathsf{ICat}_0 \longrightarrow \mathsf{Cat}$ denotes the forgetful functor defined by

$$U(\mathsf{C}, I_\mathsf{C}, \iota^\mathsf{C}) = \mathsf{C} \quad \text{and} \quad U(F) = F.$$

(3) $\mathsf{ICat}^{\mathsf{st}}$ denotes the category whose objects are small strict involutive categories and whose morphisms are strict involutive functors.
(4) $U : \mathsf{ICat}^{\mathsf{st}} \longrightarrow \mathsf{Cat}$ denotes the forgetful functor defined by

$$U(\mathsf{C}, I_\mathsf{C}) = \mathsf{C} \quad \text{and} \quad U(F) = F.$$

LEMMA 3.1.2. *A functor $F : \mathsf{C} \longrightarrow \mathsf{D}$, with $(\mathsf{C}, I_\mathsf{C}, \iota^\mathsf{C})$ and $(\mathsf{D}, I_\mathsf{D}, \iota^\mathsf{D})$ involutive categories, is a strict involutive functor if and only if*

(1) the diagram

(3.1.3)
$$\begin{array}{ccc} \mathsf{C} & \xrightarrow{F} & \mathsf{D} \\ {\scriptstyle I_\mathsf{C}}\downarrow & & \downarrow{\scriptstyle I_\mathsf{D}} \\ \mathsf{C} & \xrightarrow{F} & \mathsf{D} \end{array}$$

is commutative, and
(2) $F\iota^\mathsf{C} = \iota^\mathsf{D}_F$.

Furthermore, if C and D are strict involutive categories, then F is a strict involutive functor if and only if (1) holds. ◇

PROOF. The first assertion is true because, under the assumption $FI_\mathsf{C} = I_\mathsf{D}F$, the condition $F\iota^\mathsf{C} = \iota^\mathsf{D}_F$ is the axiom (2.2.2) for an involutive functor with identity distributive law. For the second assertion, when C and D are strict involutive categories, $\iota^{\mathsf{C}} : \mathrm{Id}_{\mathsf{C}} = I_{\mathsf{C}}^2$ and $\iota^{\mathsf{D}} : \mathrm{Id}_{\mathsf{D}} = I_{\mathsf{D}}^2$ are identity natural transformations, making condition (2) redundant. □

EXAMPLE 3.1.4. For each strict involutive category (C, I), the involution functor $I : \mathsf{C} \longrightarrow \mathsf{C}$ is a strict involutive functor. ◇

EXAMPLE 3.1.5. Suppose C is a category, and $f : G \longrightarrow H$ is a homomorphism of commutative groups. The pre-composition functor $f^* : \mathsf{C}^H \longrightarrow \mathsf{C}^G$ in Example 2.2.4 is a strict involutive functor. ◇

For an involutive category $(\mathsf{C}, I_{\mathsf{C}}, \iota^{\mathsf{C}})$, recall from Definition 2.5.1 that its category of involutive objects and morphisms is denoted by $\mathsf{Inv}(\mathsf{C}, I_{\mathsf{C}}, \iota^{\mathsf{C}})$. Also, every category C can be equipped with the trivial involutive structure $(\mathrm{Id}_{\mathsf{C}}, \mathrm{Id}_{\mathrm{Id}_{\mathsf{C}}})$. The following observation says that the category of small strict involutive categories is equal to the category of involutive objects in Cat with the trivial involutive structure.

LEMMA 3.1.6. *There is an equality of categories*

$$\mathsf{ICat}^{\mathsf{st}} = \mathsf{Inv}\big(\mathsf{Cat}, \mathrm{Id}, \mathrm{Id}_{\mathrm{Id}}\big).$$

PROOF. The objects in $\mathsf{Inv}\big(\mathsf{Cat}, \mathrm{Id}, \mathrm{Id}_{\mathrm{Id}}\big)$ are exactly small strict involutive categories, as we already mentioned in Example 2.5.6. For this category of involutive objects, the axiom (2.5.3) for a morphism is exactly the condition (3.1.3) for a strict involutive functor between strict involutive categories. □

The following observation gives explicit descriptions of the free and the cofree strict involutive categories generated by a small category.

PROPOSITION 3.1.7. *The forgetful functor $U : \mathsf{ICat}^{\mathsf{st}} \longrightarrow \mathsf{Cat}$ admits both a left adjoint $\mathcal{F}^{\mathsf{si}}$ and a right adjoint $\mathcal{F}^{\mathsf{cosi}}$*

$$\mathsf{Cat} \underset{\mathcal{F}^{\mathsf{cosi}}}{\overset{\mathcal{F}^{\mathsf{si}}}{\rightleftarrows}} \mathsf{ICat}^{\mathsf{st}} \quad (\text{with } U : \mathsf{ICat}^{\mathsf{st}} \to \mathsf{Cat} \text{ in the middle})$$

given by

$$\mathcal{F}^{\mathsf{si}}(\mathsf{C}) = \Big(\mathsf{C} \amalg \mathsf{C},\ \ \mathsf{C} \amalg \mathsf{C} \xrightarrow{\ twist\ } \mathsf{C} \amalg \mathsf{C}\Big),$$

$$\mathcal{F}^{\mathsf{cosi}}(\mathsf{C}) = \Big(\mathsf{C} \times \mathsf{C},\ \ \mathsf{C} \times \mathsf{C} \xrightarrow{\ twist\ } \mathsf{C} \times \mathsf{C}\Big)$$

for each small category C.

PROOF. To obtain the adjunctions $\mathcal{F}^{\mathsf{si}} \dashv U \dashv \mathcal{F}^{\mathsf{cosi}}$, we apply Theorem 2.5.10 to the involutive category $(\mathsf{Cat}, \mathrm{Id}, \mathrm{Id}_{\mathrm{Id}})$. This is possible because Cat has all products and coproducts. Then we use Lemma 3.1.6 to identify its category of involutive objects with the category $\mathsf{ICat}^{\mathsf{st}}$ of small strict involutive categories. □

DEFINITION 3.1.8. We call $\mathcal{F}^{\mathsf{si}}(\mathsf{C})$ and $\mathcal{F}^{\mathsf{cosi}}(\mathsf{C})$ the *free strict involutive category* and the *cofree strict involutive category* generated by C, respectively. We write

$$U\mathcal{F}^{\mathsf{si}}(\mathsf{C}) = \mathsf{C}_0 \amalg \mathsf{C}_1$$

with C_0 and C_1 indicating the two copies of C.

EXAMPLE 3.1.9. Suppose **1** is a category with one object and only the identity morphism. Then the free strict involutive category on one object $\mathcal{F}^{\mathsf{si}}(\mathbf{1})$ is a discrete category with two objects, which are interchanged by the strict involutive functor. On the other hand, the cofree strict involutive category on one object $\mathcal{F}^{\mathsf{cosi}}(\mathbf{1})$ is again a discrete category with one object. ◇

3.2 Free Involutive Categories

In Proposition 3.1.7 above, we described the free *strict* involutive category $\mathcal{F}^{\mathsf{si}}(\mathsf{C})$ generated by a category C. Recall from Definition 3.1.1 the category ICat_0 of small involutive categories and strict involutive functors between them. The purpose of this section is to describe the free involutive category generated by a category as an explicit left adjoint of the forgetful functor $U : \mathsf{ICat}_0 \longrightarrow \mathsf{Cat}$ that forgets about the involutive structure. First we define what this would-be left adjoint does on a category. We write $\mathbb{Z}_{\geq 0}$ for the set of non-negative integers.

DEFINITION 3.2.1. Suppose C is a category. Define a category $\mathcal{F}^{\mathsf{icat}}(\mathsf{C})$ as follows.

Objects: $\mathsf{Ob}(\mathcal{F}^{\mathsf{icat}}(\mathsf{C})) = \mathbb{Z}_{\geq 0} \times \mathsf{Ob}(\mathsf{C})$, so an object in $\mathcal{F}^{\mathsf{icat}}(\mathsf{C})$ is an ordered pair (m, X) with m a non-negative integer and X an object in C.

Morphisms: For objects (m, X) and (n, Y) in $\mathcal{F}^{\mathsf{icat}}(\mathsf{C})$, define the morphism set

$$\mathcal{F}^{\mathsf{icat}}(\mathsf{C})\big((m, X), (n, Y)\big) = \begin{cases} \mathsf{C}(X, Y) & \text{if } m \equiv n \pmod 2, \\ \varnothing & \text{otherwise.} \end{cases}$$

Identity: For each object (m, X) in $\mathcal{F}^{\mathsf{icat}}(\mathsf{C})$, its identity morphism is $\mathrm{Id}_X \in \mathsf{C}(X, X)$.

Composition: Suppose given morphisms

$$(m, X) \xrightarrow{f} (n, Y) \xrightarrow{g} (l, Z)$$

in $\mathcal{F}^{\mathsf{icat}}(\mathsf{C})$, so $m \equiv n \equiv l \pmod 2$, $f \in \mathsf{C}(X, Y)$, and $g \in \mathsf{C}(Y, Z)$. The composite of these two morphisms in $\mathcal{F}^{\mathsf{icat}}(\mathsf{C})$ is defined as the composite $gf \in \mathsf{C}(X, Z)$.

Involution Functor: Define a functor

$$I_{\mathsf{icat}} : \mathcal{F}^{\mathsf{icat}}(\mathsf{C}) \longrightarrow \mathcal{F}^{\mathsf{icat}}(\mathsf{C})$$

by

$$I_{\mathsf{icat}}(m, X) = (m+1, X) \quad \text{for} \quad (m, X) \in \mathcal{F}^{\mathsf{icat}}(\mathsf{C}),$$

$$I_{\mathsf{icat}}\Big((m, X) \xrightarrow{f} (n, Y) \Big) = \Big((m+1, X) \xrightarrow{f} (n+1, Y) \Big)$$

for each morphism $f \in \mathcal{F}^{\mathsf{icat}}\big((m, X), (n, Y)\big)$ with $m \equiv n \pmod 2$.

Unit of Involution: Define a natural transformation

$$\iota^{\mathsf{icat}} : \mathrm{Id} \longrightarrow I^2_{\mathsf{icat}}$$

by setting

$$\Big((m, X) \xrightarrow{\iota^{\mathsf{icat}}_{(m,X)}} (m+2, X) \Big) = \mathrm{Id}_X \in \mathsf{C}(X, X)$$

for each object $(m, X) \in \mathcal{F}^{\mathsf{icat}}(\mathsf{C})$.

LEMMA 3.2.2. *For each category* C, *the triple*

$$\big(\mathcal{F}^{\mathsf{icat}}(\mathsf{C}), I_{\mathsf{icat}}, \iota^{\mathsf{icat}}\big)$$

is an involutive category.

PROOF. The functoriality of I_{icat}, the naturality of ι^{icat}, and that ι^{icat} is a natural isomorphism, all follow directly from their definitions. The triangle identity (2.1.2) holds because both $I_{\mathsf{icat}}\iota^{\mathsf{icat}}_{(m,X)}$ and $\iota^{\mathsf{icat}}_{I_{\mathsf{icat}}(m,X)}$ are equal to Id_X. □

EXAMPLE 3.2.3. If **1** denotes a category with one object and only the identity morphism, then

$$\mathsf{Ob}\big(\mathcal{F}^{\mathsf{icat}}(\mathbf{1})\big) = \mathbb{Z}_{\geq 0}.$$

There is a unique morphism $m \longrightarrow n$ in $\mathcal{F}^{\mathsf{icat}}(\mathbf{1})$ if and only if $m \equiv n \pmod 2$, in which case the morphism is an isomorphism. The involution functor I on $\mathcal{F}^{\mathsf{icat}}(\mathbf{1})$ sends:

- an object m to the object $m+1$;
- the unique morphism $m \longrightarrow n$, when $m \equiv n \pmod 2$, to the unique morphism $m+1 \longrightarrow n+1$.

The unit $\iota : \mathrm{Id} \xrightarrow{\cong} I^2$ at an object m is the unique isomorphism $m \xrightarrow{\cong} m+2$. ◇

EXAMPLE 3.2.4. Suppose $\mathbf{2} = \{\, 0 \longrightarrow 1 \,\}$ is a category with two objects and one non-identity morphism. Then $\mathcal{F}^{\mathsf{icat}}(\mathbf{2})$ has object set

$$\mathsf{Ob}\big(\mathcal{F}^{\mathsf{icat}}(\mathbf{2})\big) = \mathbb{Z}_{\geq 0} \times \{0, 1\}$$

and morphism sets

$$\mathcal{F}^{\mathsf{icat}}(\mathbf{2})\big((m,a),(n,b)\big) = \begin{cases} \{*\} & \text{if } m \equiv n \pmod 2 \text{ and } (a,b) \neq (1,0), \\ \varnothing & \text{otherwise.} \end{cases}$$

The involution functor I on $\mathcal{F}^{\mathsf{icat}}(\mathbf{2})$ sends:

- an object $(m,a) \in \mathbb{Z}_{\geq 0} \times \{0,1\}$ to the object $(m+1,a)$;
- the unique morphism $(m,a) \longrightarrow (n,b)$, when $m \equiv n \pmod 2$ and $(a,b) \neq (1,0)$, to the unique morphism $(m+1,a) \longrightarrow (n+1,b)$.

The unit

$$\iota_{(m,a)} : (m,a) \xrightarrow{\cong} I^2(m,a) = (m+2,a)$$

is the unique isomorphism. ◇

Next we define the action of $\mathcal{F}^{\mathsf{icat}}$ on functors.

DEFINITION 3.2.5. For a functor $G : \mathsf{C} \longrightarrow \mathsf{D}$, define the functor

$$\mathcal{F}^{\mathsf{icat}}(G) : \mathcal{F}^{\mathsf{icat}}(\mathsf{C}) \longrightarrow \mathcal{F}^{\mathsf{icat}}(\mathsf{D})$$

as the assignment

$$\mathbb{Z}_{\geq 0} \times \mathsf{Ob}(\mathsf{C}) \xrightarrow{\mathrm{Id} \times G} \mathbb{Z}_{\geq 0} \times \mathsf{Ob}(\mathsf{D})$$

on objects. For a morphism $f \in \mathcal{F}^{\mathsf{icat}}(\mathsf{C})\big((m,X),(n,Y)\big)$, with $m \equiv n \pmod 2$ and $f \in \mathsf{C}(X,Y)$, define the morphism

$$\mathcal{F}^{\mathsf{icat}}(G)(f) = G(f) \in \mathsf{D}\big(G(X), G(Y)\big) = \mathcal{F}^{\mathsf{icat}}(\mathsf{D})\big((m, G(X)), (n, G(Y))\big).$$

LEMMA 3.2.6. *For each functor $G : \mathsf{C} \longrightarrow \mathsf{D}$, $\mathcal{F}^{\mathsf{icat}}(G)$ is a strict involutive functor. Moreover,*

$$\mathcal{F}^{\mathsf{icat}} : \mathsf{Cat} \longrightarrow \mathsf{ICat}_0$$

defines a functor.

PROOF. The functoriality of $\mathcal{F}^{\mathsf{icat}}(G)$ follows directly from its definition. The equality

$$\mathcal{F}^{\mathsf{icat}}(G) \circ I_{\mathsf{icat}} = I_{\mathsf{icat}} \circ \mathcal{F}^{\mathsf{icat}}(G) : \mathcal{F}^{\mathsf{icat}}(\mathsf{C}) \longrightarrow \mathcal{F}^{\mathsf{icat}}(\mathsf{D})$$

holds because both composite functors send an object $(m, X) \in \mathcal{F}^{\mathsf{icat}}(\mathsf{C})$ to the object $\big(m+1, G(X)\big) \in \mathcal{F}^{\mathsf{icat}}(\mathsf{D})$, and similar for morphisms. The equality

$$\mathcal{F}^{\mathsf{icat}}(G)\iota^{\mathsf{icat}}_{(m,X)} = \iota^{\mathsf{icat}}_{\mathcal{F}^{\mathsf{icat}}(G)(m,X)}$$

holds because both sides are given by $\mathrm{Id}_{G(X)} \in \mathsf{D}\big(G(X), G(X)\big)$. So by Lemma 3.1.2, $\mathcal{F}^{\mathsf{icat}}(G)$ is a strict involutive functor. That $\mathcal{F}^{\mathsf{icat}}$ preserves identity functors and composition of functors also follows directly from its definition. □

Next we define the morphism that will serve as the unit of the desired adjunction.

DEFINITION 3.2.7. For each category C, define the functor

$$\eta_{\mathsf{C}} : \mathsf{C} \longrightarrow U\mathcal{F}^{\mathsf{icat}}(\mathsf{C})$$

by

$$\eta_{\mathsf{C}}(X) = (0, X) \quad \text{for} \quad X \in \mathsf{Ob}(\mathsf{C}),$$
$$\eta_{\mathsf{C}}\big(f \in \mathsf{C}(X, Y)\big) = f \in \mathcal{F}^{\mathsf{icat}}(\mathsf{C})\big((0, X), (0, Y)\big) = \mathsf{C}(X, Y).$$

Here is the main result of this section, which says that $\mathcal{F}^{\mathsf{icat}}$ is the free involutive category functor.

THEOREM 3.2.8. *There is an adjunction*

$$\mathsf{Cat} \underset{U}{\overset{\mathcal{F}^{\mathsf{icat}}}{\rightleftarrows}} \mathsf{ICat}_0$$

whose unit is given in each component by η_{C} in Definition 3.2.7.

PROOF. Suppose C is a category. We will prove that η_{C} satisfies the following universal property for the unit of an adjunction:

For each functor $F : \mathsf{C} \longrightarrow \mathsf{D}$ with $(\mathsf{D}, I_{\mathsf{D}}, \iota^{\mathsf{D}})$ an involutive category, there exists a unique strict involutive functor

$$\overline{F} : (\mathcal{F}^{\mathsf{icat}}(\mathsf{C}), I_{\mathsf{icat}}, \iota^{\mathsf{icat}}) \longrightarrow (\mathsf{D}, I_{\mathsf{D}}, \iota^{\mathsf{D}})$$

such that the diagram

$$\begin{array}{ccc} & & U\mathcal{F}^{\mathsf{icat}}(\mathsf{C}) \\ & \overset{\eta_{\mathsf{C}}}{\nearrow} & \big\downarrow U\overline{F} \\ \mathsf{C} & \xrightarrow[F]{} & \mathsf{D} \end{array}$$

is commutative.

The desired functor $\overline{F}$ is defined as follows.

- For an object $(m, X) \in \mathcal{F}^{\mathsf{icat}}(\mathsf{C})$, we define

$$\overline{F}(m, X) = I_{\mathsf{D}}^m F(X) \in \mathsf{D},$$

where I_{D}^0 means Id_{D}.

- For a morphism $f \in \mathcal{F}^{\mathsf{icat}}(\mathsf{C})\big((m, X), (n, Y)\big)$, with $m \equiv n \pmod 2$ and $f \in \mathsf{C}(X, Y)$, we define the morphism

$$\overline{F}(f) \in \mathsf{D}\big(I_{\mathsf{D}}^m F(X), I_{\mathsf{D}}^n F(Y)\big)$$

as the composite

$$I_{\mathsf{D}}^m F(X) \xrightarrow{I_{\mathsf{D}}^m F(f)} I_{\mathsf{D}}^m F(Y) \xrightarrow{(\iota^{\mathsf{D}})^{(n-m)/2}} I_{\mathsf{D}}^n F(Y) \tag{3.2.9}$$

(the whole composite being $\overline{F}(f)$) in D. Here $(\iota^{\mathsf{D}})^{(n-m)/2}$ is the composite

$$I_{\mathsf{D}}^m F(Y) \xrightarrow{\iota^{\mathsf{D}}_{I_{\mathsf{D}}^m F(Y)}} I_{\mathsf{D}}^{m+2} F(Y) \xrightarrow{\iota^{\mathsf{D}}_{I_{\mathsf{D}}^{m+2} F(Y)}} \cdots \xrightarrow{\iota^{\mathsf{D}}_{I_{\mathsf{D}}^{n-2} F(Y)}} I_{\mathsf{D}}^n F(Y)$$

if $m \leq n$, and similarly with $(\iota^{\mathsf{D}})^{-1}$ if $m > n$.

It follows from the definition that $\overline{F}$ preserves identity morphisms. That $\overline{F}$ preserves compositions of morphisms follows from the naturality of the unit ι^{D}. Moreover, by construction $\overline{F}$ extends F in the sense that $U\overline{F} \circ \eta_{\mathsf{C}} = F$.

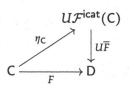

To see that $\overline{F}$ is a strict involutive functor, we first check that

$$\overline{F} I_{\mathsf{icat}} = I_{\mathsf{D}} \overline{F}.$$

For each object $(m, X) \in \mathcal{F}^{\mathsf{icat}}(\mathsf{C})$, there are equalities

$$\overline{F} I_{\mathsf{icat}}(m, X) = \overline{F}(m+1, X) = I_{\mathsf{D}}^{m+1} F(X) = I_{\mathsf{D}} \overline{F}(m, X).$$

For each morphism $f \in \mathcal{F}^{\mathsf{icat}}(\mathsf{C})\big((m, X), (n, Y)\big)$, with $m \equiv n \pmod 2$ and $f \in \mathsf{C}(X, Y)$, the desired equality

$$\overline{F} I_{\mathsf{icat}}(f) = I_{\mathsf{D}} \overline{F}(f)$$

holds if $m = n$ because both sides are equal to $I_{\mathsf{D}}^{m+1} F(f)$. If $m < n$, then the desired equality is the commutativity of the diagram

$$\begin{array}{ccccccccc}
I_{\mathsf{D}}^{m+1}F(X) & & & & & & & & \\
\big\downarrow{\scriptstyle I_{\mathsf{D}}^{m+1}F(f)} & & & & & & & & \\
I_{\mathsf{D}}^{m+1}F(Y) & \xrightarrow{\iota^{\mathsf{D}}_{I_{\mathsf{D}}^{m+1}F(Y)}} & I_{\mathsf{D}}^{m+3}F(Y) & \xrightarrow{\iota^{\mathsf{D}}_{I_{\mathsf{D}}^{m+3}F(Y)}} & \cdots & \xrightarrow{\iota^{\mathsf{D}}_{I_{\mathsf{D}}^{n-3}F(Y)}} & I_{\mathsf{D}}^{n-1}F(Y) & \xrightarrow{\iota^{\mathsf{D}}_{I_{\mathsf{D}}^{n-1}F(Y)}} & I_{\mathsf{D}}^{n+1}F(Y) \\
\| & & \| & & & & \| & & \| \\
I\big(I^m F(Y)\big) & \xrightarrow{I\iota^{\mathsf{D}}_{I^m F(Y)}} & I\big(I^{m+2} F(Y)\big) & \xrightarrow{I\iota^{\mathsf{D}}_{I^{m+2} F(Y)}} & \cdots & \xrightarrow{I\iota^{\mathsf{D}}_{I^{n-4} F(Y)}} & I\big(I^{n-2} F(Y)\big) & \xrightarrow{I\iota^{\mathsf{D}}_{I^{n-2} F(Y)}} & I\big(I^{n} F(Y)\big)
\end{array}$$

in which $I = I_{\mathsf{D}}$ in the bottom row. The bottom squares are commutative by the triangle identity (2.1.2) in D. A similar argument works when $m > n$, using $(\iota^{\mathsf{D}})^{-1}$ in place of ι^{D}.

Next, for each object $(m, X) \in \mathcal{F}^{\mathsf{icat}}(\mathsf{C})$, the equality

$$\overline{F} \iota^{\mathsf{icat}}_{(m,X)} = \iota^{\mathsf{D}}_{\overline{F}(m,X)} \in \mathsf{D}\big(I_{\mathsf{D}}^{m} F(X), I_{\mathsf{D}}^{m+2} F(X)\big)$$

holds because both sides are equal to the unit $\iota^{\mathsf{D}}_{I_{\mathsf{D}}^m F(X)}$. Therefore, by Lemma 3.1.2, $\overline{F}$ is a strict involutive functor.

In Exercise 1 in Section 3.7, the reader is asked to check the uniqueness of $\overline{F}$ as a strict involutive functor satisfying $U\overline{F} \circ \eta_{\mathsf{C}} = F$. □

DEFINITION 3.2.10. We call $\mathcal{F}^{\mathsf{icat}}(\mathsf{C})$ the *free involutive category* generated by a category C.

EXAMPLE 3.2.11. The involutive category $\mathcal{F}^{\mathsf{icat}}(\mathbf{1})$ in Example 3.2.3 is the free involutive category generated by one object. ◇

EXAMPLE 3.2.12. The involutive category $\mathcal{F}^{\mathsf{icat}}(\mathbf{2})$ in Example 3.2.4 is the free involutive category generated by one morphism. ◇

3.3 Strictification of Free Involutive Categories

For each small category C, the unit of the adjunction $\mathcal{F}^{\mathsf{si}} \dashv U$ in Proposition 3.1.7 is a functor

$$\eta^{\mathsf{si}}_{\mathsf{C}} : \mathsf{C} \longrightarrow U\mathcal{F}^{\mathsf{si}}(\mathsf{C}),$$

where $\mathcal{F}^{\mathsf{si}}(\mathsf{C})$ is the free strict involutive category generated by C. By the adjunction $\mathcal{F}^{\mathsf{icat}} \dashv U$ in Theorem 3.2.8, there exists a unique strict involutive functor

$$\mathcal{F}^{\mathsf{icat}}(\mathsf{C}) \xrightarrow{\overline{\eta^{\mathsf{si}}_{\mathsf{C}}}} \mathcal{F}^{\mathsf{si}}(\mathsf{C}) \;,$$

where $\mathcal{F}^{\mathsf{icat}}(\mathsf{C})$ is the free involutive category generated by C, such that the diagram

$$\begin{array}{ccc} & & U\mathcal{F}^{\mathsf{icat}}(\mathsf{C}) \\ & \overset{\eta_{\mathsf{C}}}{\nearrow} & \big\downarrow U\overline{\eta^{\mathsf{si}}_{\mathsf{C}}} \\ \mathsf{C} & \xrightarrow[\eta^{\mathsf{si}}_{\mathsf{C}}]{} & U\mathcal{F}^{\mathsf{si}}(\mathsf{C}) \end{array}$$

is commutative. The purpose of this section is to observe that the free involutive category $\mathcal{F}^{\mathsf{icat}}(\mathsf{C})$ and the free strict involutive category $\mathcal{F}^{\mathsf{si}}(\mathsf{C})$ generated by C are equivalent via the functor $\overline{\eta^{\mathsf{si}}_{\mathsf{C}}}$.

COROLLARY 3.3.1. *For each small category C, the strict involutive functor $\overline{\eta^{\mathsf{si}}_{\mathsf{C}}}$ is an involutive equivalence.*

PROOF. Since $\overline{\eta^{\mathsf{si}}_{\mathsf{C}}}$ is an involutive functor, to show that it is an involutive equivalence, by Corollary 2.4.2 it suffices to show that it is an equivalence of categories, i.e., fully faithful and essentially surjective. The underlying category of $\mathcal{F}^{\mathsf{si}}(\mathsf{C})$ is the coproduct $\mathsf{C}_0 \amalg \mathsf{C}_1$ of two copies of C. By the explicit construction of the functor $\overline{F}$ in the proof of Theorem 3.2.8, the functor $\overline{\eta^{\mathsf{si}}_{\mathsf{C}}}$ sends an object $(m, X) \in \mathcal{F}^{\mathsf{icat}}(\mathsf{C})$ to the object $X \in \mathsf{C}_{\epsilon(m)} \subseteq \mathcal{F}^{\mathsf{si}}(\mathsf{C})$, where

$$\epsilon(m) = \begin{cases} 0 & \text{if } m \text{ is even,} \\ 1 & \text{if } m \text{ is odd.} \end{cases}$$

So the functor $\overline{\eta^{\mathsf{si}}_{\mathsf{C}}}$ is surjective on objects.

On morphism sets, $\overline{\eta^{\mathsf{si}}_{\mathsf{C}}}$ is the identity map of

$$\begin{aligned}\mathcal{F}^{\mathsf{icat}}(\mathsf{C})\big((m,X),(n,Y)\big) &= \mathcal{F}^{\mathsf{si}}(\mathsf{C})\big(X \in \mathsf{C}_{\epsilon(m)}, Y \in \mathsf{C}_{\epsilon(n)}\big)\\ &= \begin{cases}\mathsf{C}(X,Y) & \text{if } m \equiv n \pmod 2,\\ \varnothing & \text{otherwise.}\end{cases}\end{aligned}$$

So the functor $\overline{\eta^{\mathsf{si}}_{\mathsf{C}}}$ is fully faithful. □

EXAMPLE 3.3.2. For the one-object discrete category **1**, the free strict involutive category $\mathcal{F}^{\mathsf{si}}(\mathbf{1})$ is the discrete category with two objects $\{0, 1\}$ by Example 3.1.9. The free involutive category $\mathcal{F}^{\mathsf{icat}}(\mathbf{1})$ is discussed in Example 3.2.3. The involutive equivalence

$$\overline{\eta^{\mathsf{si}}_{\mathbf{1}}} : \mathcal{F}^{\mathsf{icat}}(\mathbf{1}) \longrightarrow \mathcal{F}^{\mathsf{si}}(\mathbf{1})$$

sends an object $m \in \mathcal{F}^{\mathsf{icat}}(\mathbf{1})$ to $\epsilon(m) \in \{0, 1\}$, and the unique morphism $m \longrightarrow n$, when $m \equiv n \pmod 2$, to the identity morphism of $\epsilon(m)$. ◇

3.4 Involutive Categories as Monadic Algebras

The purpose of this section is to observe that small involutive categories are exactly the algebras over a monad in the category Cat of small categories and functors. The reader may review the definitions regarding monads in Section 1.3. We will use the adjunction $\mathcal{F}^{\mathsf{icat}} \dashv U$ in Theorem 3.2.8 with unit $\eta : \mathrm{Id}_{\mathsf{Cat}} \longrightarrow U\mathcal{F}^{\mathsf{icat}}$ in Definition 3.2.7. For a small category C, there are object sets

$$\begin{aligned}\mathsf{Ob}\big(U\mathcal{F}^{\mathsf{icat}}(\mathsf{C})\big) &= \mathbb{Z}_{\geq 0} \times \mathsf{Ob}(\mathsf{C}),\\ \mathsf{Ob}\big((U\mathcal{F}^{\mathsf{icat}})^2(\mathsf{C})\big) &= \mathbb{Z}_{\geq 0} \times \mathbb{Z}_{\geq 0} \times \mathsf{Ob}(\mathsf{C}).\end{aligned}$$

The functor $U\mathcal{F}^{\mathsf{icat}}$ and the unit η are part of our desired monad. We now define its multiplication.

DEFINITION 3.4.1. We define a natural transformation

$$\mu : (U\mathcal{F}^{\mathsf{icat}})^2 \longrightarrow U\mathcal{F}^{\mathsf{icat}}$$

as follows. For each small category C, the component

$$\mu_{\mathsf{C}} : (U\mathcal{F}^{\mathsf{icat}})^2(\mathsf{C}) \longrightarrow U\mathcal{F}^{\mathsf{icat}}(\mathsf{C})$$

is the functor that sends:

- an object $(n, m, X) \in (U\mathcal{F}^{\mathsf{icat}})^2(\mathsf{C})$, with $n, m \in \mathbb{Z}_{\geq 0}$ and $X \in \mathsf{C}$, to the object $(n+m, X) \in U\mathcal{F}^{\mathsf{icat}}(\mathsf{C})$;
- a morphism

$$f \in (U\mathcal{F}^{\mathsf{icat}})^2(\mathsf{C})\big((n, m, X), (q, p, Y)\big)$$

to the morphism

$$f \in U\mathcal{F}^{\mathsf{icat}}(\mathsf{C})\big((n+m, X), (q+p, Y)\big).$$

Recall from Definition 3.1.1 the category ICat_0 of small involutive categories and strict involutive functors.

THEOREM 3.4.2. *The tuple*

$$\big(U\mathcal{F}^{\mathsf{icat}}, \mu, \eta\big)$$

is a monad in Cat *whose category of algebras is isomorphic to the category* ICat_0.

PROOF. The existence of the monad $\big(U\mathcal{F}^{\mathsf{icat}}, \mu, \eta\big)$ follows from the adjunction in Theorem 3.2.8. For the rest of this proof, let us write T for both the functor $U\mathcal{F}^{\mathsf{icat}} : \mathsf{Cat} \longrightarrow \mathsf{Cat}$ and the monad $\big(U\mathcal{F}^{\mathsf{icat}}, \mu, \eta\big)$ in Cat.

We now identify T-algebras and their morphisms with small involutive categories and strict involutive functors, respectively. For a T-algebra $\big(\mathsf{C}, \theta : T\mathsf{C} \longrightarrow \mathsf{C}\big)$, with C a small category and θ its T-algebra structure morphism, the associativity and unity axioms in (1.3.4) mean the equalities

$$\begin{aligned} \theta(n+m, X) &= \theta\big(n, \theta(m, X)\big), \\ \theta(0, X) &= X \end{aligned}$$

for $n, m \in \mathbb{Z}_{\geq 0}$ and $X \in \mathsf{Ob}(\mathsf{C})$, and similarly for morphisms. We now define a functor $I_{\mathsf{C}} : \mathsf{C} \longrightarrow \mathsf{C}$ by

$$I_{\mathsf{C}}(X) = \theta(1, X) \quad \text{for} \quad X \in \mathsf{Ob}(\mathsf{C}),$$

$$I_{\mathsf{C}}(f) = \theta\Big((1, X) \xrightarrow{f} (1, Y) \Big) \quad \text{for} \quad f \in \mathsf{C}(X, Y).$$

Then we define a natural transformation $\iota^{\mathsf{C}} : \mathrm{Id}_{\mathsf{C}} \longrightarrow I_{\mathsf{C}}^2$ whose component

$$\theta(0, X) = X \xrightarrow{\iota^{\mathsf{C}}_X} I_{\mathsf{C}}^2(X) = \theta\big(1, \theta(1, X)\big) = \theta(2, X)$$

for $X \in \mathsf{Ob(C)}$ is the morphism

$$\theta\Big((0, X) \xrightarrow{\ \mathrm{Id}_X\ } (2, X) \Big).$$

An inspection shows that $(\mathsf{C}, I_{\mathsf{C}}, \iota^{\mathsf{C}})$ is an involutive category. In other words, ι^{C} is a natural isomorphism that satisfies the equality

$$I_{\mathsf{C}}\iota^{\mathsf{C}} = \iota^{\mathsf{C}}_{I_{\mathsf{C}}} : I_{\mathsf{C}} \xrightarrow{\ \cong\ } I^3_{\mathsf{C}}.$$

If $G : (\mathsf{C}, \theta^{\mathsf{C}}) \longrightarrow (\mathsf{D}, \theta^{\mathsf{D}})$ is a morphism of T-algebras, then $G : \mathsf{C} \longrightarrow \mathsf{D}$ is a functor that strictly commutes with the structure morphisms in the sense that

$$G \circ \theta^{\mathsf{C}} = \theta^{\mathsf{D}} \circ T(G) : T\mathsf{C} \longrightarrow \mathsf{D}.$$

Using Lemma 3.1.2, this implies that

$$G : (\mathsf{C}, I_{\mathsf{C}}, \iota^{\mathsf{C}}) \longrightarrow (\mathsf{D}, I_{\mathsf{D}}, \iota^{\mathsf{D}})$$

is a strict involutive functor.

Conversely, suppose $(\mathsf{D}, I_{\mathsf{D}}, \iota^{\mathsf{D}})$ is a small involutive category. We define a structure morphism $\phi^{\mathsf{D}} : T\mathsf{D} \longrightarrow \mathsf{D}$ as the functor that sends:

- an object $(m, X) \in T\mathsf{D}$ to the object $I^m_{\mathsf{D}} X \in \mathsf{D}$ with $I^0_{\mathsf{D}} = \mathrm{Id}_{\mathsf{D}}$;
- a morphism $f : (m, X) \longrightarrow (n, Y) \in T\mathsf{D}$ to the composite

$$\begin{array}{ccc} I^m_{\mathsf{D}} X & \xrightarrow{\ \phi^{\mathsf{D}}(f) = \overline{\mathrm{Id}_{\mathsf{D}}(f)}\ } & I^n_{\mathsf{D}} Y \\ {\scriptstyle I^m_{\mathsf{D}}(f)}\downarrow & & \| \\ I^m_{\mathsf{D}} Y & \xrightarrow{\ (\iota^{\mathsf{D}})^{(n-m)/2}\ } & I^n_{\mathsf{D}} Y \end{array}$$

in (3.2.9) with $F = \mathrm{Id}_{\mathsf{D}}$.

We need to check the associativity and unity axioms (1.3.4) for $(\mathsf{D}, \phi^{\mathsf{D}})$. The associativity axiom on objects follows from the equalities

$$\begin{aligned} \phi^{\mathsf{D}}\mu(n, m, X) &= \phi^{\mathsf{D}}(n + m, X) \\ &= I^{n+m}_{\mathsf{D}} X \\ &= I^n_{\mathsf{D}}(I^m_{\mathsf{D}} X) \\ &= \phi^{\mathsf{D}}(n, \phi^{\mathsf{D}}(m, X)), \end{aligned}$$

and similarly for morphisms. The unity axiom is immediate from the definition. Therefore, $(\mathsf{D}, \phi^{\mathsf{D}})$ is a T-algebra.

If $G : (\mathsf{C}, I_{\mathsf{C}}, \iota^{\mathsf{C}}) \longrightarrow (\mathsf{D}, I_{\mathsf{D}}, \iota^{\mathsf{D}})$ is a strict involutive functor between small involutive categories, then Lemma 3.1.2 implies the equality

$$G \circ \phi^{\mathsf{C}} = \phi^{\mathsf{D}} \circ T(G) : T\mathsf{C} \longrightarrow \mathsf{D}.$$

So $G : (\mathsf{C}, \phi^{\mathsf{C}}) \longrightarrow (\mathsf{D}, \phi^{\mathsf{D}})$ is a morphism of T-algebras.

An inspection of the above constructions

$$(\mathsf{C}, \theta) \longmapsto (\mathsf{C}, I_{\mathsf{C}}, \iota) \quad \text{and} \quad (\mathsf{D}, I_{\mathsf{D}}, \iota^{\mathsf{D}}) \longmapsto (\mathsf{D}, \phi^{\mathsf{D}})$$

shows that they are both functorial, and that they are inverses of each other. □

3.5 Presentability of Involutive Categories

The purpose of this section is to observe that the category ICat_0 of small involutive categories and strict involutive functors is locally finitely presentable. Our references for locally finitely presentable categories are [AR94, Bor94a, Bor94b]. We first recall some relevant definitions regarding directed colimits, finitely presentable objects, and generators.

DEFINITION 3.5.1. A partially ordered set is called *directed* if each pair of elements has an upper bound.

(1) A *directed category* is a directed partially ordered set, regarded as a small category.
(2) A *directed colimit* in a category C is a colimit of a functor $\mathsf{D} \longrightarrow \mathsf{C}$ with D a directed category.
(3) For a set $\mathcal{X}$ of objects in a category C, a *directed colimit of objects in* $\mathcal{X}$ is a directed colimit $G : \mathsf{D} \longrightarrow \mathsf{C}$ with $G(d) \in \mathcal{X}$ for each object $d \in \mathsf{D}$.

DEFINITION 3.5.2. Suppose C is a category, and $X \in \mathsf{C}$ is an object.

(1) X is called *finitely presentable* if the functor

$$\mathsf{C}(X, -) : \mathsf{C} \longrightarrow \mathsf{Set}$$

preserves directed colimits.
(2) C is called *locally finitely presentable* if it satisfies the following two conditions.

- C has all small colimits.
- C has a set $\mathcal{X}$ of finitely presentable objects such that every object in C is a directed colimit of objects in $\mathcal{X}$.

(3) Two monomorphisms $f : R \longrightarrow X$ and $g : S \longrightarrow X$ in C are *equivalent* if there exists an isomorphism $h : R \xrightarrow{\cong} S$ such that $f = gh$.
(4) A *sub-object of* X is an equivalence class of monomorphisms with codomain X. The sub-object represented by a monomorphism $f : R \longrightarrow X$ is also denoted by the domain R.
(5) A sub-object of X is *proper* if it has a representing monomorphism $f : R \longrightarrow X$ that is not an isomorphism.
(6) A *generator* of C is a set of objects $\mathcal{X} = \{X_j\}_{j\in J}$ in C such that, for each pair of distinct morphisms $f, g : A \longrightarrow B$ in C, there exists a morphism

$$h_j : X_j \longrightarrow A$$

for some index $j \in J$ such that

$$fh_j \neq gh_j : X_j \longrightarrow B.$$

(7) A *strong generator* of C is a generator $\mathcal{X} = \{X_j\}_{j\in J}$ such that, for each object B and each proper sub-object A of B, there exists a morphism

$$h_j : X_j \longrightarrow B$$

for some index $j \in J$ that does not factor through A.

To show that ICat_0 is locally finitely presentable, we will need the following preliminary observation. Recall the free-forgetful adjunction $\mathcal{F}^{\mathsf{icat}} \dashv U$ in Theorem 3.2.8.

LEMMA 3.5.3. *The forgetful functor* $U : \mathsf{ICat}_0 \longrightarrow \mathsf{Cat}$ *preserves monomorphisms and proper sub-objects.*

PROOF. Suppose $F : \mathsf{C} \longrightarrow \mathsf{D}$ is a monomorphism in ICat_0. To show that the functor $UF : U\mathsf{C} \longrightarrow U\mathsf{D}$ is a monomorphism, suppose $H_1, H_2 : \mathsf{A} \rightrightarrows U\mathsf{C}$ are functors such that

$$(UF)H_1 = (UF)H_2 : \mathsf{A} \longrightarrow U\mathsf{D}.$$

We must show that $H_1 = H_2$. By the adjunction in Theorem 3.2.8 for each $i = 1, 2$ the functor H_i has a unique extension to a strict involutive functor $H_i' : \mathcal{F}^{\mathsf{icat}}(\mathsf{A}) \longrightarrow \mathsf{C}$. Then

$$FH_1' = FH_2' : \mathcal{F}^{\mathsf{icat}}(\mathsf{A}) \longrightarrow \mathsf{D}$$

because:

- Their underlying functors are equal by adjunction, since their restrictions to A are equal as functors.

- The distributive laws of F, H_1', and H_2' are identities.

Since F is a monomorphism, it follows that $H_1' = H_2'$. By adjunction this implies that $H_1 = H_2$.

Each proper sub-object in ICat_0 is represented by a monomorphism $F : \mathsf{C} \longrightarrow \mathsf{D}$ that is not an isomorphism. We already showed that UF is a monomorphism in Cat. If UF is an isomorphism in Cat, then the underlying functor of F, which is UF, is an isomorphism. But then F is an isomorphism, since its distributive law is the identity. Since this violates our assumption on F, we conclude that the monomorphism UF is not an isomorphism in Cat. In other words, UF represents a proper sub-object in Cat. □

The involutive categories in the following definition were described in Examples 3.2.3 and 3.2.4.

DEFINITION 3.5.4. Define the set

$$S = \{\mathcal{F}^{\mathsf{icat}}(\mathbf{1}), \mathcal{F}^{\mathsf{icat}}(\mathbf{2})\}$$

that consists of two small involutive categories. Here:

- $\mathbf{1}$ is the discrete category with one object 0.
- $\mathbf{2} = \{0 \longrightarrow 1\}$ is the category with two objects and one non-identity morphism.

LEMMA 3.5.5. $S = \{\mathcal{F}^{\mathsf{icat}}(\mathbf{1}), \mathcal{F}^{\mathsf{icat}}(\mathbf{2})\}$ *is a strong generator of* ICat_0.

PROOF. To show that S is a generator of ICat_0, suppose $F, G : \mathsf{C} \rightrightarrows \mathsf{D}$ are distinct strict involutive functors between small involutive categories. Then F and G are distinct as functors, since both of them have the identity as the distributive law.

- If $X \in \mathsf{C}$ is an object such that $F(X) \neq G(X)$, then the functor

$$\mathbf{1} \xrightarrow{H} U\mathsf{C} \quad \text{defined by} \quad H(0) = X$$

satisfies

$$(UF)H \neq (UG)H.$$

By the adjunction in Theorem 3.2.8, H extends uniquely to a strict involutive functor $\mathcal{F}^{\mathsf{icat}}(\mathbf{1}) \longrightarrow \mathsf{C}$, whose composites with F and G are different.
- On the other hand, if $f \in \mathsf{C}$ is a morphism such that $F(f) \neq G(f)$, then the functor

$$\mathbf{2} \xrightarrow{H} U\mathsf{C} \quad \text{defined by} \quad H(0 \to 1) = f$$

satisfies

$$(UF)H \neq (UG)H.$$

By the adjunction in Theorem 3.2.8, H extends uniquely to a strict involutive functor $\mathcal{F}^{\mathsf{icat}}(\mathbf{2}) \longrightarrow \mathsf{C}$, whose composites with F and G are different.

Therefore, S is a generator of ICat_0.

To show that S is a strong generator, suppose $F : \mathsf{C} \longrightarrow \mathsf{D}$ is a monomorphism in ICat_0 that represents a proper sub-object of D. By Lemma 3.5.3 the functor $UF : U\mathsf{C} \longrightarrow U\mathsf{D}$ is a monomorphism that represents a proper sub-object of $U\mathsf{D}$ in Cat. Since UF is a monomorphism but not an isomorphism, at least one of the following two statements holds:

(1) There exists an object $X \in U\mathsf{D}$ that is not in the image of UF.
(2) There exists a morphism $f \in U\mathsf{D}$ that is not in the image of UF.

In the first case, the unique extension $H' : \mathcal{F}^{\mathsf{icat}}(\mathbf{1}) \longrightarrow \mathsf{D}$ of the functor

$$\mathbf{1} \xrightarrow{H} U\mathsf{D} \quad \text{defined by} \quad H(0) = X$$

does not factor through F. In the second case, the functor

$$\mathbf{2} \xrightarrow{H} U\mathsf{D} \quad \text{defined by} \quad H(0 \to 1) = f$$

has a unique extension $H' : \mathcal{F}^{\mathsf{icat}}(\mathbf{2}) \longrightarrow \mathsf{D}$ that does not factor through F. □

LEMMA 3.5.6. *The involutive categories $\mathcal{F}^{\mathsf{icat}}(\mathbf{1})$ and $\mathcal{F}^{\mathsf{icat}}(\mathbf{2})$ are finitely presentable objects in ICat_0.*

PROOF. The functor $\mathsf{ICat}_0(\mathcal{F}^{\mathsf{icat}}(\mathbf{1}), -)$ preserves directed colimits because a strict involutive functor $\mathcal{F}^{\mathsf{icat}}(\mathbf{1}) \longrightarrow \mathsf{C}$ in ICat_0 is uniquely determined by the image of the object $0 \in \mathbf{1}$ by the adjunction in Theorem 3.2.8. Similarly, a strict involutive functor $\mathcal{F}^{\mathsf{icat}}(\mathbf{2}) \longrightarrow \mathsf{C}$ in ICat_0 is uniquely determined by the image of the non-identity morphism $0 \longrightarrow 1$ in $\mathbf{2}$ by adjunction. □

PROPOSITION 3.5.7. *The category ICat_0 has all small limits and colimits, and they are created by the forgetful functor $U : \mathsf{ICat}_0 \longrightarrow \mathsf{Cat}$*

PROOF. Now suppose $G : \mathsf{D} \longrightarrow \mathsf{ICat}_0$ is a functor with D a small category. For the composite functor $UG : \mathsf{D} \longrightarrow \mathsf{Cat}$, the colimit

$$\mathsf{C} = \mathrm{colim}_{\mathsf{D}}\, UG \in \mathsf{Cat}$$

exists by Theorem 1.4.9. Its involution functor I_{C} is defined by the commutative diagrams

$$\begin{array}{ccc} UG(d) & \xrightarrow{\text{natural}} & \mathsf{C} \\ {\scriptstyle UI_{G(d)}}\downarrow & & \downarrow{\scriptstyle I_{\mathsf{C}}} \\ UG(d) & \xrightarrow{\text{natural}} & \mathsf{C} \end{array}$$

for $d \in \mathsf{D}$. The required natural isomorphism $\iota^{\mathsf{C}} : \mathrm{Id}_{\mathsf{C}} \xrightarrow{\cong} I^2_{\mathsf{C}}$ is also uniquely defined by those in the involutive categories $G(d)$ for $d \in \mathsf{D}$, since every object in C is represented by an object in some category $UG(d)$. The triangle identity (2.1.2) for $(I_{\mathsf{C}}, \iota^{\mathsf{C}})$ follows from those for $(I_{G(d)}, \iota^{G(d)})$.

By the construction of I_{C}, each natural functor $UG(d) \longrightarrow \mathsf{C}$ for $d \in \mathsf{D}$ yields a strict involutive functor

$$\bigl(G(d), I_{G(d)}, \iota^{G(d)}\bigr) \xrightarrow{\text{natural}} (\mathsf{C}, I_{\mathsf{C}}, \iota^{\mathsf{C}}) \ .$$

That these natural functors and $(\mathsf{C}, I_{\mathsf{C}}, \iota^{\mathsf{C}})$ form a colimit of G follows from:

- the definition of C as a colimit of UG;
- the fact that the involutive structure on C is uniquely defined by those in $G(d)$ for $d \in \mathsf{D}$.

A similar argument proves the existence of limits in ICat_0. □

THEOREM 3.5.8. *The category* ICat_0 *is locally finitely presentable.*

PROOF. By Theorem 1.11 in [AR94], a category is locally finitely presentable if and only if it is cocomplete, and has a strong generator consisting of finitely presentable objects. The category ICat_0 is cocomplete by Proposition 3.5.7. It has a strong generator consisting of finitely presentable objects by Lemma 3.5.5 and Lemma 3.5.6. □

There is a generalization of Theorem 3.5.8 for the category ICat of small involutive categories and involutive functors. First we need a few definitions. The reader may consult [Pin14, SV02] for discussion of cardinals.

DEFINITION 3.5.9. A *regular cardinal* is an infinite cardinal that cannot be expressed as a sum of a smaller number of smaller cardinals. Suppose λ is a regular cardinal, and C is a category.

(1) A partially ordered set $(\mathsf{D}, \leq)$ is called *λ-directed* if every subset of D of cardinality smaller than λ has an upper bound.
(2) A *λ-directed colimit* in C is a colimit of a functor $\mathsf{D} \longrightarrow \mathsf{C}$ with D a λ-directed partially ordered set, regarded as a small category.
(3) For a set $\mathcal{X}$ of objects in C, a *λ-directed colimit of objects in $\mathcal{X}$* is a λ-directed colimit $G : \mathsf{D} \longrightarrow \mathsf{C}$ with $G(d) \in \mathcal{X}$ for each object $d \in \mathsf{D}$.

(4) An object $X \in \mathsf{C}$ is called *λ-presentable* if the functor $\mathsf{C}(X, -)$ preserves λ-directed colimits.
(5) C is called *locally λ-presentable* if it has all small colimits, and has a set $\mathcal{X}$ of λ-presentable objects such that every object in C is a λ-directed colimit of objects in $\mathcal{X}$.
(6) C is called *locally presentable* if it is locally β-presentable for some regular cardinal β.

The proof of the following result is similar to that of Theorem 3.5.8. The reader is asked to supply the proof in Exercise 4 below.

THEOREM 3.5.10. *The category* ICat *is locally presentable.*

3.6 Right Adjoint of the Forgetful Functor

Recall from Proposition 3.1.7 that the forgetful functor $U : \mathsf{ICat}^{\mathsf{st}} \longrightarrow \mathsf{Cat}$, from the category of small strict involutive categories and strict involutive functors to the category of small categories, admits both a left adjoint and a right adjoint. For the category ICat_0 of small involutive categories and strict involutive functors, the forgetful functor $U : \mathsf{ICat}_0 \longrightarrow \mathsf{Cat}$ has a left adjoint by Theorem 3.2.8. The purpose of this section is to observe that this second forgetful functor also admits a right adjoint. First we need some definitions from [Bor94a] Chapter 4.1. Recall from Definition 3.5.2 the concepts of a sub-object and of a generator.

DEFINITION 3.6.1. Suppose C is a category.

(1) C is called *well-powered* if the sub-objects of each object constitute a set.
(2) C is called *cowell-powered* if its opposite category C^{op} is well-powered.
(3) C has a *cogenerator* if C^{op} has a generator.

THEOREM 3.6.2. *The forgetful functor*

$$U : \mathsf{ICat}_0 \longrightarrow \mathsf{Cat}$$

admits a right adjoint.

PROOF. We use the Special Adjoint Functor Theorem in [Bor94a] (Corollary 3.3.5):

> Suppose $G : \mathsf{C} \longrightarrow \mathsf{D}$ is a small-limit-preserving functor such that C is complete and well-powered, and has a cogenerator. Then G has a left adjoint.

To show that U admits a right adjoint, it suffices to show that the functor

$$U^{\mathsf{op}} : \mathsf{ICat}_0^{\mathsf{op}} \longrightarrow \mathsf{Cat}^{\mathsf{op}}$$

satisfies the hypotheses of the Special Adjoint Functor Theorem. We observed in Proposition 3.5.7 that ICat_0 is cocomplete and that U preserves small colimits.

By Lemma 3.5.5 ICat_0 has a strong generator. A highly non-trivial result, [AR94] Theorem 1.58, says that every locally presentable category is cowell-powered. So by Theorem 3.5.8, ICat_0 is cowell-powered. Therefore, U^{op} admits a left adjoint, so U admits a right adjoint. □

3.7 Exercises

(1) Finish the proof of Theorem 3.2.8 by proving that the strict involutive functor $\overline{F}$ satisfying $U\overline{F} \circ \eta_{\mathsf{C}} = F$ is unique. Hint: To check the uniqueness of $\overline{F}$ on morphisms $f \in \mathcal{F}^{\mathsf{icat}}(\mathsf{C})\big((m, X), (n, Y)\big)$, first observe that f factors as the composite

$$\overbrace{(m,X) = I^m_{\mathsf{icat}}(0,X) \xrightarrow{I^m_{\mathsf{icat}} f} I^m_{\mathsf{icat}}(0,Y) = (m,Y) \xrightarrow{\mathrm{Id}_Y} (n,Y)}^{f}.$$

(2) In the proof of Theorem 3.4.2, check that:
- μ in Definition 3.4.1 is a well-defined natural transformation.
- $\big(U\mathcal{F}^{\mathsf{icat}}, \mu, \eta\big)$ is a monad in Cat.
- $(\mathsf{C}, I_{\mathsf{C}}, \iota^{\mathsf{C}})$ is an involutive category.
- (D, ϕ) satisfies the associativity axiom (1.3.4) on morphisms.
- the constructions

$$(\mathsf{C}, \theta) \longmapsto (\mathsf{C}, I_{\mathsf{C}}, \iota) \quad \text{and} \quad (\mathsf{D}, I_{\mathsf{D}}, \iota^{\mathsf{D}}) \longmapsto (\mathsf{D}, \phi^{\mathsf{D}})$$

are inverse isomorphisms

$$\mathsf{Alg}\big(U\mathcal{F}^{\mathsf{icat}}, \mu, \eta\big) \cong \mathsf{ICat}_0.$$

(3) Prove that there is a monad on Cat whose category of algebras is isomorphic to $\mathsf{ICat}^{\mathsf{st}}$, the category of small strict involutive categories and strict involutive functors.

(4) Prove Theorem 3.5.10: ICat is locally presentable. Hint: It is possible for two distinct involutive functors

$$(F, \nu^F) \neq (G, \nu^G) : \mathsf{C} \rightrightarrows \mathsf{D}$$

to be equal as functors, as long as $\nu^F \neq \nu^G$. To construct a strong generator of ICat, consider the set $S = \big\{\mathcal{F}^{\mathsf{icat}}(\mathbf{1}), \mathcal{F}^{\mathsf{icat}}(\mathbf{2})\big\}$ in Definition 3.5.4. Then use [AR94] Theorem 1.20, which states that a category is locally λ-presentable if and only if it is cocomplete, and has a strong generator consisting of λ-presentable objects.

Chapter 4
Involutive Monoidal Categories

In this chapter, we develop the involutive versions of (symmetric) monoidal categories, (commutative) monoids, monads, and algebras over a monad. In an involutive symmetric monoidal category, in addition to involutive commutative monoids, there is also a notion of a reversing involutive monoid, which incorporates the symmetry isomorphism in a way that is different from a commutative monoid. Reversing involutive monoids are important because they include complex unital $*$-algebras and their differential graded analogues as examples.

4.1 Definitions and Examples

The purposes of this section are (i) to define an involutive monoidal category, (ii) to prove a few of their basic properties, and (iii) to provide some examples.

Recall from Section 1.2 that a monoidal category is a category C equipped with

- a monoidal product $\otimes : \mathsf{C} \times \mathsf{C} \longrightarrow \mathsf{C}$,
- a monoidal unit $\mathbb{1} \in \mathsf{C}$,
- an associativity isomorphism α, and
- left/right unit isomorphisms λ and ρ,

that satisfies the unity axioms and the pentagon axiom. Recall from Definition 1.2.9 that a monoidal functor $(F, F_2, F_0) : \mathsf{C} \longrightarrow \mathsf{D}$ between two monoidal categories is a functor $F : \mathsf{C} \longrightarrow \mathsf{D}$ equipped with a morphism $F_0 : \mathbb{1}^{\mathsf{D}} \longrightarrow F\mathbb{1}^{\mathsf{C}}$ and a natural transformation

$$FX \otimes FY \xrightarrow{F_2} F(X \otimes Y)$$

© The Author(s), under exclusive license to Springer Nature Switzerland AG 2020
D. Yau, *Involutive Category Theory*, Lecture Notes in Mathematics 2279,
https://doi.org/10.1007/978-3-030-61203-0_4

for objects $X, Y \in \mathsf{C}$, that commute with α, λ, and ρ in the sense of (1.2.12)–(1.2.14). Also recall from Definition 2.1.1 the concept of an involutive category.

DEFINITION 4.1.1. An *involutive monoidal category* is a tuple

$$\big((\mathsf{C}, \otimes, \mathbb{1}, \alpha, \lambda, \rho), I, \iota\big)$$

in which:

- $(\mathsf{C}, \otimes, \mathbb{1}, \alpha, \lambda, \rho)$ is a monoidal category.
- (C, I, ι) is an involutive category.
- The involution functor $I : \mathsf{C} \longrightarrow \mathsf{C}$ is equipped with the structure of a monoidal functor.
- The unit $\iota : \mathrm{Id}_{\mathsf{C}} \xrightarrow{\cong} I^2$ is a monoidal natural transformation.

If there is no danger of confusion, we will abbreviate such an involutive monoidal category to C. We also define the following.

(1) An *involutive symmetric monoidal category* is an involutive monoidal category whose underlying monoidal category is equipped with a symmetric monoidal structure such that I is a symmetric monoidal functor.
(2) An *involutive strict (symmetric) monoidal category* is an involutive (symmetric) monoidal category whose underlying monoidal category is strict.
(3) A *strict involutive (strict) monoidal category* is an involutive (strict) monoidal category whose underlying involutive category is strict.
(4) A *strict involutive (strict) symmetric monoidal category* is an involutive (strict) symmetric monoidal category whose underlying involutive category is strict.

EXPLANATION 4.1.2. Let us explain Definition 4.1.1 in more details. Suppose C is a category that is both a monoidal category and an involutive category.

(1) That the involution functor $I : \mathsf{C} \longrightarrow \mathsf{C}$ is a monoidal functor means that it is part of a triple (I, I_2, I_0) with

   - a natural transformation

$$I(X) \otimes I(Y) \xrightarrow{(I_2)_{X,Y}} I(X \otimes Y) \quad \in \mathsf{C}$$

   for objects $X, Y \in \mathsf{C}$, and
   - a morphism $I_0 : \mathbb{1} \longrightarrow I\mathbb{1}$

   that satisfies the associativity axiom

(4.1.3)
$$\begin{array}{ccc}
\big(I(X)\otimes I(Y)\big)\otimes I(Z) & \xrightarrow{\alpha_{I(X),I(Y),I(Z)}} & I(X)\otimes\big(I(Y)\otimes I(Z)\big)\\
\Big\downarrow{\scriptstyle (I_2)_{X,Y}\otimes I(Z)} & & \Big\downarrow{\scriptstyle I(X)\otimes(I_2)_{Y,Z}}\\
I(X\otimes Y)\otimes I(Z) & & I(X)\otimes I(Y\otimes Z)\\
\Big\downarrow{\scriptstyle (I_2)_{X\otimes Y,Z}} & & \Big\downarrow{\scriptstyle (I_2)_{X,Y\otimes Z}}\\
I\big((X\otimes Y)\otimes Z\big) & \xrightarrow{I(\alpha_{X,Y,Z})} & I\big(X\otimes(Y\otimes Z)\big)
\end{array}$$

and the left and the right unity axioms

(4.1.4)
$$\begin{array}{ccc}
\mathbb{1}\otimes I(X) & \xrightarrow{\lambda_{I(X)}} & I(X)\\
\Big\downarrow{\scriptstyle I_0\otimes I(X)} & & \Big\uparrow{\scriptstyle I(\lambda_X)}\\
I(\mathbb{1})\otimes I(X) & \xrightarrow{(I_2)_{\mathbb{1},X}} & I(\mathbb{1}\otimes X)
\end{array}
\qquad
\begin{array}{ccc}
I(X)\otimes\mathbb{1} & \xrightarrow{\rho_{I(X)}} & I(X)\\
\Big\downarrow{\scriptstyle I(X)\otimes I_0} & & \Big\uparrow{\scriptstyle I(\rho_X)}\\
I(X)\otimes I(\mathbb{1}) & \xrightarrow{(I_2)_{X,\mathbb{1}}} & I(X\otimes\mathbb{1})
\end{array}$$

in Definition 1.2.9 for objects $X, Y, Z \in \mathsf{C}$.

(2) That the unit $\iota : \mathrm{Id}_{\mathsf{C}} \xrightarrow{\cong} I^2$ is a monoidal natural transformation means the diagrams (1.2.18) and (1.2.19) are commutative. In the case of ι, these diagrams are

(4.1.5)
$$\begin{array}{ccc}
X\otimes Y & \xrightarrow{\iota_{X\otimes Y}} & I^2(X\otimes Y)\\
\Big\downarrow{\scriptstyle \iota_X\otimes\iota_Y} & & \Big\uparrow{\scriptstyle I(I_2)_{X,Y}}\\
I^2(X)\otimes I^2(Y) & \xrightarrow{(I_2)_{I(X),I(Y)}} & I\big(I(X)\otimes I(Y)\big)
\end{array}$$

and

(4.1.6)
$$\begin{array}{ccc}
\mathbb{1} & \xrightarrow{\iota_{\mathbb{1}}} & I^2(\mathbb{1})\\
\Big\downarrow{\scriptstyle I_0} & & \Big\|\\
I(\mathbb{1}) & \xrightarrow{I(I_0)} & I^2(\mathbb{1})
\end{array}$$

for objects $X, Y \in \mathsf{C}$.

If C is furthermore a symmetric monoidal category with symmetry isomorphism ξ, then the axiom (1.2.29) for (I, I_2, I_0) to be a symmetric monoidal functor is the commutative diagram

(4.1.7)
$$\begin{array}{ccc}
I(X)\otimes I(Y) & \xrightarrow[\cong]{\xi_{I(X),I(Y)}} & I(Y)\otimes I(X)\\
\Big\downarrow{\scriptstyle (I_2)_{X,Y}} & & \Big\downarrow{\scriptstyle (I_2)_{Y,X}}\\
I(X\otimes Y) & \xrightarrow[\cong]{I(\xi_{X,Y})} & I(Y\otimes X)
\end{array}$$

for objects $X, Y \in \mathsf{C}$. ◇

The next observation says that in an involutive monoidal category, the domain and the codomain of the associativity isomorphism α, the domain of the left unit isomorphism λ, and the domain of the right unit isomorphism ρ are all involutive functors.

PROPOSITION 4.1.8. *Suppose C is an involutive monoidal category as in Definition 4.1.1. Then the following statements hold:*

(1) The functor

$$(\mathrm{Id} \otimes \mathrm{Id}) \otimes \mathrm{Id} : \mathsf{C} \times \mathsf{C} \times \mathsf{C} \longrightarrow \mathsf{C}$$

is an involutive functor with distributive law the composite

$$\begin{array}{ccc} \big(I(X) \otimes I(Y)\big) \otimes I(Z) & \xrightarrow{\quad \nu \quad} & I\big((X \otimes Y) \otimes Z\big) \\ {\scriptstyle (I_2)_{X,Y} \otimes I(Z)} \searrow & & \nearrow {\scriptstyle (I_2)_{X \otimes Y, Z}} \\ & I(X \otimes Y) \otimes I(Z) & \end{array}$$

(2) The functor

$$\mathrm{Id} \otimes (\mathrm{Id} \otimes \mathrm{Id}) : \mathsf{C} \times \mathsf{C} \times \mathsf{C} \longrightarrow \mathsf{C}$$

is an involutive functor with distributive law the composite

$$\begin{array}{ccc} I(X) \otimes \big(I(Y) \otimes I(Z)\big) & \xrightarrow{\quad \nu \quad} & I\big(X \otimes (Y \otimes Z)\big) \\ {\scriptstyle I(X) \otimes (I_2)_{Y,Z}} \searrow & & \nearrow {\scriptstyle (I_2)_{X, Y \otimes Z}} \\ & I(X) \otimes I(Y \otimes Z) & \end{array}$$

(3) The functor

$$\mathbb{1} \otimes \mathrm{Id} : \mathsf{C} \longrightarrow \mathsf{C}$$

is an involutive functor with distributive law the composite

$$\mathbb{1} \otimes I(X) \xrightarrow{\;I_0 \otimes I(X)\;} I(\mathbb{1}) \otimes I(X) \xrightarrow{\;(I_2)_{\mathbb{1},X}\;} I(\mathbb{1} \otimes X).$$

(4) The functor

$$\mathrm{Id} \otimes \mathbb{1} : \mathsf{C} \longrightarrow \mathsf{C}$$

is an involutive functor with distributive law the composite

$$I(X) \otimes \mathbb{1} \xrightarrow{I(X)\otimes I_0} I(X) \otimes I(\mathbb{1}) \xrightarrow{(I_2)_{X,\mathbb{1}}} I(X \otimes \mathbb{1}).$$

In the above statements, $X, Y, Z \in \mathsf{C}$ *are objects.*

PROOF. We will prove (1). The proofs for the other assertions are similar. To prove that the functor in (1) is an involutive functor, we need to check the axiom (2.2.2), which is the outer-most diagram below.

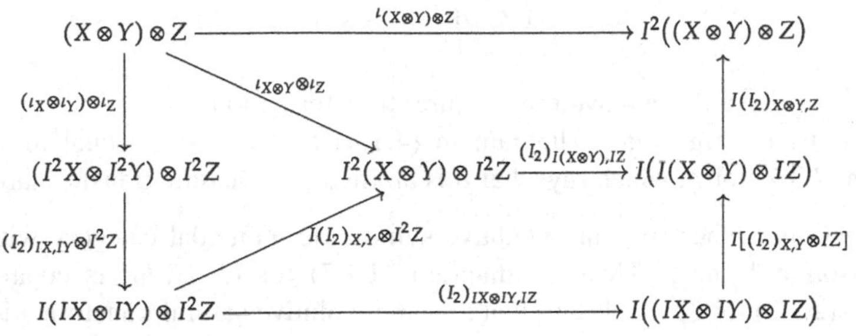

The left triangle is the monoidal product of the commutative diagram (4.1.5) with ι_Z, so it is commutative. The top right trapezoid is commutative by (4.1.5). The bottom right trapezoid is commutative by the naturality of I_2. □

In Definition 4.1.1, an involutive monoidal category is defined as a category that is both involutive and monoidal in which the involutive structure (I, ι) is monoidal. The next observation states that this last condition can be replaced by the requirement that the monoidal structure be involutive. The involutive functor structures in Proposition 4.1.8 will be used below.

PROPOSITION 4.1.9. *An involutive (symmetric) monoidal category is exactly a (symmetric) monoidal category* $(\mathsf{C}, \otimes, \mathbb{1}, \alpha, \lambda, \rho)$ *(with symmetry isomorphism* ξ*) that is equipped with an involutive category structure* (I, ι) *such that the following statements hold:*

- *The monoidal unit* $\mathbb{1}$ *is equipped with the structure of an involutive object* $(\mathbb{1}, I_0)$.
- *The monoidal product* $\otimes : \mathsf{C} \times \mathsf{C} \longrightarrow \mathsf{C}$ *is equipped with the structure of an involutive functor* $(\otimes, I_2)$.
- *The monoidal category structure natural isomorphisms* α, λ, ρ, *and* ξ *in the case that* C *is symmetric, are involutive natural isomorphisms.*

In this case, (I, I_2, I_0) *is the monoidal functor structure on* I *in Definition 4.1.1.*

PROOF. For an involutive monoidal category C as in Definition 4.1.1:

- The diagram (4.1.6) is equal to the diagram (2.5.2) that says that $(\mathbb{1}, I_0)$ is an involutive object.
- The diagram (4.1.5) is equal to the diagram (2.2.2) for $(\otimes, I_2)$, which says that $(\otimes, I_2)$ is an involutive functor.

- The associativity diagram (4.1.3) for I_2 is equal to the diagram (2.3.2) for

$$(\mathrm{Id} \otimes \mathrm{Id}) \otimes \mathrm{Id} \xrightarrow{\alpha} \mathrm{Id} \otimes (\mathrm{Id} \otimes \mathrm{Id}) \ ,$$

  which says that α is an involutive natural transformation.
- The left unity diagram in (4.1.4) for (I_2, I_0) is equal to the diagram (2.3.2) for

$$\mathbb{1} \otimes \mathrm{Id} \xrightarrow{\lambda} \mathrm{Id} \ ,$$

  which says that λ is an involutive natural transformation.
- Similarly, the right unity diagram in (4.1.4) for (I_2, I_0) is equal to the diagram (2.3.2) for ρ, which says that ρ is an involutive natural transformation.

Suppose C is furthermore an involutive symmetric monoidal category with symmetry isomorphism ξ. Then the diagram (4.1.7) for (I, I_2, I_0) is equal to the diagram (2.3.2) for ξ, which says that ξ is an involutive natural transformation.

The converse uses the same identifications. □

Therefore, an involutive (symmetric) monoidal category is a category that is both involutive and (symmetric) monoidal whose involutive structure is (symmetric) monoidal, or equivalently whose (symmetric) monoidal structure is involutive.

Recall that a monoidal functor (F, F_2, F_0) is called *strong* if the structure morphisms F_2 and F_0 are isomorphisms. In an involutive monoidal category, the involution functor is assumed to be a monoidal functor. In fact, it must be a strong monoidal functor.

COROLLARY 4.1.10. *In an involutive monoidal category, the involution functor is a strong monoidal functor.*

PROOF. For an involutive monoidal category with involution functor I, whose monoidal functor structure is (I_2, I_0), we must show that I_0 and I_2 are isomorphisms. By Proposition 4.1.9 $(\mathbb{1}, I_0)$ is an involutive object, so by Proposition 2.5.4(1) the involutive structure map I_0 is an isomorphism. Similarly, by Proposition 4.1.9 $(\otimes, I_2)$ is an involutive functor, so by Proposition 2.2.5 the distributive law I_2 is a natural isomorphism. □

EXAMPLE 4.1.11 (Trivial Involution). Suppose C is a (symmetric) monoidal category. When equipped with the trivial involutive structure $\big(\mathrm{Id}_{\mathsf{C}}, \mathrm{Id}_{\mathrm{Id}_{\mathsf{C}}}\big)$ in Example 2.1.8 and the identity monoidal functor structure on Id_{C}, C becomes an involutive (symmetric) monoidal category. ◇

EXAMPLE 4.1.12 (Conjugate Vector Spaces). As discussed in Example 2.1.10, the symmetric monoidal category $(\mathsf{Vect}(\mathbb{C}), \otimes, \mathbb{C})$ of complex vector spaces has an involutive structure $I = \overline{(-)}$ that sends each complex vector space V to its conjugate $\overline{V}$. Then $\mathsf{Vect}(\mathbb{C})$ becomes an involutive symmetric monoidal category, in which the structure morphism $I_0 : \mathbb{C} \longrightarrow \overline{\mathbb{C}}$ sends each $z \in \mathbb{C}$ to its complex conjugate. The structure morphism

$$I_2 : \overline{V} \otimes \overline{W} \longrightarrow \overline{V \otimes W}$$

is the identity function of the underlying additive groups. Similarly:

- The symmetric monoidal category $(\mathsf{Chain}(\mathbb{C}), \otimes, \mathbb{C})$ of chain complexes of $\mathbb{C}$-vector spaces, when equipped with the conjugate involution structure, is an involutive symmetric monoidal category.
- The symmetric monoidal category $(\mathsf{Hilb}, \widehat{\otimes}, \mathbb{C})$ of complex Hilbert spaces and bounded linear maps, with the completed tensor product $\widehat{\otimes}$, when equipped with the conjugate involution structure, is an involutive symmetric monoidal category.

◇

EXAMPLE 4.1.13 (Opposite Categories). As we discussed in Example 2.1.11, the symmetric monoidal category $(\mathsf{Cat}, \times, \mathbf{1})$ of small categories has an involutive structure $I = (-)^{\mathsf{op}}$ given by taking the opposite category. Then Cat becomes an involutive symmetric monoidal category, in which the structure morphism I_0 is the identity functor on $\mathbf{1}$. The structure morphism

$$I_2 : \mathsf{C}^{\mathsf{op}} \times \mathsf{D}^{\mathsf{op}} \longrightarrow (\mathsf{C} \times \mathsf{D})^{\mathsf{op}}$$

is the identity function on objects, and sends $(f^{\mathsf{op}}, g^{\mathsf{op}})$ to $(f, g)^{\mathsf{op}}$ on morphisms. Furthermore, restricted to the subcategory Poset of partially ordered sets, this gives Poset the structure of an involutive symmetric monoidal category. ◇

4.2 Rigidity of Involutive Monoidal Categories

The purpose of this section is to show that an involutive monoidal category is actually determined by less data than stated in Definition 4.1.1. We begin with the following rigidity result.

LEMMA 4.2.1. *Suppose $(\mathsf{C}, \otimes, \mathbb{1}, \alpha, \lambda, \rho)$ is a monoidal category that is equipped with*

- *a functor $I : \mathsf{C} \longrightarrow \mathsf{C}$,*
- *an isomorphism $\iota_{\mathbb{1}} : \mathbb{1} \xrightarrow{\cong} I^2\mathbb{1}$,*
- *an isomorphism $I_2 : I^2\mathbb{1} \otimes I\mathbb{1} \xrightarrow{\cong} I(I\mathbb{1} \otimes \mathbb{1})$, and*
- *a morphism $I_0 : \mathbb{1} \longrightarrow I\mathbb{1}$ such that the diagram*

(4.2.2)
$$\begin{array}{ccc} I^2\mathbb{1}\otimes\mathbb{1} & \xrightarrow{I^2\mathbb{1}\otimes I_0} & I^2\mathbb{1}\otimes I\mathbb{1} \\ {\scriptstyle \rho_{I^2\mathbb{1}}}\downarrow & & \downarrow{\scriptstyle I_2} \\ I^2\mathbb{1} & \xleftarrow{I(\rho_{I\mathbb{1}})} & I(I\mathbb{1}\otimes\mathbb{1}) \end{array}$$

is commutative.

Then I_0 is equal to the composite

(4.2.3)
$$\begin{array}{ccccccc} \mathbb{1} & & \xrightarrow{\qquad\qquad I_0 \qquad\qquad} & & & & I\mathbb{1} \\ {\scriptstyle \iota_{\mathbb{1}}}\downarrow & & & & & & \uparrow{\scriptstyle \lambda_{I\mathbb{1}}} \\ I^2\mathbb{1} & \xrightarrow{I(\rho^{-1}_{I\mathbb{1}})} & I(I\mathbb{1}\otimes\mathbb{1}) & \xrightarrow{I_2^{-1}} & I^2\mathbb{1}\otimes I\mathbb{1} & \xrightarrow{\iota_{\mathbb{1}}^{-1}\otimes I\mathbb{1}} & \mathbb{1}\otimes I\mathbb{1} \end{array}$$

which is an isomorphism.

PROOF. If I_0 is the composite in (4.2.3), then it must be an isomorphism because every one of the five constituent morphisms is an isomorphism. It remains to prove the commutativity of the diagram (4.2.3), which is the outer-most diagram below.

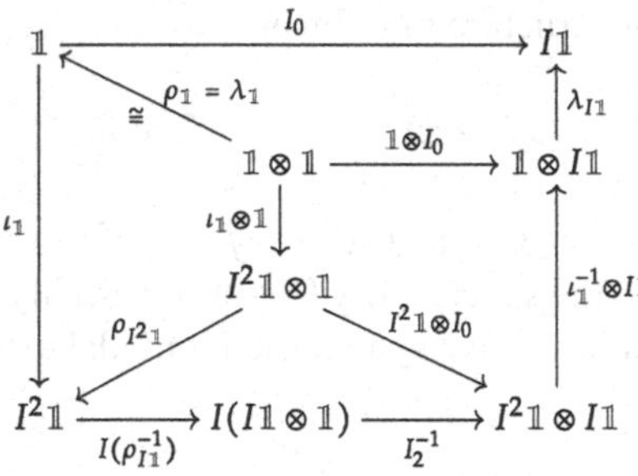

In the above diagram, the top right trapezoid, the left trapezoid, and the lower right trapezoid are commutative by the naturality of λ, the naturality of ρ, and the functoriality of $\otimes$, respectively. The bottom triangle is commutative by (4.2.2). □

PROPOSITION 4.2.4. *In an involutive monoidal category* C *as in Definition 4.1.1, the morphism* $I_0 : \mathbb{1} \longrightarrow I\mathbb{1}$ *is equal to the composite in* (4.2.3).

PROOF. In an involutive monoidal category C, the involution functor I is part of a strong monoidal functor (I, I_2, I_0) by Corollary 4.1.10, so I_2 is a natural isomorphism. The diagram (4.2.2) is the right unity diagram (4.1.4) for $X = I\mathbb{1}$, which is commutative in C. So we can apply Lemma 4.2.1 to finish the proof. □

Recall the explicit description of an involutive monoidal category in Explanation 4.1.2. We now observe that most of the unity axioms (4.1.4) are consequences of the other structures.

THEOREM 4.2.5. *Suppose* $(\mathsf{C}, \otimes, \mathbb{1}, \alpha, \lambda, \rho)$ *is a (symmetric) monoidal category that is equipped with:*

- *an involutive structure* (I, ι)*;*
- *a natural isomorphism*

$$I(X) \otimes I(Y) \xrightarrow[\cong]{I_2} I(X \otimes Y) \quad \text{for} \quad X, Y \in C$$

such that (4.1.3) *and* (4.1.5) *(and also* (4.1.7) *if* C *is symmetric) are commutative;*
- *a morphism* $I_0 : \mathbb{1} \longrightarrow I\mathbb{1}$ *such that* (4.1.6) *and* (4.2.2) *are commutative.*

Then this data defines an involutive (symmetric) monoidal category C*.*

PROOF. By Explanation 4.1.2 we must prove the unity axioms in (4.1.4). By Lemma 4.2.1 the morphism I_0 is equal to the composite in (4.2.3), for which the left unity axiom in (4.1.4) is the outer-most diagram below.

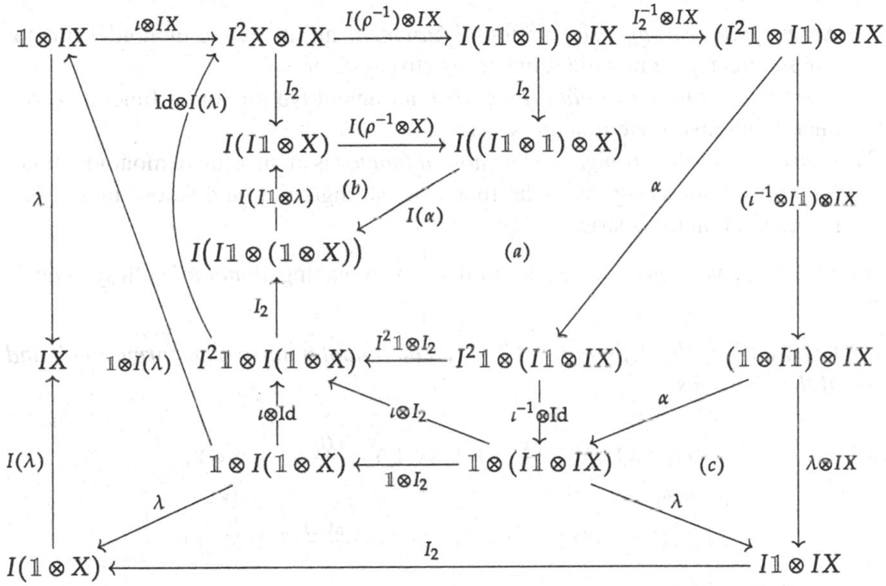

In the above diagram:

- The sub-diagram (a) is commutative by the associativity axiom (4.1.3).
- The triangle (b) is commutative by the unity axiom (1.2.4) of a monoidal category.
- The triangle (c) is commutative by the first unity diagram in (1.2.6).
- All the other sub-diagrams are commutative by the naturality of λ, α, and I_2, and the functoriality of $\otimes$.

The proof for the right unity axiom is similar. □

4.3 Involutive Monoidal Functors

Recall from Definition 1.2.9 that a monoidal functor (F, F_2, F_0) is called *strong* (resp., *strict*) if F_2 and F_0 are isomorphisms (resp., identity morphisms). Also recall from Definition 3.1.1 that an involutive functor (F, ν) is called *strict* if the distributive law ν is the identity natural transformation.

DEFINITION 4.3.1. Suppose C and D are involutive (symmetric) monoidal categories. An *involutive (symmetric) monoidal functor* $F : \mathsf{C} \longrightarrow \mathsf{D}$ is a functor that is equipped with both

- an involutive functor structure (F, ν) and
- a (symmetric) monoidal functor structure (F, F_2, F_0)

such that the distributive law $\nu : FI_{\mathsf{C}} \longrightarrow I_{\mathsf{D}}F$ is a monoidal natural transformation. Moreover:

(1) An *involutive strong/strict monoidal functor* is an involutive monoidal functor whose underlying monoidal functor is strong/strict.
(2) A *strict involutive monoidal functor* is an involutive monoidal functor whose underlying involutive functor is strict.
(3) A *strict involutive strong/strict monoidal functor* is an involutive monoidal functor whose underlying monoidal functor is strong/strict, and whose underlying involutive functor is strict.

The symmetric versions are defined similarly by replacing *monoidal* with *symmetric monoidal*.

LEMMA 4.3.2. *In Definition 4.3.1, ν is a monoidal natural transformation if and only if the diagrams*

(4.3.3)
$$\begin{array}{ccccc} FI_{\mathsf{C}}X \otimes FI_{\mathsf{C}}Y & \xrightarrow{F_2} & F(I_{\mathsf{C}}X \otimes I_{\mathsf{C}}Y) & \xrightarrow{F(I_2)} & FI_{\mathsf{C}}(X \otimes Y) \\ \downarrow{\scriptstyle \nu_X \otimes \nu_Y} & & & & \downarrow{\scriptstyle \nu_{X \otimes Y}} \\ I_{\mathsf{D}}F(X) \otimes I_{\mathsf{D}}F(Y) & \xrightarrow{I_2} & I_{\mathsf{D}}(FX \otimes FY) & \xrightarrow{I_{\mathsf{D}}(F_2)} & I_{\mathsf{D}}F(X \otimes Y) \end{array}$$

for objects $X, Y \in \mathsf{C}$ and

(4.3.4)
$$\begin{array}{ccccc} \mathbb{1}^{\mathsf{D}} & \xrightarrow{F_0} & F(\mathbb{1}^{\mathsf{C}}) & \xrightarrow{F(I_0)} & FI_{\mathsf{C}}(\mathbb{1}^{\mathsf{C}}) \\ \| & & & & \downarrow{\scriptstyle \nu_{\mathbb{1}^{\mathsf{C}}}} \\ \mathbb{1}^{\mathsf{D}} & \xrightarrow{I_0} & I_{\mathsf{D}}(\mathbb{1}^{\mathsf{D}}) & \xrightarrow{I_{\mathsf{D}}(F_0)} & I_{\mathsf{D}}F(\mathbb{1}^{\mathsf{C}}) \end{array}$$

are commutative.

PROOF. Since C and D are involutive (symmetric) monoidal categories, their involution functors I_{C} and I_{D} are (symmetric) monoidal functors. Therefore, so are the composite functors FI_{C} and $I_{\mathsf{D}}F$. The diagrams (4.3.4) and (4.3.3) are the diagrams in Definition 1.2.17 for $\nu : FI_{\mathsf{C}} \longrightarrow I_{\mathsf{D}}F$ to be a monoidal natural transformation. □

Recall from Proposition 2.2.6 that the composite of two involutive functors is an involutive functor. The following observation ensures that involutive monoidal functors can be composed.

LEMMA 4.3.5. *Suppose*

$$\mathsf{C} \xrightarrow{F} \mathsf{D} \xrightarrow{G} \mathsf{E}$$

are two involutive monoidal functors between involutive monoidal categories. Then the composite $GF : \mathsf{C} \longrightarrow \mathsf{E}$ is an involutive monoidal functor. The corresponding statements for all other variants in Definition 4.3.1 are also true.

PROOF. The composite GF is a monoidal functor and an involutive functor by Proposition 2.2.6. So we need to check the commutativity of the two diagrams (4.3.3) and (4.3.4) for GF. The diagram (4.3.3) for GF is the outer-most diagram below.

$$\begin{array}{ccccc}
GFI_{\mathsf{C}}X \otimes GFI_{\mathsf{C}}Y & \xrightarrow{G_2} & G(FI_{\mathsf{C}}X \otimes FI_{\mathsf{C}}Y) & \xrightarrow{G(F_2)} & GF(I_{\mathsf{C}}X \otimes I_{\mathsf{C}}Y) \\
\downarrow{\scriptstyle G(\nu^F_X)\otimes G(\nu^F_Y)} & & \downarrow{\scriptstyle G(\nu^F_X \otimes \nu^F_Y)} & & \downarrow{\scriptstyle GF(I_2)} \\
GI_{\mathsf{D}}FX \otimes GI_{\mathsf{D}}FY & \xrightarrow{G_2} & G(I_{\mathsf{D}}FX \otimes I_{\mathsf{D}}FY) & & GFI_{\mathsf{C}}(X \otimes Y) \\
\downarrow{\scriptstyle \nu^G_{FX}\otimes \nu^G_{FY}} & & \downarrow{\scriptstyle G(I_2)} & & \downarrow{\scriptstyle G(\nu^F_{X\otimes Y})} \\
I_{\mathsf{E}}GF(X) \otimes I_{\mathsf{E}}GF(Y) & & GI_{\mathsf{D}}(FX \otimes FY) & \xrightarrow{GI_{\mathsf{D}}(F_2)} & GI_{\mathsf{D}}F(X \otimes Y) \\
\downarrow{\scriptstyle I_2} & & \downarrow{\scriptstyle \nu^G_{FX\otimes FY}} & & \downarrow{\scriptstyle \nu^G_{F(X\otimes Y)}} \\
I_{\mathsf{E}}(GFX \otimes GFY) & \xrightarrow{I_{\mathsf{E}}(G_2)} & I_{\mathsf{E}}G(FX \otimes FY) & \xrightarrow{I_{\mathsf{E}}G(F_2)} & I_{\mathsf{E}}GF(X \otimes Y)
\end{array}$$

The upper left rectangle and the lower right rectangle are commutative by the naturality of G_2 and the naturality of ν^G, respectively. The lower left rectangle and the upper right rectangle are commutative by (4.3.3) for G and F, respectively.

Similarly, the diagram (4.3.4) for GF factors into three sub-diagrams, which are commutative by (4.3.4) for G and F, and by the naturality of ν^G. This shows that GF is an involutive monoidal functor.

The corresponding statements for all other variants in Definition 4.3.1 are true because strong/strict monoidal functors (resp., strict involutive functors) are closed under composition, and similarly for their symmetric versions. □

DEFINITION 4.3.6. Define the following categories.

(1) IMCat is the category whose objects are small involutive monoidal categories and whose morphisms are involutive monoidal functors.

(2) IMCat_0 is the category whose objects are small involutive monoidal categories and whose morphisms are strict involutive strict monoidal functors.
(3) $\mathsf{ISymCat}$ and $\mathsf{ISymCat}_0$ are the symmetric versions obtained from IMCat and IMCat_0, respectively, by replacing *monoidal* with *symmetric monoidal*.

In the definition of an involutive monoidal functor, instead of asking that the distributive law be a monoidal natural transformation, we could also require that the monoidal functor structure (F, F_2, F_0) be involutive.

PROPOSITION 4.3.7. *Suppose $(F, \nu) : \mathsf{C} \longrightarrow \mathsf{D}$ is an involutive functor with C and D involutive monoidal categories. Then the following statements hold.*

(1) The functor

$$\mathsf{C} \times \mathsf{C} \xrightarrow{\otimes \circ (F,F)} \mathsf{D}$$

is an involutive functor whose distributive law is the composite

$$FI_{\mathsf{C}}(X) \otimes FI_{\mathsf{C}}(Y) \xrightarrow{\nu_X \otimes \nu_Y} I_{\mathsf{D}}F(X) \otimes I_{\mathsf{D}}F(Y) \xrightarrow{I_2} I_{\mathsf{D}}(FX \otimes FY)$$

for $X, Y \in \mathsf{C}$.
(2) The functor

$$\mathsf{C} \times \mathsf{C} \xrightarrow{F \circ \otimes} \mathsf{D}$$

is an involutive functor whose distributive law is the composite

$$F(I_{\mathsf{C}}X \otimes I_{\mathsf{C}}Y) \xrightarrow{F(I_2)} FI_{\mathsf{C}}(X \otimes Y) \xrightarrow{\nu} I_{\mathsf{D}}F(X \otimes Y)$$

for $X, Y \in \mathsf{C}$.
(3) Suppose F is also equipped with a monoidal functor structure (F, F_2, F_0). Then F is an involutive monoidal functor if and only if:

- *$F_2 : \otimes \circ (F, F) \longrightarrow F \circ \otimes$ is an involutive natural transformation, and*
- *$F_0 : (\mathbb{1}^{\mathsf{D}}, I_0) \longrightarrow \big(F\mathbb{1}^{\mathsf{C}}, \nu_{\mathbb{1}} \circ F(I_0)\big)$ is a morphism of involutive objects.*

PROOF. We will omit the superscripts and subscripts C and D to simplify the notations. For the first assertion, the involutive functor axiom (2.2.2) for $\otimes \circ (F, F)$ is the outer-most diagram below.

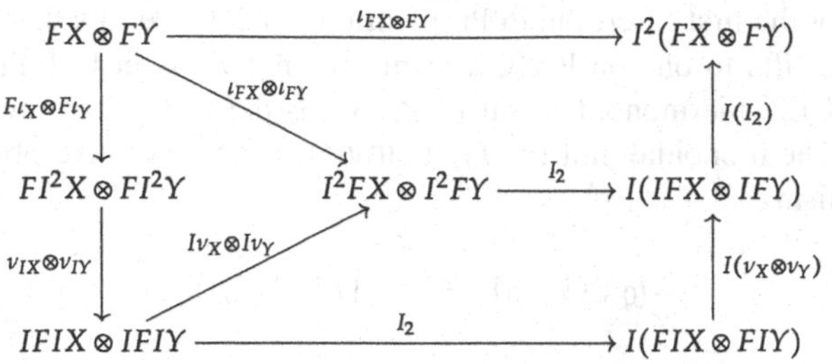

The left triangle is commutative by the axiom (2.2.2) for (F, ν). The top right trapezoid is commutative because the unit ι is a monoidal natural transformation (4.1.5). The bottom right trapezoid is commutative by the naturality of I_2. The second assertion is proved similarly.

For the last assertion, we observe that the diagram (4.3.3) is equal to the diagram (2.3.2) for F_2, which says that it is an involutive natural transformation. Similarly, the diagram (4.3.4) is equal to the diagram (2.5.3) for F_0, which says that it is a morphism of involutive objects. Note that since $(\mathbb{1}, I_0)$ is an involutive object in C, $\big(F\mathbb{1}, \nu_{\mathbb{1}} \circ F(I_0)\big)$ is an involutive object in D as in (2.6.4). □

In summary, an involutive monoidal functor is a functor that is both involutive and monoidal such that (i) the involutive structure is monoidal, or equivalently that (ii) the monoidal structure is involutive.

4.4 Involutive Monoidal Category of Involutive Objects

We now observe that if C is an involutive (symmetric) monoidal category, then so is its category $\mathsf{Inv}(\mathsf{C})$ of involutive objects. Recall from Definition 1.2.33 that a symmetric monoidal category is *closed* if each functor $- \otimes X$, with X an object, admits a right adjoint.

THEOREM 4.4.1. *Suppose C is an involutive (symmetric) monoidal category. Then:*

(1) $\mathsf{Inv}(\mathsf{C})$ inherits from C the structure of an involutive (symmetric) monoidal category.
(2) The forgetful functor $U : \mathsf{Inv}(\mathsf{C}) \longrightarrow \mathsf{C}$ is an involutive (symmetric) monoidal functor.
(3) Each involutive (symmetric) monoidal functor $F : \mathsf{C} \longrightarrow \mathsf{D}$ lifts to an involutive (symmetric) monoidal functor

$$\mathsf{Inv}(F) : \mathsf{Inv}(\mathsf{C}) \longrightarrow \mathsf{Inv}(\mathsf{D}).$$

(4) If C is closed, then $\mathsf{Inv}(\mathsf{C})$ inherits a closed structure from C.

PROOF. For the first assertion, in Proposition 2.6.2(1) we saw that the involutive structure on C lifts to one on $\mathsf{Inv}(\mathsf{C})$. Using the notations in Definition 4.1.1 and Explanation 4.1.2, the monoidal unit in $\mathsf{Inv}(\mathsf{C})$ is defined as the involutive object $(\mathbb{1}, I_0)$ in C. The monoidal unit $(\mathbb{1}, I_0) \in \mathsf{Inv}(\mathsf{C})$ is an involutive object in $\mathsf{Inv}(\mathsf{C})$ via the morphism

$$I_0 : (\mathbb{1}, I_0) \longrightarrow \left(I\mathbb{1}, I(I_0)\right).$$

By Proposition 4.1.9

$$(\otimes, I_2) : \mathsf{C} \times \mathsf{C} \longrightarrow \mathsf{C}$$

is an involutive functor. By the last statement in Proposition 2.5.4 and Proposition 2.6.3, this lifts to an involutive functor

$$(\otimes, I_2) : \mathsf{Inv}(\mathsf{C}) \times \mathsf{Inv}(\mathsf{C}) \longrightarrow \mathsf{Inv}(\mathsf{C})$$

given by

$$(X, j^X) \otimes (Y, j^Y) = \left(X \otimes Y, \; \overbrace{X \otimes Y \xrightarrow{\; j^X \otimes j^Y \;} IX \otimes IY \xrightarrow{\; I_2 \;} I(X \otimes Y)}^{j^{X \otimes Y}}\right)$$

on objects. It is the same as $\otimes : \mathsf{C} \times \mathsf{C} \longrightarrow \mathsf{C}$ on morphisms.

Moreover, by Proposition 4.1.9 the monoidal structure natural isomorphisms in C–i.e., α, λ, ρ, and ξ when C is symmetric–are involutive natural isomorphisms. By Proposition 2.6.5(1), each of them lifts to an involutive natural isomorphism at the level of involutive objects. The axioms for a (symmetric) monoidal category are satisfied in $\mathsf{Inv}(\mathsf{C})$ because they are satisfied in C. Therefore, Proposition 4.1.9 applies once again to show that $\mathsf{Inv}(\mathsf{C})$ is an involutive (symmetric) monoidal category.

For assertion (2), we observed in Proposition 2.6.2(2) that the forgetful functor U is an involutive functor with identity distributive law. By definition U is a strict monoidal functor, for which the two diagrams in Lemma 4.3.2 are commutative. So U is an involutive (symmetric) monoidal functor.

Assertion (3) follows similarly from Proposition 2.6.3, Proposition 2.6.5(1), and Proposition 4.3.7(3).

For assertion (4), we are assuming that C is an involutive symmetric monoidal category that is also closed. By assertion (1), $\mathsf{Inv}(\mathsf{C})$ inherits from C the structure of an involutive symmetric monoidal category. We must show that for each object $(X, j^X) \in \mathsf{Inv}(\mathsf{C})$, the functor

$$- \otimes (X, j^X) : \mathsf{Inv}(\mathsf{C}) \longrightarrow \mathsf{Inv}(\mathsf{C})$$

admits a right adjoint. Since C is closed, the functor $- \otimes X : \mathsf{C} \longrightarrow \mathsf{C}$ admits a right adjoint $[X, -] : \mathsf{C} \longrightarrow \mathsf{C}$. By Proposition 2.6.5(2), it suffices to show that

$$- \otimes X \dashv [X, -]$$

is an involutive adjunction on C. Moreover, by Theorem 2.3.8 it suffices to show that $- \otimes X$ is an involutive functor.

The distributive law of $- \otimes X$ is defined as the composite

$$IY \otimes X \xrightarrow{IY \otimes j^X} IY \otimes IX \xrightarrow{I_2} I(Y \otimes X)$$
$$\overbrace{\phantom{IY \otimes X \xrightarrow{IY \otimes j^X} IY \otimes IX \xrightarrow{I_2}}}^{v_Y^{-\otimes X}}$$

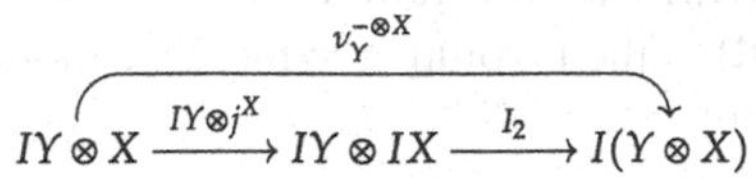

for each object $Y \in \mathsf{C}$. The involutive functor axiom (2.2.2) is the outer-most diagram below.

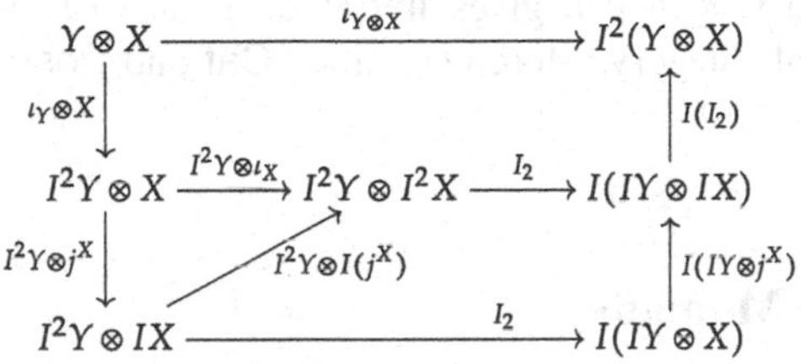

The top rectangle is commutative because ι is a monoidal natural transformation (4.1.5). The lower-left triangle is commutative because (X, j^X) is an involutive object (2.5.2). The lower-right trapezoid is commutative by the naturality of I_2. So $(- \otimes X, v^{-\otimes X})$ is an involutive functor. □

Theorem 4.4.1 applies in the following examples.

EXAMPLE 4.4.2 (Trivial Involution). Suppose C is a (symmetric) monoidal category, regarded as an involutive (symmetric) monoidal category with the trivial involutive structure $(\mathrm{Id}_\mathsf{C}, \mathrm{Id}_{\mathrm{Id}_\mathsf{C}})$ in Example 2.1.8 and the identity monoidal functor structure on Id_C. Then $\mathsf{Inv}(\mathsf{C})$ is the involutive (symmetric) monoidal category in which an object (X, j) consists of

- an object $X \in \mathsf{C}$ and
- a morphism $j : X \longrightarrow X$

such that $j \circ j = \mathrm{Id}_X$.   ◇

EXAMPLE 4.4.3 (Conjugate Vector Spaces). Suppose the symmetric monoidal category $(\mathsf{Vect}(\mathbb{C}), \otimes, \mathbb{C})$ of complex vector spaces is equipped with the involutive symmetric monoidal category structure in Example 4.1.12, with involution functor $I = \overline{(-)}$ given by conjugate vector spaces. Then $\mathsf{Inv}(\mathsf{Vect}(\mathbb{C}))$ is the involutive symmetric monoidal category in which an object (X, j) consists of

- a $\mathbb{C}$-vector space X and
- a $\mathbb{C}$-linear map $j : X \longrightarrow \overline{X}$

such that $\overline{j} \circ j = \mathrm{Id}_X$. A similar discussion applies to the involutive symmetric monoidal categories $(\mathsf{Chain}(\mathbb{C}), \otimes, \mathbb{C})$ and $(\mathsf{Hilb}, \widehat{\otimes}, \mathbb{C})$. Moreover, both $\mathsf{Vect}(\mathbb{C})$ and $\mathsf{Chain}(\mathbb{C})$ are closed, so $\mathsf{Inv}(\mathsf{Vect}(\mathbb{C}))$ and $\mathsf{Inv}(\mathsf{Chain}(\mathbb{C}))$ are also closed. $\diamond$

EXAMPLE 4.4.4 (Opposite Categories). Suppose the symmetric monoidal category $(\mathsf{Cat}, \times, \mathbf{1})$ is equipped with the involutive symmetric monoidal category structure in Example 4.1.13, with involution functor $I = (-)^{\mathsf{op}}$ given by opposite categories. Then $\mathsf{Inv}(\mathsf{Cat})$ is the involutive symmetric monoidal category in which an object (C, j) consists of

- a small category C and
- a functor $j : \mathsf{C} \longrightarrow \mathsf{C}^{\mathsf{op}}$

such that $j^{\mathsf{op}} \circ j = \mathrm{Id}_{\mathsf{C}}$. The same discussion applies to the subcategory Poset of partially ordered sets, which gives $\mathsf{Inv}(\mathsf{Poset})$ the structure of an involutive symmetric monoidal category. Moreover, since Cat and Poset are closed, so are $\mathsf{Inv}(\mathsf{Cat})$ and $\mathsf{Inv}(\mathsf{Poset})$. $\diamond$

4.5 Involutive Monoids

The purpose of this section is to define involutive monoids in an involutive monoidal category. An important class of involutive monoids will be considered in Definition 8.3.5, where we define involutive operads as involutive monoids in an involutive monoidal category of symmetric sequences.

To define involutive monoids, we first observe that the category $\mathsf{Mon}(\mathsf{C})$ of monoids in C inherits an involutive structure from C. We use the notations in Definition 4.1.1 and Explanation 4.1.2 below.

THEOREM 4.5.1. *Suppose*

$$\big((\mathsf{C}, \otimes, \mathbb{1}, \alpha, \lambda, \rho), (I, I_2, I_0), \iota\big)$$

is an involutive monoidal category. Then the category $\mathsf{Mon}(\mathsf{C})$ *inherits an involutive structure* $\big(I_{\mathsf{mon}}, \iota^{\mathsf{mon}}\big)$ *given by*

$$I_{\mathsf{mon}}(X, \mu, \mathsf{1}) = \Big(IX, I\mu \circ I_2, I\mathsf{1} \circ I_0\Big),$$

$$\iota^{\mathsf{mon}}_{(X,\mu,\mathsf{1})} = \iota_X$$

for $(X, \mu, \mathsf{1}) \in \mathsf{Mon}(\mathsf{C})$.

PROOF. First we check that $I_{\mathsf{mon}}(X, \mu, 1)$ is a monoid in C. The associativity of its multiplication $I\mu \circ I_2$ is the outer-most diagram below.

(4.5.2)

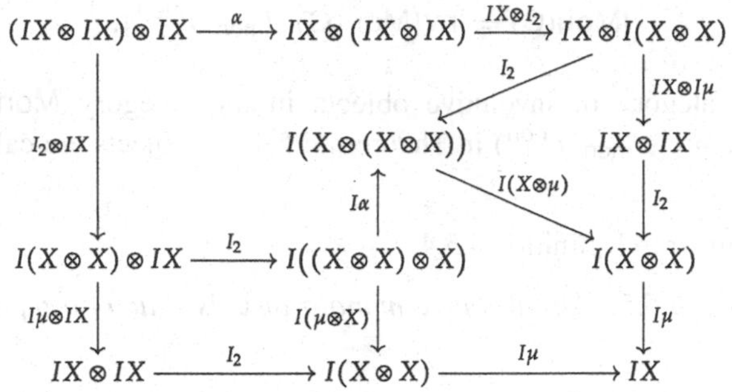

$(IX \otimes IX) \otimes IX$
α
$IX \otimes (IX \otimes IX)$
$IX \otimes I_2$
$IX \otimes I(X \otimes X)$
I_2
$IX \otimes I\mu$
$I_2 \otimes IX$
$I(X \otimes (X \otimes X))$
$IX \otimes IX$
$I(X \otimes \mu)$
$I\alpha$
I_2
$I(X \otimes X) \otimes IX$
I_2
$I((X \otimes X) \otimes X)$
$I(X \otimes X)$
$I\mu \otimes IX$
$I(\mu \otimes X)$
$I\mu$
$IX \otimes IX$
I_2
$I(X \otimes X)$
$I\mu$
IX

Starting at the top left and going counter-clockwise, the four sub-diagrams above are commutative by (1.2.12), the naturality of I_2, the associativity of μ, and the naturality of I_2, respectively. The unity of $I(1) \circ I_0$ is checked similarly. For a morphism $f \in \mathsf{Mon}(\mathsf{C})$, we define

$$I_{\mathsf{mon}}(f) = I(f).$$

This defines a functor

$$I_{\mathsf{mon}} : \mathsf{Mon}(\mathsf{C}) \longrightarrow \mathsf{Mon}(\mathsf{C}).$$

That

$$\iota^{\mathsf{mon}}_{(X,\mu,1)} : (X, \mu, 1) \longrightarrow I^2_{\mathsf{mon}}(X, \mu, 1)$$

is compatible with the multiplications is the outer-most diagram below.

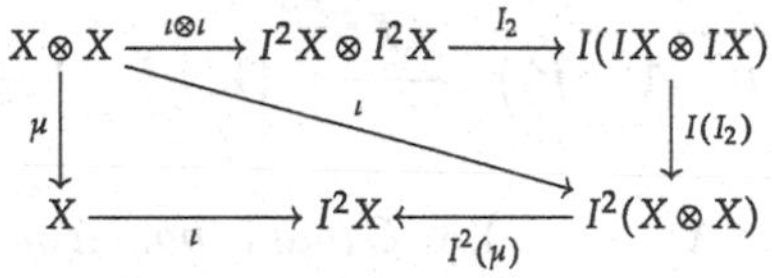

$X \otimes X$
$\iota \otimes \iota$
$I^2X \otimes I^2X$
I_2
$I(IX \otimes IX)$
ι
μ
$I(I_2)$
X
ι
I^2X
$I^2(\mu)$
$I^2(X \otimes X)$

The lower left triangle is commutative by the naturality of ι. The upper right triangle is commutative by (4.1.5). That $\iota^{\mathsf{mon}}_{(X,\mu,1)}$ is compatible with the units is checked similarly. The naturality of ι^{mon} follows from the naturality of ι. Finally, $(I_{\mathsf{mon}}, \iota^{\mathsf{mon}})$ satisfies the triangle identity (2.1.2) because (I, ι) does. □

REMARK 4.5.3. The existence of the induced functor I_{mon} in Theorem 4.5.1 also follows from the general fact that each monoidal functor induces a functor between the respective categories of monoids. ◇

DEFINITION 4.5.4. Suppose C is an involutive monoidal category. Define the category

$$\mathsf{IMon}(\mathsf{C}) = \mathsf{Inv}\big(\mathsf{Mon}(\mathsf{C}), I_{\mathsf{mon}}, \iota^{\mathsf{mon}}\big),$$

which is the category of involutive objects in the category $\mathsf{Mon}(\mathsf{C})$ with the involutive structure $(I_{\mathsf{mon}}, \iota^{\mathsf{mon}})$ in Theorem 4.5.1. Its objects are called *involutive monoids* in C.

Let us first unwrap Definition 4.5.4.

PROPOSITION 4.5.5. *An involutive monoid in C is exactly a tuple $(X, \mu, \mathsf{1}, j)$ with*

- *$(X, \mu, \mathsf{1})$ a monoid in C, and*
- *$(X, j : X \longrightarrow IX)$ an involutive object in C*

such that the diagrams

(4.5.6)
$$\begin{CD} \mathbb{1} @>{\mathsf{1}}>> X \\ @V{I_0}VV @VV{j}V \\ I\mathbb{1} @>{I\mathsf{1}}>> IX \end{CD}$$

and

(4.5.7)
$$\begin{array}{ccccc} X \otimes X & & \xrightarrow{\quad\quad\mu\quad\quad} & & X \\ {\scriptstyle j\otimes j}\big\downarrow & & & & \big\downarrow{\scriptstyle j} \\ IX \otimes IX & \xrightarrow{I_2} & I(X \otimes X) & \xrightarrow{I\mu} & IX \end{array}$$

are commutative. Moreover, a morphism of involutive monoids

$$\big(X, \mu^X, \mathsf{1}^X, j^X\big) \xrightarrow{\quad f \quad} \big(Y, \mu^Y, \mathsf{1}^Y, j^Y\big)$$

is exactly a morphism $f : X \longrightarrow Y \in \mathsf{C}$ that is both a morphism of monoids and a morphism of involutive objects.

PROOF. With respect to the involutive structure $(I_{\mathsf{mon}}, \iota^{\mathsf{mon}})$, an involutive object in $\mathsf{Mon}(\mathsf{C})$ consists of

- an object $(X, \mu, \mathsf{1}) \in \mathsf{Mon}(\mathsf{C})$, and
- a morphism

$$j : (X, \mu, \mathsf{1}) \longrightarrow I_{\mathsf{mon}}(X, \mu, \mathsf{1})$$

in $\mathsf{Mon(C)}$ such that

$$I_{\mathsf{mon}}(j) \circ j = \iota^{\mathsf{mon}}_{(X,\mu,1)}.$$

This equality means

$$Ij \circ j = \iota_X,$$

so (X, j) is an involutive object in C. The diagrams (4.5.6) and (4.5.7) mean exactly that the morphism j preserves the units and the multiplications, i.e., $j \in \mathsf{Mon(C)}$. The assertion about morphisms follows from the definition of the category $\mathsf{IMon(C)}$. □

An involutive monoid is defined as an involutive object in the category of monoids in C. There is another description of an involutive monoid in which the roles of the involutive structure and the monoid structure are switched.

PROPOSITION 4.5.8. *For each involutive monoidal category C, there is an isomorphism*

$$\mathsf{IMon}(\mathsf{C}) \cong \mathsf{Mon}\big(\mathsf{Inv}(\mathsf{C})\big)$$

between

- *the category of involutive monoids in C and*
- *the category of monoids in $\mathsf{Inv}(\mathsf{C})$.*

PROOF. By Theorem 4.4.1(1), the category $\mathsf{Inv(C)}$ of involutive objects inherits an involutive monoidal category structure from C. So the category $\mathsf{Mon}\big(\mathsf{Inv(C)}\big)$ of monoids in $\mathsf{Inv(C)}$ is defined. A monoid in $\mathsf{Inv(C)}$ consists of

- an object $(X, j) \in \mathsf{Inv(C)}$,
- a multiplication

$$(X, j) \otimes (X, j) \xrightarrow{\mu} (X, j) \quad \in \mathsf{Inv(C)},$$

and
- a unit

$$(\mathbb{1}, I_0) \xrightarrow{1} (X, j) \quad \in \mathsf{Inv(C)}$$

such that $\big((X, j), \mu, 1\big)$ is associative and unital. The associativity and unity of (X, j) mean exactly that $(X, \mu, 1)$ is a monoid in C. That 1 and μ are morphisms of involutive objects in C means exactly the diagrams (4.5.6) and (4.5.7). The identification of the morphisms in the two categories is similar. □

In summary, for an involutive monoidal category C, there is a commutative diagram

(4.5.9)
$$\begin{array}{ccc} \mathsf{Inv}(\mathsf{Mon}(\mathsf{C})) = \mathsf{IMon}(\mathsf{C}) \cong \mathsf{Mon}(\mathsf{Inv}(\mathsf{C})) & \longrightarrow & \mathsf{Inv}(\mathsf{C}) \\ \downarrow & & \downarrow \\ \mathsf{Mon}(\mathsf{C}) & \longrightarrow & \mathsf{C} \end{array}$$

of forgetful functors. However, even if C is small, this square is *not* a pullback square in Cat in general. The reason is that by Proposition 4.5.5, an involutive monoid is more than just an object equipped with both a monoid structure and an involutive structure. To be an involutive monoid, the diagrams (4.5.6) and (4.5.7) must also be commutative.

PROPOSITION 4.5.10. *Suppose $(F, \nu) : C \longrightarrow D$ is an involutive monoidal functor between involutive monoidal categories. Then the induced functor in Proposition 1.2.16 lifts further to an involutive functor*

$$\mathsf{IMon}(C) \xrightarrow{\;F_*\;} \mathsf{IMon}(D)\ .$$

PROOF. Using Proposition 2.6.3, it suffices to show that the induced functor

$$F : \mathsf{Mon}(\mathsf{C}) \longrightarrow \mathsf{Mon}(\mathsf{D})$$

in Proposition 1.2.16 is an involutive functor, where $\mathsf{Mon}(\mathsf{C})$ and $\mathsf{Mon}(\mathsf{D})$ have the inherited involutive structures in Theorem 4.5.1. The distributive law on the induced functor F is componentwise the one for the original involutive functor $F : \mathsf{C} \longrightarrow \mathsf{D}$. For a monoid $(X, \mu, 1)$ in C, we first check that

$$FIX \xrightarrow{\;\nu^X\;} IFX$$

is a morphism of monoids in D.

The compatibility of ν^X with the multiplications is the outer-most diagram below.

$$\begin{array}{ccccc} FIX \otimes FIX & \xrightarrow{(\nu^X, \nu^X)} & IFX \otimes IFX & \xrightarrow{I_2} & I(FX \otimes FX) \\ \downarrow{\scriptstyle F_2} & & & & \downarrow{\scriptstyle IF_2} \\ F(IX \otimes IX) & \xrightarrow{FI_2} & FI(X \otimes X) & \xrightarrow{\nu^{X \otimes X}} & IF(X \otimes X) \\ \downarrow{\scriptstyle FI_2} & & \downarrow{\scriptstyle FI\mu} & & \downarrow{\scriptstyle IF\mu} \\ FI(X \otimes X) & \xrightarrow{FI\mu} & FIX & \xrightarrow{\nu^X} & IFX \end{array}$$

The top rectangle is the commutative diagram (4.3.3). The bottom left square is commutative by definition. The bottom right square is commutative by the naturality of ν. The compatibility of ν^X with the units are proved similarly. The involutive functor axiom (2.2.2) for the induced functor F follows from that for the original involutive functor F. □

EXAMPLE 4.5.11 (Trivial Involution). Suppose C is a monoidal category, regarded as an involutive monoidal category with the trivial involutive structure $(\mathrm{Id}_{\mathsf{C}}, \mathrm{Id}_{\mathrm{Id}_{\mathsf{C}}})$. An involutive monoid in C is a tuple $(X, \mu, \mathbb{1}, j)$ such that:

- $(X, \mu, \mathbb{1})$ is a monoid in C.
- $j : X \longrightarrow X$ is a morphism in C such that $j \circ j = \mathrm{Id}_X$.
- j is compatible with $\mathbb{1}$ and μ in the sense that the diagrams (4.5.6) and (4.5.7) are commutative. ◇

EXAMPLE 4.5.12 (Almost Star-Algebras). Consider the involutive symmetric monoidal category $(\mathsf{Vect}(\mathbb{C}), \otimes, \mathbb{C})$ of complex vector spaces with involution functor $I = \overline{(-)}$ given by conjugate vector spaces. An involutive monoid in $\mathsf{Vect}(\mathbb{C})$ is a unital $\mathbb{C}$-algebra $(X, \mu, \mathbb{1})$ equipped with a $\mathbb{C}$-linear map $j : X \longrightarrow \overline{X}$ such that

- $\overline{j} \circ j = \mathrm{Id}_X$ and
- the diagrams (4.5.6) and (4.5.7) are commutative.

In terms of elements $a, b \in X$, the commutative diagram (4.5.7) means the equality

$$j(ab) = j(a) \cdot j(b).$$

This is different from the usual concept of a complex unital $*$-algebra, which reverses the order of the multiplication in the sense that

$$(ab)^* = b^* \cdot a^*.$$

In Example 4.7.6 we will obtain $*$-algebras as examples of a different kind of involutive monoids. A similar discussion applies to the involutive symmetric monoidal categories $(\mathsf{Chain}(\mathbb{C}), \otimes, \mathbb{C})$ and $(\mathsf{Hilb}, \widehat{\otimes}, \mathbb{C})$. ◇

4.6 Involutive Commutative Monoids

The purpose of this section is to define the commutative version of involutive monoids. For a symmetric monoidal category C, its category of commutative monoids is denoted by $\mathsf{CMon}(\mathsf{C})$. Theorem 4.5.1 has the following commutative variation.

PROPOSITION 4.6.1. *Suppose*

$$\big((\mathsf{C}, \otimes, \mathbb{1}, \alpha, \lambda, \rho, \xi), (I, I_2, I_0), \iota\big)$$

is an involutive symmetric monoidal category. Then the category $\mathsf{CMon}(\mathsf{C})$ *inherits an involutive structure from* C *defined as in Theorem 4.5.1.*

PROOF. Reusing most of the proof of Theorem 4.5.1, we just need to check that, for a commutative monoid $(X, \mu, 1)$ in C, the multiplication in $I_{\mathsf{mon}}(X, \mu, 1)$ is commutative. This follows from the naturality of the symmetry isomorphism ξ and the commutativity of the multiplication μ. □

DEFINITION 4.6.2. Suppose C is an involutive symmetric monoidal category. Define the category

$$\mathsf{ICMon}(\mathsf{C}) = \mathsf{Inv}\big(\mathsf{CMon}(\mathsf{C}), I_{\mathsf{mon}}, \iota^{\mathsf{mon}}\big),$$

which is the category of involutive objects in the category $\mathsf{CMon}(\mathsf{C})$ with the involutive structure $(I_{\mathsf{mon}}, \iota^{\mathsf{mon}})$ in Proposition 4.6.1. Its objects are called *involutive commutative monoids* in C.

Unraveling Definition 4.6.2 we obtain the following explicit description of an involutive commutative monoid. Its proof is almost identical to that of Proposition 4.5.5.

PROPOSITION 4.6.3. *An involutive commutative monoid in an involutive symmetric monoidal category* C *is exactly a tuple* $(X, \mu, 1, j)$ *with*

- $(X, \mu, 1)$ *a commutative monoid in* C*, and*
- $(X, j : X \longrightarrow IX)$ *an involutive object in* C

such that the diagrams (4.5.6) *and* (4.5.7) *are commutative. Moreover, a morphism of involutive commutative monoids*

$$\big(X, \mu^X, 1^X, j^X\big) \xrightarrow{\;f\;} \big(Y, \mu^Y, 1^Y, j^Y\big)$$

is exactly a morphism $f : X \longrightarrow Y \in \mathsf{C}$ *that is both a morphism of monoids and a morphism of involutive objects.*

By Theorem 4.4.1(1), the category $\mathsf{Inv}(\mathsf{C})$ of involutive objects inherits an involutive symmetric monoidal structure from C. So the category $\mathsf{CMon}\big(\mathsf{Inv}(\mathsf{C})\big)$ of commutative monoids in $\mathsf{Inv}(\mathsf{C})$ is defined. Next is the commutative version of Proposition 4.5.8, whose proof can be reused here essentially without change.

PROPOSITION 4.6.4. *For each involutive symmetric monoidal category* C*, there is an isomorphism*

$$\mathsf{ICMon}(\mathsf{C}) \cong \mathsf{CMon}\big(\mathsf{Inv}(\mathsf{C})\big)$$

between

- *the category of involutive commutative monoids in* C *and*
- *the category of commutative monoids in* Inv(C).

In summary, for an involutive symmetric monoidal category C, there is a commutative diagram

$$\begin{array}{ccc} \mathsf{Inv}(\mathsf{CMon}(\mathsf{C})) = \mathsf{ICMon}(\mathsf{C}) \cong \mathsf{CMon}(\mathsf{Inv}(\mathsf{C})) & \longrightarrow & \mathsf{Inv}(\mathsf{C}) \\ \downarrow & & \downarrow \\ \mathsf{CMon}(\mathsf{C}) & \longrightarrow & \mathsf{C} \end{array}$$

of forgetful functors. As in (4.5.9) this square is in general not a pullback square because by Proposition 4.6.3, an involutive commutative monoid is more than just an object equipped with a commutative monoid structure and an involutive structure.

Next is the symmetric version of Proposition 4.5.10.

PROPOSITION 4.6.5. *Suppose* $(F, \nu) : C \longrightarrow D$ *is an involutive symmetric monoidal functor between involutive symmetric monoidal categories. Then the induced functor in Proposition 1.2.30 lifts further to an involutive functor*

$$\mathsf{ICMon}(C) \xrightarrow{F_*} \mathsf{ICMon}(D) .$$

PROOF. We reuse the proof of Proposition 4.5.10, replacing Proposition 1.2.16 and Theorem 4.5.1 by their symmetric versions, namely Propositions 1.2.30 and 4.6.1. □

EXAMPLE 4.6.6 (Commutative Star-Algebras). Continuing Example 4.5.12, an involutive commutative monoid in Vect($\mathbb{C}$) is a complex unital commutative $*$-algebra, which satisfies

$$(ab)^* = a^* \cdot b^*.$$

A similar discussion applies to Chain($\mathbb{C}$) and Hilb. ◇

4.7 Reversing Involutive Monoids

For an involutive monoidal category C, we observed in Theorem 4.5.1 that the category Mon(C) of monoids in C inherits an involutive structure from C. Involutive monoids are defined as involutive objects in Mon(C). Equivalently, involutive monoids are monoids in the category Inv(C) of involutive objects in C by Proposition 4.5.8. In this section, we observe that for an involutive symmetric monoidal category C, there is another involutive structure on Mon(C) that incorporates the symmetry isomorphism, leading to the concept of a *reversing involutive monoid.* These objects are important because they include complex unital $*$-algebras and their differential graded analogues as examples. In Theorem 8.5.4 we will show that reversing involutive monoids are the involutive algebras over the involutive associative operad.

PROPOSITION 4.7.1. *Suppose*

$$\big((C, \otimes, \mathbb{1}, \alpha, \lambda, \rho, \xi), (I, I_2, I_0), \iota\big)$$

is an involutive symmetric monoidal category. Then the category Mon(*C*) *inherits an involutive structure* $\big(I_{\mathsf{rmon}}, \iota^{\mathsf{rmon}}\big)$ *given by*

$$I_{\mathsf{rmon}}(X, \mu, \mathsf{1}) = \Big(IX, I\mu \circ I_2 \circ \xi, I\mathsf{1} \circ I_0\Big),$$
$$\iota^{\mathsf{rmon}}_{(X,\mu,\mathsf{1})} = \iota_X$$

for $(X, \mu, \mathsf{1}) \in$ Mon(*C*).

PROOF. First we check that $I_{\mathsf{rmon}}(X, \mu, \mathsf{1})$ is a monoid in C. The associativity of its multiplication $I\mu \circ I_2 \circ \xi$ is the outer-most diagram below.

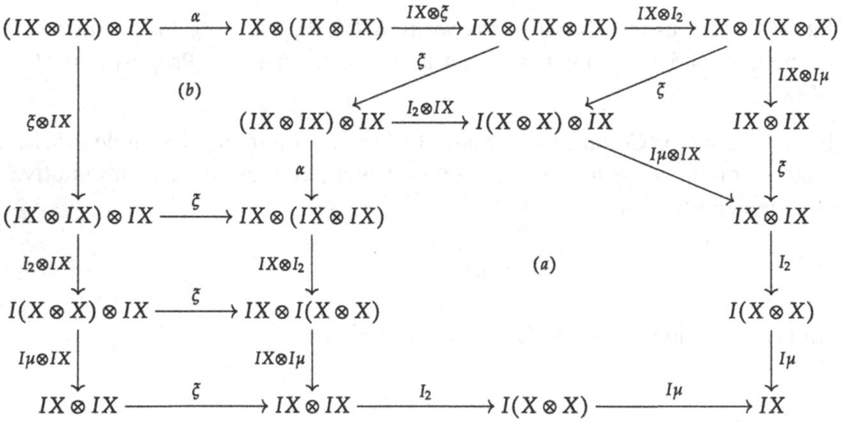

In this diagram:

- The sub-diagram (a) is the outer-most diagram in (4.5.2), which was proved to be commutative.
- The sub-diagram (b) is commutative by Lemma 1.2.26.
- The other four sub-diagrams are commutative by the naturality of the symmetry isomorphism ξ.

Therefore, the multiplication in $I_{\mathsf{rmon}}(X, \mu, \mathbb{1})$ is associative. The rest of the proof is almost identical to the proof of Theorem 4.5.1 after the diagram (4.5.2). □

DEFINITION 4.7.2. Suppose C is an involutive symmetric monoidal category. Define the category

$$\mathsf{RIMon}(\mathsf{C}) = \mathsf{Inv}\big(\mathsf{Mon}(\mathsf{C}), I_{\mathsf{rmon}}, \iota^{\mathsf{rmon}}\big),$$

which is the category of involutive objects in the category $\mathsf{Mon}(\mathsf{C})$ with the involutive structure $(I_{\mathsf{rmon}}, \iota^{\mathsf{rmon}})$ in Proposition 4.7.1. Its objects are called *reversing involutive monoids* in C.

Let us unwrap Definition 4.7.2.

PROPOSITION 4.7.3. *A reversing involutive monoid in an involutive symmetric monoidal category C is exactly a tuple $(X, \mu, \mathbb{1}, j)$ with*

- $(X, \mu, \mathbb{1})$ *a monoid in C, and*
- $(X, j : X \longrightarrow IX)$ *an involutive object in C*

such that the diagrams (4.5.6) *and*

(4.7.4)
$$\begin{array}{ccccccc} X\otimes X & & \xrightarrow{\quad\quad\quad\quad\mu\quad\quad\quad\quad} & & & & X \\ \big\downarrow{\scriptstyle j\otimes j} & & & & & & \big\downarrow{\scriptstyle j} \\ IX\otimes IX & \xrightarrow{\xi} & IX\otimes IX & \xrightarrow{I_2} & I(X\otimes X) & \xrightarrow{I\mu} & IX \end{array}$$

are commutative. Moreover, a morphism of reversing involutive monoids

$$\big(X, \mu^X, \mathbb{1}^X, j^X\big) \xrightarrow{\quad f \quad} \big(Y, \mu^Y, \mathbb{1}^Y, j^Y\big)$$

is exactly a morphism $f : X \longrightarrow Y \in \mathsf{C}$ that is both a morphism of monoids and a morphism of involutive objects.

PROOF. We reuse the proof of Proposition 4.5.5, except that here the diagram (4.7.4) states that j preserves the multiplications in $(X, \mu, \mathbb{1})$ and $I_{\mathsf{rmon}}(X, \mu, \mathbb{1})$. □

Next is the reversing analogue of Proposition 4.5.10.

PROPOSITION 4.7.5. *Suppose* $(F, \nu) : \mathsf{C} \longrightarrow \mathsf{D}$ *is an involutive symmetric monoidal functor between involutive symmetric monoidal categories. Then the induced functor in Proposition 1.2.16 lifts further to an involutive functor*

$$\mathsf{RIMon}(\mathsf{C}) \xrightarrow{F_*} \mathsf{RIMon}(\mathsf{D}) .$$

PROOF. We reuse the proof of Proposition 4.5.10, and replace Theorem 4.5.1 by Proposition 4.7.1. □

EXAMPLE 4.7.6 (Star-Algebras). Continuing Example 4.5.12, a reversing involutive monoid in the involutive symmetric monoidal category $\mathsf{Vect}(\mathbb{C})$ is a unital $\mathbb{C}$-algebra $(X, \mu, 1)$ equipped with a $\mathbb{C}$-linear map $j : X \longrightarrow \overline{X}$ such that

- $\overline{j} \circ j = \mathrm{Id}_X$ and
- that the diagrams (4.5.6) and (4.7.4) are commutative.

In terms of elements $a, b \in X$, the commutative diagram (4.7.4) means the equality

$$j(ab) = j(b) \cdot j(a).$$

So a reversing involutive monoid in $\mathsf{Vect}(\mathbb{C})$ is exactly a complex unital $*$-algebra in the usual sense. Similarly, a reversing involutive monoid in $\mathsf{Chain}(\mathbb{C})$ is exactly a complex unital differential graded $*$-algebra in the usual sense. ◇

Recall from Proposition 2.5.4(2) that an involutive object is equivalent to a pair

$$\bigl(X, j : IX \longrightarrow X\bigr)$$

such that

$$j \circ Ij = \iota_X^{-1} : I^2X \xrightarrow{\cong} X.$$

PROPOSITION 4.7.7. *A reversing involutive monoid in an involutive symmetric monoidal category* C *is exactly a tuple* $(X, \mu, 1, j)$ *with*

- $(X, \mu, 1)$ *a monoid in* C*, and*
- $(X, j : IX \longrightarrow X)$ *an involutive object in* C

such that the diagrams

$$(4.7.8)\qquad \begin{array}{ccc} \mathbb{1} & \xrightarrow{1} & X \\ {\scriptstyle I_0}\downarrow & & \uparrow{\scriptstyle j} \\ I\mathbb{1} & \xrightarrow{I1} & IX \end{array}$$

and

(4.7.9)
$$\begin{array}{ccccccc} IX \otimes IX & \xrightarrow{\xi} & IX \otimes IX & \xrightarrow{I_2} & I(X \otimes X) & \xrightarrow{I\mu} & IX \\ \downarrow{\scriptstyle j\otimes j} & & & & & & \downarrow{\scriptstyle j} \\ X \otimes X & & & \xrightarrow{\mu} & & & X \end{array}$$

are commutative.

PROOF. We again reuse the proof of Proposition 4.5.5, along with the equivalent description of an involutive object in Proposition 2.5.4(2). □

4.8 Involutive Monads

The purpose of this section is to define the involutive versions of a monad and of an algebra over a monad. In Theorem 8.4.5 we will see that every involutive operad yields an involutive monad.

Recall the concepts of a monad from Definition 1.3.1, of an involutive functor from Definition 2.2.1, and of an involutive natural transformation from Definition 2.3.1.

DEFINITION 4.8.1. Suppose (C, I, ι) is an involutive category. An *involutive monad* in C is a tuple

$$(T, \mu, \epsilon, \nu)$$

such that:

- (T, μ, ϵ) is a monad in C.
- (T, ν) is an involutive functor.
- μ and ϵ are involutive natural transformations.

If there is no danger of confusion, we abbreviate (T, μ, ϵ, ν) to T.

Let us make explicit the last condition of an involutive monad.

PROPOSITION 4.8.2. *In Definition 4.8.1:*

(1) $\mu : T^2 \longrightarrow T$ *is an involutive natural transformation if and only if the diagram*

(4.8.3)
$$\begin{array}{ccccc} T^2I & & \xrightarrow{\mu_I} & & TI \\ \downarrow{\scriptstyle T\nu} & & & & \downarrow{\scriptstyle \nu} \\ TIT & \xrightarrow{\nu_T} & IT^2 & \xrightarrow{I\mu} & IT \end{array}$$

is commutative

(2) $\epsilon : \mathrm{Id}_{\mathsf{C}} \longrightarrow T$ *is an involutive natural transformation if and only if the diagram*

(4.8.4)
$$\begin{array}{ccc} I & \xrightarrow{\epsilon_I} & TI \\ \Vert & & \downarrow{\scriptstyle \nu} \\ I & \xrightarrow{I\epsilon} & IT \end{array}$$

is commutative.

Recall from Proposition 1.3.2 that monads in C are exactly monoids in the strict monoidal category $\mathsf{End}(\mathsf{C})$ of functors on C. There is a similar description of involutive monads as monoids using the strict monoidal category $\mathsf{IEnd}(\mathsf{C})$ in Example 2.3.4.

PROPOSITION 4.8.5. *An involutive monad in a small involutive category (C, I, ι) is exactly a monoid in the strict monoidal category $\mathsf{IEnd}(\mathsf{C})$.*

PROOF. This follows from Definition 4.8.1 and Proposition 4.8.2. □

Recall from Definition 1.3.3 that each monad T has an associated category $\mathsf{Alg}(T)$ of T-algebras. For an involutive monad, its category of algebras has an extra structure.

PROPOSITION 4.8.6. *Suppose (T, μ, ϵ, ν) is an involutive monad in an involutive category (C, I, ι). Then the category $\mathsf{Alg}(T)$ inherits an involutive structure (I_T, ι^T) from C defined by*

$$I_T(X, \theta) = \big(IX, I\theta \circ \nu_X\big),$$
$$\iota^T_{(X,\theta)} = \iota_X$$

for $(X, \theta) \in \mathsf{Alg}(T)$.

PROOF. First we check that $I_T(X, \theta)$ is a T-algebra. Its associativity is the outermost diagram below.

$$\begin{array}{ccccc} T^2IX & \xrightarrow{T\nu_X} & TITX & \xrightarrow{TI\theta} & TIX \\ & & \downarrow{\scriptstyle \nu_{TX}} & & \downarrow{\scriptstyle \nu_X} \\ \downarrow{\scriptstyle \mu_{IX}} & & IT^2X & \xrightarrow{IT\theta} & ITX \\ & & \downarrow{\scriptstyle I\mu_X} & & \downarrow{\scriptstyle I\theta} \\ TIX & \xrightarrow{\nu_X} & ITX & \xrightarrow{I\theta} & IX \end{array}$$

The left rectangle is commutative by (4.8.3). The top right square is commutative by the naturality of ν. The bottom right square is commutative by the associativity of the T-algebra (X, θ). The unity of $I_T(X, \theta)$ is checked similarly. For a morphism $f : (X, \theta^X) \longrightarrow (Y, \theta^Y)$ of T-algebras, we define

$$I_T(f) = I(f) : I_T(X, \theta^X) \longrightarrow I_T(Y, \theta^Y),$$

which is a morphism of T-algebras by the naturality of ν. This defines a functor

$$I_T : \mathsf{Alg}(T) \longrightarrow \mathsf{Alg}(T).$$

That

$$\iota^T_{(X,\theta)} : (X, \theta) \longrightarrow I^2_T(X, \theta)$$

is compatible with the T-actions is the outer-most diagram below.

$$
\begin{array}{ccc}
TX \xrightarrow{T\iota_X} TI^2X \xrightarrow{\nu_{IX}} ITIX \\
\theta \downarrow \quad \searrow^{\iota_{TX}} \quad \downarrow^{I\nu_X} \\
X \xrightarrow{\iota_X} I^2X \xleftarrow{I^2\theta} I^2TX
\end{array}
$$

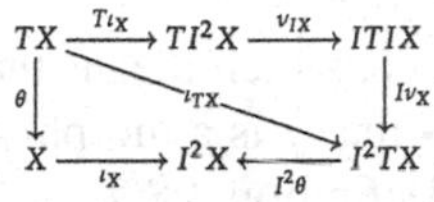

The lower left triangle is commutative by the naturality of ι. The upper right triangle is commutative because (T, ν) is an involutive functor (2.2.2). The naturality of ι^T follows from the naturality of ι. Finally, (I_T, ι^T) satisfies the triangle identity (2.1.2) because (I, ι) does. $\qquad\qquad\square$

DEFINITION 4.8.7. Suppose (T, μ, ϵ, ν) is an involutive monad in an involutive category (C, I, ι). Define the category

$$\mathsf{IAlg}(T, \mu, \epsilon, \nu) = \mathsf{Inv}(\mathsf{Alg}(T), I_T, \iota^T),$$

which is the category of involutive objects in the category $\mathsf{Alg}(T)$ of T-algebras with the involutive structure (I_T, ι^T) in Proposition 4.8.6. Its objects are called *involutive T-algebras*.

Let us unwrap Definition 4.8.7.

PROPOSITION 4.8.8. *For an involutive monad (T, μ, ϵ, ν) in an involutive category (C, I, ι), an involutive T-algebra is exactly a tuple (X, θ, j) with*

- *(X, θ) a T-algebra and*
- *(X, j) an involutive object in C*

such that the diagram

(4.8.9)
$$
\begin{array}{ccc}
TX & \xrightarrow{Tj} & TIX & \xrightarrow{\nu_X} & ITX \\
\theta \downarrow & & & & \downarrow I\theta \\
X & \xrightarrow{\qquad j \qquad} & & & IX
\end{array}
$$

is commutative. Moreover, a morphism of involutive T-algebras

$$f : (X, \theta^X, j^X) \longrightarrow (Y, \theta^Y, j^Y)$$

is exactly a morphism $f : X \longrightarrow Y \in \mathsf{C}$ *that is both a morphism of* T*-algebras and a morphism of involutive objects in* C.

PROOF. An involutive object in $\big(\mathsf{Alg}(T), I_T, \iota^T\big)$ consists of

- an object $(X, \theta) \in \mathsf{Alg}(T)$ and
- a morphism $j : (X, \theta) \longrightarrow I_T(X, \theta)$ in $\mathsf{Alg}(T)$

such that

$$I_T(j) \circ j = \iota^T_{(X,\theta)}.$$

This equality means $I(j) \circ j = \iota_X$, which means that (X, j) is an involutive object in C. The diagram (4.8.9) means that j is a morphism of T-algebras. The assertion about morphisms follows from Definition 4.8.7. □

There is an equivalent description of an involutive T-algebra in which the roles of the involutive structure and the T-algebra structure are switched. To make this statement precise, first we observe that an involutive monad induces a monad on the category of involutive objects. Recall from Proposition 2.6.2 that $\mathsf{Inv}(\mathsf{C})$ inherits an involutive structure from C.

LEMMA 4.8.10. *Suppose* (T, μ, ϵ, ν) *is an involutive monad in an involutive category* (C, I, ι)*. Then there is an induced monad*

$$\big(T_{\mathsf{inv}}, \mu^{\mathsf{inv}}, \epsilon^{\mathsf{inv}}\big)$$

in $\mathsf{Inv}(\mathsf{C})$ *given by*

- $T_{\mathsf{inv}}(X, j) = \big(TX, \nu_X \circ Tj\big)$,
- $\mu^{\mathsf{inv}}_{(X,j)} = \mu_X$, *and*
- $\epsilon^{\mathsf{inv}}_{(X,j)} = \epsilon_X$

for $(X, j) \in \mathsf{Inv}(\mathsf{C})$.

PROOF. The involutive object axiom (2.5.2) for $T_{\mathsf{inv}}(X, j)$ is the outer-most diagram below.

$$\begin{array}{ccccc}
TX & \xrightarrow{Tj} & TIX & \xrightarrow{\nu_X} & ITX \\
\| & & \downarrow{\scriptstyle TIj} & & \downarrow{\scriptstyle ITj} \\
TX & \xrightarrow{T\iota_X} & TI^2X & \xrightarrow{\nu_{IX}} & ITIX \\
{\scriptstyle \iota_{TX}}\downarrow & & & & \downarrow{\scriptstyle I\nu_X} \\
I^2TX & = & = & = & I^2TX
\end{array}$$

The top left square is commutative because (X, j) is an involutive object in C. The top right square is commutative by the naturality of ν. The bottom rectangle is commutative because (T, ν) is an involutive functor (2.2.2). For a morphism f in

$\mathsf{Inv}(\mathsf{C})$, we define

$$T_{\mathsf{inv}}(f) = Tf,$$

which is a morphism in $\mathsf{Inv}(\mathsf{C})$ by the naturality of ν. This defines a functor

$$T_{\mathsf{inv}} : \mathsf{Inv}(\mathsf{C}) \longrightarrow \mathsf{Inv}(\mathsf{C}).$$

That

$$\mu^{\mathsf{inv}}_{(X,j)} : T^2_{\mathsf{inv}}(X, j) \longrightarrow T_{\mathsf{inv}}(X, j)$$

is a morphism of involutive objects in C is the outer-most diagram below.

$$\begin{array}{ccccc}
T^2X & & \xrightarrow{\quad\mu_X\quad} & & TX \\
{\scriptstyle T^2j}\downarrow & & & & \downarrow{\scriptstyle Tj} \\
T^2IX & & \xrightarrow{\quad\mu_{IX}\quad} & & TIX \\
{\scriptstyle T\nu_X}\downarrow & & & & \downarrow{\scriptstyle \nu_X} \\
TITX & \xrightarrow{\nu_{TX}} & IT^2X & \xrightarrow{I\mu_X} & ITX
\end{array}$$

The top rectangle is commutative by the naturality of $\mu : T^2 \longrightarrow T$. The bottom rectangle is commutative by (4.8.3). The naturality and associativity of μ^{inv} follow from those of μ, and similarly for ϵ^{inv}. □

PROPOSITION 4.8.11. *Suppose (T, μ, ϵ, ν) is an involutive monad in an involutive category (C, I, ι). Then there is an isomorphism*

$$\mathsf{IAlg}(T) \cong \mathsf{Alg}\big(T_{\mathsf{inv}}, \mu^{\mathsf{inv}}, \epsilon^{\mathsf{inv}}\big)$$

between

- *the category of involutive T-algebras and*
- *the category of T_{inv}-algebras in $\mathsf{Inv}(\mathsf{C})$.*

PROOF. A T_{inv}-algebra consists of an object $(X, j) \in \mathsf{Inv}(\mathsf{C})$ and a morphism

$$\theta : T_{\mathsf{inv}}(X, j) \longrightarrow (X, j) \in \mathsf{Inv}(\mathsf{C})$$

such that

$$\theta \circ \mu^{\mathsf{inv}}_{(X,j)} = \theta \circ T_{\mathsf{inv}}\theta \quad \text{and} \quad \mathrm{Id}_{(X,\theta)} = \theta \circ \epsilon^{\mathsf{inv}}_{(X,\theta)}.$$

These two equalities mean that (X, θ) is a T-algebra. The diagram (4.8.9) is equivalent to the condition that θ is a morphism of involutive objects. So the desired isomorphism on objects follows from Proposition 4.8.8.

A morphism between two T_{inv}-algebras is a morphism in $\mathsf{Inv}(\mathsf{C})$ that is also compatible with the T_{inv}-algebra structures. By Lemma 4.8.10 this compatibility condition is the same thing as being compatible with the underlying T-algebra structures. Therefore, a morphism between two T_{inv}-algebras is the same thing as a morphism between the underlying objects in C that is both a morphism of T-algebras and a morphism of involutive objects in C. By Proposition 4.8.8 this is the same thing as a morphism of involutive T-algebras. □

In summary, for an involutive monad (T, μ, ϵ, ν) in an involutive category (C, I, ι), there is a commutative diagram

$$\begin{array}{ccc} \mathsf{Inv}(\mathsf{Alg}(T)) = \mathsf{IAlg}(T) \cong \mathsf{Alg}(T_{\mathsf{inv}}) & \longrightarrow & \mathsf{Inv}(\mathsf{C}) \\ \downarrow & & \downarrow \\ \mathsf{Alg}(T) & \longrightarrow & \mathsf{C} \end{array}$$

of forgetful functors. This diagram is in general not a pullback square because by Proposition 4.8.8, an involutive T-algebra is more than just an object equipped with a T-algebra structure and an involutive structure. To be an involutive T-algebra, the diagram (4.8.9) must also be commutative.

EXAMPLE 4.8.12 (Involutive Monoids). Suppose C is an involutive monoidal category as in Definition 4.1.1 that has all countable coproducts, which are furthermore preserved by the monoidal product on each side. There is a monad $(T_{\mathsf{mon}}, \mu, \epsilon)$ in C whose category of algebras is isomorphic to the category $\mathsf{Mon}(\mathsf{C})$ of monoids in C; see, e.g., [Mac98] (VII.3 Theorem 2). In more details:

- The free monoid functor T_{mon} sends an object $X \in \mathsf{C}$ to the countable coproduct

$$T_{\mathsf{mon}}X = \coprod_{n \geq 0} X^{\otimes n}$$

  with

$$X^{\otimes 0} = \mathbb{1}, \quad X^{\otimes 1} = X, \quad \text{and} \quad X^{\otimes n+1} = (X^{\otimes n}) \otimes X$$

  for $n \geq 1$.
- The unit

$$\epsilon_X : X \longrightarrow T_{\mathsf{mon}}X$$

  corresponds to the $n = 1$ summand.

- The multiplication

$$\mu_X : T_{\mathsf{mon}}^2 X \longrightarrow T_{\mathsf{mon}} X$$

is given by the unique isomorphisms

$$\Big(\cdots (X^{\otimes m_1} \otimes X^{\otimes m_2}) \otimes \cdots\Big) \otimes X^{\otimes m_n} \cong X^{\otimes m_1 + \cdots + m_n}$$

in Mac Lane's Coherence Theorem [Mac98] (VII.2 Corollary).

Since the involution functor I preserves countable coproducts, (T_{mon}, ν) is an involutive functor whose distributive law ν is given by the composite

$$\begin{array}{ccccc} T_{\mathsf{mon}} IX & & \xrightarrow{\quad \nu_X \quad} & & IT_{\mathsf{mon}} X \\ \| & & & & \| \\ \coprod_{n \geq 0} (IX)^{\otimes n} & \xrightarrow{\cong} & \coprod_{n \geq 0} I(X^{\otimes n}) & \xrightarrow{\cong} & I\Big(\coprod_{n \geq 0} X^{\otimes n}\Big) \end{array}$$

for each object $X \in \mathsf{C}$. The bottom left isomorphism uses Corollary 4.1.10, which says that the involution functor in an involutive monoidal category is a strong monoidal functor. Furthermore, the diagrams (4.8.3) and (4.8.4) are commutative, so $(T_{\mathsf{mon}}, \mu, \epsilon, \nu)$ is an involutive monad.

Under the isomorphism

$$\mathsf{Mon(C)} \cong \mathsf{Alg}(T_{\mathsf{mon}}),$$

the involutive structure on $\mathsf{Mon(C)}$ in Theorem 4.5.1 and the involutive structure on $\mathsf{Alg}(T_{\mathsf{mon}})$ in Proposition 4.8.6 coincide, so their categories of involutive objects also coincide. In other words, there is an isomorphism

$$\begin{aligned} \mathsf{IMon(C)} &= \mathsf{Inv}\big(\mathsf{Mon(C)}\big) \\ &\cong \mathsf{Inv}\big(\mathsf{Alg}(T_{\mathsf{mon}})\big) = \mathsf{IAlg}(T_{\mathsf{mon}}) \end{aligned}$$

between the category of involutive monoids in C and the category of involutive T_{mon}-algebras. ◇

4.9 Exercises and Notes

(1) Prove (2), (3), and (4) in Proposition 4.1.8.
(2) In the proof of Theorem 4.2.5, prove the right unity axiom in (4.1.4).
(3) In the proof of Lemma 4.3.5, check the diagram (4.3.4) for GF.
(4) Prove (2) in Proposition 4.3.7.

(5) In the proof of Theorem 4.5.1, check:

- The unity axiom for $I_{\mathsf{mon}}(X, \mu, 1)$.
- That ι_X is compatible with the units of $(X, \mu, 1)$ and $I^2_{\mathsf{mon}}(X, \mu, 1)$.

(6) In the proof of Proposition 4.5.10, check that ν^X is compatible with the units in FIX and IFX.
(7) Fill in the rest of the proof of Proposition 4.7.1.
(8) Give a detailed proof of Proposition 4.7.5.
(9) Suppose C is an involutive monoidal category with countable coproducts, which are preserved by the monoidal product on each side. Give an explicit construction of a left adjoint of the forgetful functor $\mathsf{IMon(C)} \longrightarrow \mathsf{C}$.
(10) Suppose C is a cocomplete involutive symmetric monoidal category in which the monoidal product commutes with colimits on each side. Give an explicit construction of a left adjoint of each of the forgetful functors

$$\mathsf{ICMon(C)} \longrightarrow \mathsf{C} \longleftarrow \mathsf{RIMon(C)} .$$

(11) In the proof of Proposition 4.8.6, check the unity of $I_T(X, \theta)$.
(12) In Lemma 4.8.10, show that $(T_{\mathsf{inv}}, \mu^{\mathsf{inv}}, \epsilon^{\mathsf{inv}})$ is part of an involutive monad in the involutive category $\mathsf{Inv(C)}$, whose distributive law is given by that of T.
(13) Suppose T is an involutive monad in an involutive category C with coproducts. Give an explicit construction of a left adjoint of the forgetful functor $\mathsf{IAlg}(T) \longrightarrow \mathsf{C}$.
(14) Work out the commutative version of Example 4.8.12 for involutive commutative monoids in an involutive symmetric monoidal category.
(15) For an involutive monad T in an involutive category C, prove that the Eilenberg-Moore adjunctions

$$\mathsf{C} \rightleftarrows \mathsf{Alg}(T) \quad \text{and} \quad \mathsf{Inv(C)} \rightleftarrows \mathsf{IAlg}(T)$$

are involutive adjunctions.

Notes. Most of the concepts in this chapter are due to Jacobs [Jac12], the main exceptions being 4.1.9, 4.5.1, 4.5.10, 4.6.1, 4.6.5, 4.7.1, and 4.7.5.

Slightly different concepts of involutive monoidal categories are considered by Beggs-Majid [BM09] and Egger [Egg11]. In an involutive monoidal category in the sense of [Egg11] (Definition 2.1), the involution functor I is not a monoidal functor. Instead, it is equipped with a natural isomorphism

$$I(X) \otimes I(Y) \xrightarrow{\cong} I(Y \otimes X)$$

that reverses the order of the monoidal product. For the discussion of involutive operads in Chapter 8, the concept of an involutive monoidal category in

Definition 4.1.1 is more suitable. The reason is that involutive operads, as in Definition 8.3.5, are involutive monoids in an involutive monoidal category $\mathsf{SSeq}_{\mathsf{C}}^{\mathfrak{C}}$ that is *not* symmetric.

The rigidity result Lemma 4.2.1 is similar to Lemma 2.3 in [Egg11]. Involutive monoidal categories in [Egg11] are equal to strong bar categories in [BM09]. A symmetric involutive monoidal category in the sense of [Egg11] (Remark 4.6) is equal to an involutive symmetric monoidal category in Definition 4.1.1.

For a small involutive category C, Lemma 4.8.10 can also be obtained from general 2-categorical facts. An involutive monad in a small involutive category C is the same thing as a monad in the 2-category ICat acting on the object C. Since $\mathsf{Inv}(-) : \mathsf{ICat} \longrightarrow \mathsf{ICat}$ is a 2-functor by Exercise 2.7(4), it preserves monads. For more discussion of monads in 2-categories and bicategories, the reader is referred to Chapter 6 in [JY21].

The concept of an involutive monoidal category is a special case of the notion of a monoidal category equipped with a group action. The group in this case is the cyclic group of order 2. For discussion of monoidal categories with group action, the reader is referred to [BNY18, DGNO10, Gal17, Tam01].

Chapter 5
Coherence of Involutive Monoidal Categories

The purpose of this chapter is to study coherence of involutive monoidal categories. The next chapter will deal with the symmetric case. In Section 5.1 and Section 5.2, we give explicit constructions of the free involutive monoidal category and of the free involutive strict monoidal category generated by a category. We observe that they are equivalent via a strict involutive strict monoidal functor. In Section 5.3 we show that in a small involutive monoidal category, every formal diagram is commutative. Here a formal diagram is defined as in the case of monoidal categories, but with the involutive structure also taken into account. In Section 5.4 we show that every involutive monoidal category can be strictified to an involutive strict monoidal category via an involutive adjoint equivalence involving involutive strong monoidal functors. The remaining two sections contain explicit constructions of the free involutive (strict) monoidal category generated by an involutive category.

5.1 Free Involutive Monoidal Categories

The purpose of this section is to give an explicit description of the free involutive monoidal category generated by a category. It involves the following definitions.

DEFINITION 5.1.1. Suppose S is a class, and $\mathbb{1}$, I, and $\otimes$ are symbols that do not appear in S.

(1) The class $\mathsf{W}_{\mathsf{im}}(S)$ of *involutive monoidal words generated by* S is defined recursively by the following three conditions:

   (i) $S \subseteq \mathsf{W}_{\mathsf{im}}(S)$ and $\mathbb{1} \in \mathsf{W}_{\mathsf{im}}(S)$.
   (ii) If $X \in \mathsf{W}_{\mathsf{im}}(S)$, then $IX \in \mathsf{W}_{\mathsf{im}}(S)$.
   (iii) If $X, Y \in \mathsf{W}_{\mathsf{im}}(S)$, then $X \otimes Y \in \mathsf{W}_{\mathsf{im}}(S)$. The operation $\otimes$ is not associative by definition.

© The Author(s), under exclusive license to Springer Nature Switzerland AG 2020
D. Yau, *Involutive Category Theory*, Lecture Notes in Mathematics 2279,
https://doi.org/10.1007/978-3-030-61203-0_5

(2) For $X \in \mathsf{W}_{\mathsf{im}}(S)$, its *underlying word*

$$\mathsf{word}(X) \in \coprod_{k\geq 0} (\mathbb{Z}_{\geq 0} \times S)^{\times k}$$

is the possibly-empty finite sequence in $\mathbb{Z}_{\geq 0} \times S$ defined recursively by the following four conditions:

(i) $\mathsf{word}(\mathbb{1}) = \varnothing$, the empty sequence.
(ii) $\mathsf{word}(x) = (0, x)$ for $x \in S$.
(iii) If $X \in \mathsf{W}_{\mathsf{im}}(S)$ with

$$\mathsf{word}(X) = (n_1, x_1) \cdots (n_k, x_k) \in (\mathbb{Z}_{\geq 0} \times S)^{\times k},$$

then

$$\mathsf{word}(IX) = \begin{cases} \varnothing & \text{if } \mathsf{word}(X) = \varnothing, \\ (n_1 + 1, x_1) \cdots (n_k + 1, x_k) & \text{if } k > 0. \end{cases}$$

(iv) If $X, Y \in \mathsf{W}_{\mathsf{im}}(S)$ with $\mathsf{word}(X)$ and $\mathsf{word}(Y)$ already defined, then

$$\mathsf{word}(X \otimes Y) = \mathsf{word}(X)\mathsf{word}(Y),$$

the concatenation of $\mathsf{word}(X)$ and $\mathsf{word}(Y)$. Concatenation is associative by definition.

If $\mathsf{word}(X) \in (\mathbb{Z}_{\geq 0} \times S)^{\times k}$, then we call k the *length* of X, and write $|X| = k$.

(3) For $X \in \mathsf{W}_{\mathsf{im}}(S)$, its *underlying integer sequence*

$$\mathsf{word}_1(X) \in \coprod_{k\geq 0} (\mathbb{Z}_{\geq 0})^{\times k}$$

is obtained from $\mathsf{word}(X)$ by dropping the S-entries.

(4) Define an equivalence relation on $\coprod_{k\geq 0} (\mathbb{Z}_{\geq 0})^{\times k}$ as follows. For

$$N = (n_1, \ldots, n_q),\ M = (m_1, \ldots, m_p) \in \coprod_{k\geq 0} (\mathbb{Z}_{\geq 0})^{\times k},$$

we define

$$N \equiv M \pmod 2$$

if

(i) $p = q$ and
(ii) $n_i \equiv m_i \pmod 2$ for $1 \leq i \leq p$.

(5) For $X, Y \in \mathsf{W}_{\mathsf{im}}(S)$, we write

$$X \equiv Y \pmod 2 \quad \text{if} \quad \mathsf{word}_1(X) \equiv \mathsf{word}_1(Y) \pmod 2.$$

EXAMPLE 5.1.2. Suppose S is a class with $x, y \in S$. Then one involutive monoidal word generated by S is

$$X = I^4\big(I^2(I\mathbb{1} \otimes I^3x) \otimes I^6y\big) \in \mathsf{W}_{\mathsf{im}}(S),$$

where $I^n = I \cdots I$ with $n \geq 1$ copies of I. Its underlying word and underlying integer sequence are

$$\begin{aligned}\mathsf{word}(X) &= (9, x)(10, y) \in (\mathbb{Z}_{\geq 0} \times S)^{\times 2},\\ \mathsf{word}_1(X) &= (9, 10) \in \mathbb{Z}^{\times 2}_{\geq 0}.\end{aligned}$$

Suppose further that $w, z \in S$. Then the involutive monoidal word

$$W = Iw \otimes I^3(Iz \otimes I^5\mathbb{1}) \in \mathsf{W}_{\mathsf{im}}(S)$$

has

$$\begin{aligned}\mathsf{word}(W) &= (1, w)(4, z) \in (\mathbb{Z}_{\geq 0} \times S)^{\times 2},\\ \mathsf{word}_1(W) &= (1, 4) \in \mathbb{Z}^{\times 2}_{\geq 0}.\end{aligned}$$

It follows that

$$X \equiv W \pmod 2,$$

although $X \neq W$ as involutive monoidal words. ◇

We now define a category that will be shown to be the free involutive monoidal category.

DEFINITION 5.1.3. Suppose C is a category. Define the category $\mathcal{F}_{\mathsf{im}}(\mathsf{C})$ with extra structures as follows.

Objects: $\mathsf{Ob}\big(\mathcal{F}_{\mathsf{im}}(\mathsf{C})\big) = \mathsf{W}_{\mathsf{im}}\big(\mathsf{Ob}(\mathsf{C})\big)$, the class of involutive monoidal words generated by $\mathsf{Ob}(\mathsf{C})$.

Morphisms: For $X, Y \in \mathsf{Ob}\big(\mathcal{F}_{\mathsf{im}}(\mathsf{C})\big)$ with

(5.1.4)
$$\mathsf{word}(X) = (m_1, x_1) \cdots (m_j, x_j) \quad \text{and} \quad \mathsf{word}(Y) = (n_1, y_1) \cdots (n_k, y_k),$$

it has the morphism set

$$\mathcal{F}_{\mathsf{im}}(\mathsf{C})(X; Y) = \begin{cases} \prod_{i=1}^{j} \mathsf{C}(x_i, y_i) & \text{if } X \equiv Y \pmod 2, \\ \varnothing & \text{otherwise,} \end{cases}$$

where an empty product, for the case $j = 0$, is taken as a one-point set.

Identities: For X as above, its identity morphism is

$$\mathrm{Id}_X = \{\mathrm{Id}_{x_i}\}_{i=1}^{j} \in \prod_{i=1}^{j} \mathsf{C}(x_i, x_i).$$

Composition: Categorical composition is defined entrywise in C.

Monoidal Product: For objects $X, Y \in \mathsf{Ob}(\mathcal{F}_{\mathsf{im}}(\mathsf{C}))$, their monoidal product $X \otimes Y$ is defined as part of the definition of the class $\mathsf{W}_{\mathsf{im}}(\mathsf{Ob}(\mathsf{C}))$. The monoidal product on morphisms in $\mathcal{F}_{\mathsf{im}}(\mathsf{C})$ is defined as concatenation of finite sequences of morphisms in C.

Monoidal Unit: The monoidal unit in $\mathcal{F}_{\mathsf{im}}(\mathsf{C})$ is $\mathbb{1} \in \mathsf{W}_{\mathsf{im}}(\mathsf{Ob}(\mathsf{C}))$.

Associativity Isomorphisms: For X, Y as above and $Z \in \mathsf{Ob}(\mathcal{F}_{\mathsf{im}}(\mathsf{C}))$ with

$$\mathsf{word}(Z) = (p_1, z_1) \cdots (p_l, z_l),$$

the associativity isomorphism

$$(X \otimes Y) \otimes Z \xrightarrow[\cong]{\alpha_{X,Y,Z}} X \otimes (Y \otimes Z) \quad \in \mathcal{F}_{\mathsf{im}}(\mathsf{C})$$

is defined as

$$\left\{\{\mathrm{Id}_{x_i}\}_{i=1}^{j}, \{\mathrm{Id}_{y_i}\}_{i=1}^{k}, \{\mathrm{Id}_{z_i}\}_{i=1}^{l}\right\} \in \prod_{i=1}^{j} \mathsf{C}(x_i, x_i) \times \prod_{i=1}^{k} \mathsf{C}(y_i, y_i) \times \prod_{i=1}^{l} \mathsf{C}(z_i, z_i).$$

Left and Right Units: For X as above, the left and the right unit isomorphisms

$$\mathbb{1} \otimes X \xrightarrow[\cong]{\lambda_X} X \xleftarrow[\cong]{\rho_X} X \otimes \mathbb{1} \quad \in \mathcal{F}_{\mathsf{im}}(\mathsf{C})$$

are both given by $\{\mathrm{Id}_{x_i}\}_{i=1}^{j}$.

Involution Functor: Define a functor

$$I : \mathcal{F}_{\mathsf{im}}(\mathsf{C}) \longrightarrow \mathcal{F}_{\mathsf{im}}(\mathsf{C})$$

by the assignment $X \longmapsto IX$ on objects, and the identity function

$$\mathcal{F}_{\mathsf{im}}(X;Y) \xrightarrow{\ \mathrm{Id}\ } \mathcal{F}_{\mathsf{im}}(IX;IY)$$

on morphism sets. This is well-defined because

$$X \equiv Y \pmod 2 \quad \text{if and only if} \quad IX \equiv IY \pmod 2,$$

and I only changes the integer entries.

Unit: Define a natural isomorphism

$$\iota : \mathrm{Id}_{\mathcal{F}_{\mathsf{im}}(\mathsf{C})} \xrightarrow{\ \cong\ } I^2$$

by

$$\iota_X = \{\mathrm{Id}_{x_i}\}_{i=1}^{j} : X \xrightarrow{\ \cong\ } I^2X \in \mathcal{F}_{\mathsf{im}}(\mathsf{C})(X; I^2X) = \prod_{i=1}^{j} \mathsf{C}(x_i, x_i)$$

for each object $X \in \mathcal{F}_{\mathsf{im}}(\mathsf{C})$ as above. This is well-defined because

$$\mathsf{word}(I^2X) = (m_1+2, x_1)\cdots(m_j+2, x_j).$$

Involutive Monoidal Structure: Define the morphism

$$I_0 : \mathbb{1} \longrightarrow I\mathbb{1}$$

as the unique isomorphism in

$$\mathcal{F}_{\mathsf{im}}(\mathsf{C})(\mathbb{1}; I\mathbb{1}) = \{*\}.$$

This is well-defined because

$$\mathsf{word}(\mathbb{1}) = \mathsf{word}(I\mathbb{1}) = \varnothing.$$

Finally, for objects $X, Y \in \mathcal{F}_{\mathsf{im}}(\mathsf{C})$ as above, define the isomorphism

$$IX \otimes IY \xrightarrow[\cong]{(I_2)_{X,Y}} I(X \otimes Y) \ \in \mathcal{F}_{\mathsf{im}}(\mathsf{C})\big(IX \otimes IY, I(X \otimes Y)\big)$$

as

$$\Big\{\{\mathrm{Id}_{x_i}\}_{i=1}^{j}, \{\mathrm{Id}_{y_i}\}_{i=1}^{k}\Big\} \in \prod_{i=1}^{j} \mathsf{C}(x_i, x_i) \times \prod_{i=1}^{k} \mathsf{C}(y_i, y_i).$$

This is well-defined because

$$\begin{aligned}\mathsf{word}\big(IX \otimes IY\big) &= \mathsf{word}\big(I(X \otimes Y)\big)\\ &= (m_1+1, x_1)\cdots(m_j+1, x_j)(n_1+1, y_1)\cdots(n_k+1, y_k).\end{aligned}$$

This finishes the definition of $\mathcal{F}_{\mathsf{im}}(\mathsf{C})$.

LEMMA 5.1.5. *For each category* C*, the tuple*

$$\big((\mathcal{F}_{\mathsf{im}}(\mathsf{C}), \otimes, \mathbb{1}, \alpha, \lambda, \rho), (I, I_2, I_0), \iota\big)$$

in Definition 5.1.3 is an involutive monoidal category.

PROOF. An inspection shows that $\big(\mathcal{F}_{\mathsf{im}}(\mathsf{C}), \otimes, \mathbb{1}, \alpha, \lambda, \rho\big)$ is a monoidal category, that $\big(\mathcal{F}_{\mathsf{im}}(\mathsf{C}), I, \iota\big)$ is an involutive category, and that the diagrams (4.1.3), (4.1.4), (4.1.5), and (4.1.6) in Explanation 4.1.2 are commutative. □

REMARK 5.1.6. For each category C, the involutive monoidal category $\mathcal{F}_{\mathsf{im}}(\mathsf{C})$ is neither strict involutive (since $\mathbb{1} \neq I^2\mathbb{1}$) nor strict monoidal (since $\mathbb{1} \otimes \mathbb{1} \neq \mathbb{1}$). ◇

To show that $\mathcal{F}_{\mathsf{im}}(\mathsf{C})$ has the desired universal property, we will use the following reduced objects.

DEFINITION 5.1.7. Suppose C is a category.

(1) An object in $\mathcal{F}_{\mathsf{im}}(\mathsf{C})$ is called *reduced* if either

 (i) it is $\mathbb{1}$, or
 (ii) it has the form

$$\big(\cdots(I^{\epsilon_1}x_1 \otimes I^{\epsilon_2}x_2) \otimes \cdots\big) \otimes I^{\epsilon_k}x_k \tag{5.1.8}$$

 for some $k > 0$ and objects $x_1, \ldots, x_k \in \mathsf{C}$, with each $\epsilon_i \in \{0, 1\}$. Here $I^0 = \mathrm{Id}$, and the left half of each pair of parentheses in (5.1.8) is at the front.

(2) The collection of *basic isomorphisms* in $\mathcal{F}_{\mathsf{im}}(\mathsf{C})$ is defined recursively as follows:

- The identity morphisms of $\mathbb{1}$ and of objects in C are basic isomorphisms.
- I_0, each component of I_2, ι, α, λ, and ρ, and their inverses are basic isomorphisms.
- Basic isomorphisms are closed under compositions, $\otimes$, and I.

EXAMPLE 5.1.9. For objects x, y in a category C, the composite δ_X,

$$\begin{array}{ccc} X = I(I^4x \otimes I^2(I^3y \otimes I^9\mathbb{1})) & \xrightarrow{\delta_X} & Ix \otimes y = X' \\ \Big\downarrow{\scriptstyle I_2^{-1}} & & \\ I^5x \otimes I^3(I^3y \otimes I^9\mathbb{1}) & & \Big\uparrow{\scriptstyle Ix\otimes\rho} \\ \Big\downarrow{\scriptstyle I^5x\otimes(I_2^{-1})^{\circ 3}} & & \\ I^5x \otimes (I^6y \otimes I^{12}\mathbb{1}) & \xrightarrow{(\iota^{-1})^{\circ 2}\otimes[(\iota^{-1})^{\circ 3}\otimes(I_0^{-1})^{\circ 12}]} & Ix \otimes (y \otimes \mathbb{1}) \end{array}$$

is a basic isomorphism in $\mathcal{F}_{\mathsf{im}}(\mathsf{C})$ from an object X that is not reduced to a reduced object X'. The morphism $(I_0^{-1})^{\circ 12}$ denotes the composite

$$I^{12}\mathbb{1} \xrightarrow{I^{11}(I_0^{-1})} I^{11}\mathbb{1} \xrightarrow{I^{10}(I_0^{-1})} \cdots \xrightarrow{I^2(I_0^{-1})} I^2\mathbb{1} \xrightarrow{I(I_0^{-1})} I\mathbb{1} \xrightarrow{I_0^{-1}} \mathbb{1}\,, \qquad \text{(total: } (I_0^{-1})^{\circ 12}\text{)}$$

The morphism $(\iota^{-1})^{\circ 2}$ denotes the composite

$$I^5x \xrightarrow{\iota_{I^3x}^{-1}} I^3x \xrightarrow{\iota_{Ix}^{-1}} Ix\,, \qquad \text{(total: } (\iota^{-1})^{\circ 2}\text{)}$$

and similarly for $(\iota^{-1})^{\circ 3}$. ◇

The next observation is an ingredient in proving that $\mathcal{F}_{\mathsf{im}}(\mathsf{C})$ is the free involutive monoidal category generated by C.

LEMMA 5.1.10. *Suppose* C *is a category. For each object* $X \in \mathcal{F}_{\mathsf{im}}(\mathsf{C})$, *there exists a canonical choice of a basic isomorphism*

$$X \xrightarrow[\cong]{\delta_X} X' \in \mathcal{F}_{\mathsf{im}}(\mathsf{C})$$

with X' *reduced.*

PROOF. If X is reduced, then its identity morphism is a basic isomorphism. So we may assume that X is not reduced.

The basic isomorphism δ_X is defined as the composite

$$X \xrightarrow{\delta_1} X_1 \xrightarrow{\delta_2} X_2 \xrightarrow{\delta_3} X_3 \xrightarrow{\delta_4} X' \qquad \text{(total: } \delta_X\text{)}$$

of the following basic isomorphisms in $\mathcal{F}_{\mathsf{im}}(\mathsf{C})$:

(1) The basic isomorphism

$$\delta_1 : X \xrightarrow{\cong} X_1$$

is a composite of monoidal products of identity morphisms with components of

$$I_2^{-1} : I(- \otimes -) \xrightarrow{\cong} I(-) \otimes I(-),$$

that brings X to the form

$$X_1 = I^{m_1} x_1 \otimes \cdots \otimes I^{m_j} x_j$$

with the same parenthesization as in X, for some objects $x_i \in \mathsf{Ob}(\mathsf{C}) \amalg \{\mathbb{1}\}$ and integers $m_i \geq 0$ for $1 \leq i \leq j$. Counting from the left, the ith factor in X_1 is $I^{m_i} x_i$. The objects x_i are the same ones that appear in X, and the object X_1 is uniquely determined by X. Moreover, as stated in [Mac98] (XI.2 Eq. (7)), the naturality of I_2 implies that δ_1 is unique.

(2) The basic isomorphism

$$\delta_2 : X_1 \xrightarrow{\cong} X_2$$

is the monoidal product

$$\delta_2 = \delta_2^1 \otimes \cdots \otimes \delta_2^j$$

with the same parenthesization as above, in which for each $1 \leq i \leq j$:

- $\delta_2^i : I^{m_i} x_i \xrightarrow{\cong} I x_i$ is a composite of components of ι^{-1} if $x_i \in \mathsf{C}$ and m_i is odd;
- $\delta_2^i : I^{m_i} x_i \xrightarrow{\cong} x_i$ is a composite of components of ι^{-1} if $x_i \in \mathsf{C}$ and m_i is even;
- $\delta_2^i : I^{m_i} \mathbb{1} \xrightarrow{\cong} \mathbb{1}$ is a composite of $I^?(I_0^{-1})$ if $x_i = \mathbb{1}$.

So

$$X_2 = I^{\epsilon_1} x_1 \otimes \cdots \otimes I^{\epsilon_j} x_j$$

with the same parenthesization as above, $\epsilon_i \in \{0, 1\}$ if $x_i \in \mathsf{C}$, and $\epsilon_i = 0$ if $x_i = \mathbb{1}$. The object X_2 is uniquely determined by X_1, hence by X as well. The definition of δ_2 involves no choices at all.

(3) The basic isomorphism

$$\delta_3 : X_2 \xrightarrow{\cong} X_3$$

is a composite of monoidal products of identity morphisms with components of λ and ρ, that removes all occurrences of $\mathbb{1}$ in X_2. So

$$X_3 = I^{\epsilon_1} x_1' \otimes \cdots \otimes I^{\epsilon_k} x_k'$$

for some parenthesization, with each $x_i' \in \mathsf{C}$ and $\epsilon_i \in \{0, 1\}$. The objects $x_i' \in \mathsf{C}$ are still exactly the ones that appear in X, and X_3 is uniquely determined by X_2, hence also by X. Moreover, δ_3 is unique by the naturality of λ and ρ, and the monoidal category axiom $\lambda_{\mathbb{1}} = \rho_{\mathbb{1}}$.

(4) The basic isomorphism

$$\delta_4 : X_3 \xrightarrow{\cong} X'$$

is a composite of monoidal products of identity morphisms with components of α^{-1}, that moves the parentheses in X_3 to the left as in (5.1.8). The object X' is uniquely determined by X_3, hence also by X. Moreover, δ_4 is unique by Mac Lane's Coherence Theorem for monoidal categories [Mac98] (VII.2 Corollary).

By construction

$$\delta_X = \delta_4 \circ \delta_3 \circ \delta_2 \circ \delta_1 : X \xrightarrow{\cong} X'$$

is a composite of basic isomorphisms, hence a basic isomorphism, with X' reduced. □

LEMMA 5.1.11. *In the context of Lemma 5.1.10, if*

$$\mathsf{word}(X) = (m_1, x_1) \cdots (m_j, x_j) \quad \textit{and} \quad \mathsf{word}(X') = (n_1, x_1') \cdots (n_k, x_k'),$$

then

- $j = k$,
- $x_i = x_i' \in \mathsf{C}$ *for each* $1 \leq i \leq j$, *and*
- $m_i \equiv n_i \pmod 2$ *for each* $1 \leq i \leq j$.

PROOF. In the decomposition $\delta_X = \delta_4\delta_3\delta_2\delta_1$, the isomorphisms δ_1, δ_3, and δ_4 do not change the underlying words of the objects involved. Moreover, copies of $\mathbb{1}$ do not contribute to the underlying words, and $\iota^{-1} : I^2 \xrightarrow{\cong} \mathrm{Id}$ reduces the exponent of I by 2. Therefore, the isomorphism δ_2 only changes the integer components by multiples of 2. □

DEFINITION 5.1.12. For each category C, denote by

$$\mathsf{C} \xrightarrow{\varepsilon_{\mathsf{C}}} \mathcal{F}_{\mathsf{im}}(\mathsf{C})$$

the full and faithful embedding given by

- the inclusion $\mathsf{Ob}(\mathsf{C}) \subseteq \mathsf{W}_{\mathsf{im}}(\mathsf{Ob}(\mathsf{C}))$ on objects, and
- the identity function

$$\mathsf{C}(x, y) \xrightarrow{\text{Id}} \mathcal{F}_{\mathsf{im}}(\mathsf{C})(x; y)$$

on morphism sets for objects $x, y \in \mathsf{C}$.

The following observation, which uses Definition 4.3.1, says that $\mathcal{F}_{\mathsf{im}}(\mathsf{C})$ is the free involutive monoidal category generated by C.

THEOREM 5.1.13. *Suppose $F : C \longrightarrow D$ is a functor with C a category and D an involutive monoidal category. Then there exists a strict involutive strict monoidal functor*

$$\mathcal{F}_{\mathsf{im}}(C) \xrightarrow{\overline{F}} D$$

such that the following three statements hold.

(1) $\overline{F}$ extends F in the sense that $\overline{F} \circ \varepsilon_C = F$.
(2) As a strict involutive strict monoidal functor extending F, $\overline{F}$ is unique.
(3) As an involutive strong monoidal functor extending F, $\overline{F}$ is unique up to an involutive monoidal natural isomorphism.

PROOF. On the objects in $\mathcal{F}_{\mathsf{im}}(\mathsf{C})$, $\overline{F}$ is defined recursively as follows:

- $\overline{F}(x) = F(x)$ for $x \in \mathsf{Ob}(\mathsf{C}) \subseteq \mathsf{Ob}(\mathcal{F}_{\mathsf{im}}(\mathsf{C}))$.
- $\overline{F}(\mathbb{1}) = \mathbb{1}^{\mathsf{D}}$, the monoidal unit in D.
- For $X, Y \in \mathsf{Ob}(\mathcal{F}_{\mathsf{im}}(\mathsf{C}))$ with $\overline{F}(X), \overline{F}(Y) \in \mathsf{D}$ already defined, we set

$$\overline{F}(X \otimes Y) = \overline{F}(X) \otimes \overline{F}(Y),$$
$$\overline{F}(IX) = I_{\mathsf{D}}\overline{F}(X).$$

For example, we have

$$\overline{F}\Big(I^4(I^2(I\mathbb{1} \otimes I^3x) \otimes I^6y)\Big) = I_{\mathsf{D}}^4(I_{\mathsf{D}}^2(I_{\mathsf{D}}\mathbb{1}^{\mathsf{D}} \otimes I_{\mathsf{D}}^3F(x)) \otimes I_{\mathsf{D}}^6F(y)).$$

To define $\overline{F}$ on morphisms, suppose $X, Y \in \mathcal{F}_{\mathsf{im}}(\mathsf{C})$ are objects with underlying words as in (5.1.4). Lemma 5.1.11 implies the equality

$$\mathcal{F}_{\mathsf{im}}(\mathsf{C})(X; Y) = \mathcal{F}_{\mathsf{im}}(\mathsf{C})(X'; Y')$$

with X' and Y' reduced objects. For a morphism

$$f = \{f_i\}_{i=1}^{j} \in \mathcal{F}_{\mathsf{im}}(\mathsf{C})(X; Y) = \prod_{i=1}^{j} \mathsf{C}(x_i, y_i) = \mathcal{F}_{\mathsf{im}}(\mathsf{C})(X'; Y'), \tag{5.1.14}$$

we define $\overline{F}(f) \in \mathsf{D}\big(\overline{F}(X), \overline{F}(Y)\big)$ as the composite

$$\overline{F}(X) \xrightarrow{\overline{F}(\delta_X)} \overline{F}(X') \xrightarrow{\overline{F}(f')} \overline{F}(Y') \xrightarrow{\overline{F}(\delta_Y)^{-1}} \overline{F}(Y), \qquad \overline{F}(f) \text{ the composite} \tag{5.1.15}$$

in D as follows:

- Using the basic isomorphism $\delta_X = \delta_4\delta_3\delta_2\delta_1$ in Lemma 5.1.10, we define

$$\overline{F}(\delta_X) = \overline{F}(\delta_4) \circ \overline{F}(\delta_3) \circ \overline{F}(\delta_2) \circ \overline{F}(\delta_1) \in \mathsf{D}$$

  as the composite $\delta_4\delta_3\delta_2\delta_1$ interpreted in the involutive monoidal category D, starting from the object $\overline{F}(X)$. The same definition applies to $\overline{F}(\delta_Y)$. For example, for the basic isomorphism δ_X in Example 5.1.9, $\overline{F}(\delta_X)$ is the composite

$$\begin{array}{ccc}
\overline{F}(X) = I_{\mathsf{D}}\big(I_{\mathsf{D}}^{4}Fx \otimes I_{\mathsf{D}}^{2}(I_{\mathsf{D}}^{3}Fy \otimes I_{\mathsf{D}}^{9}\mathbb{1}^{\mathsf{D}})\big) & \xrightarrow{\overline{F}(\delta_X)} & I_{\mathsf{D}}Fx \otimes Fy = \overline{F}(X') \\
\Big\downarrow {\scriptstyle I_2^{-1}} & & \Big\uparrow {\scriptstyle I_{\mathsf{D}}Fx \otimes \rho} \\
I_{\mathsf{D}}^{5}Fx \otimes I_{\mathsf{D}}^{3}(I_{\mathsf{D}}^{3}Fy \otimes I_{\mathsf{D}}^{9}\mathbb{1}^{\mathsf{D}}) & & \\
\Big\downarrow {\scriptstyle I_{\mathsf{D}}^{5}Fx \otimes (I_2^{-1})^{\circ 3}} & & \\
I_{\mathsf{D}}^{5}Fx \otimes (I_{\mathsf{D}}^{6}Fy \otimes I_{\mathsf{D}}^{12}\mathbb{1}^{\mathsf{D}}) & \xrightarrow{(\iota^{-1})^{\circ 2} \otimes [(\iota^{-1})^{\circ 3} \otimes (I_0^{-1})^{\circ 12}]} & I_{\mathsf{D}}Fx \otimes (Fy \otimes \mathbb{1}^{\mathsf{D}})
\end{array}$$

  in D, in which $\otimes$, $\mathbb{1}^{\mathsf{D}}$, I_{D}, I_2, I_0, ι, and ρ are part of the involutive monoidal structure of D.
- If $X' = \mathbb{1}$, then also $Y' = \mathbb{1}$ because $\mathbb{1}$ is the only reduced object whose underlying word is empty. In this case, $j = 0$, and $\overline{F}(f') = \mathrm{Id}_{\mathbb{1}^{\mathsf{D}}}$.
- If

$$X' = I^{\epsilon_1}x_1 \otimes \cdots \otimes I^{\epsilon_j}x_j$$

  with the parenthesization as in (5.1.8), then

$$Y' = I^{\epsilon_1}y_1 \otimes \cdots \otimes I^{\epsilon_j}y_j$$

  with the same parenthesization, and $\overline{F}(f')$ is the following monoidal product.

$$\begin{array}{c} \overline{F}(X') = I_{\mathsf{D}}^{\epsilon_1} F(x_1) \otimes \cdots \otimes I_{\mathsf{D}}^{\epsilon_j} F(x_j) \\ \overline{F}(f') = \Big\downarrow I_{\mathsf{D}}^{\epsilon_1} F(f_1) \otimes \cdots \otimes I_{\mathsf{D}}^{\epsilon_j} F(f_j) \\ \overline{F}(Y') = I_{\mathsf{D}}^{\epsilon_1} F(y_1) \otimes \cdots \otimes I_{\mathsf{D}}^{\epsilon_j} F(y_j) \end{array}$$

An inspection shows that $\overline{F} : \mathcal{F}_{\mathsf{im}}(\mathsf{C}) \longrightarrow \mathsf{D}$ is a functor such that $\overline{F} \circ \varepsilon_{\mathsf{C}} = F$.

We assign to $\overline{F}$:

- the identity distributive law

$$\overline{F}(IX) \xrightarrow[=]{\nu_X^{\overline{F}}} I_{\mathsf{D}} \overline{F}(X) \ ;$$

- the strict monoidal functor structure

$$\mathbb{1}^{\mathsf{D}} \xrightarrow[=]{\overline{F}_0} \overline{F}(\mathbb{1}) \quad \text{and} \quad \overline{F}(X) \otimes \overline{F}(Y) \xrightarrow[=]{\overline{F}_2} \overline{F}(X \otimes Y)$$

for objects X, Y in $\mathcal{F}_{\mathsf{im}}(\mathsf{C})$.

Using the criteria in Lemma 4.3.2, another inspection shows that $\overline{F}$ with the above structure is a strict involutive strict monoidal functor.

To see that $\overline{F}$ is unique as a strict involutive strict monoidal functor extending F, first observe that $\overline{F}$ on objects is uniquely determined as in the first paragraph of this proof. To see that $\overline{F}$ is uniquely determined on morphisms as well, note that for a morphism f as in (5.1.14), the same entries define a morphism

$$f' = \{f_i\}_{i=1}^{j} \in \mathcal{F}_{\mathsf{im}}(\mathsf{C})(X'; Y').$$

Moreover, the original morphism f factors as follows.

$$X \xrightarrow{\delta_X} X' \xrightarrow{f'} Y' \xrightarrow{\delta_Y^{-1}} Y \qquad (\text{composite } f: X \to Y)$$

The strict involutive strict monoidal functor requirement then forces $\overline{F}(f)$ to be equal to the composite $\overline{F}(\delta_Y)^{-1} \circ \overline{F}(f') \circ \overline{F}(\delta_X)$, as defined above.

To see that $\overline{F}$, as an involutive strong monoidal functor extending F, is unique up to an involutive monoidal natural isomorphism, suppose

$$G : \mathcal{F}_{\mathsf{im}}(\mathsf{C}) \longrightarrow \mathsf{D}$$

is an involutive strong monoidal functor extending F with distributive law ν^G and strong monoidal structure (G_2, G_0). We define a natural transformation

$$\sigma : \overline{F} \longrightarrow G$$

as follows:

- For an object $x \in \mathsf{C}$, there are equalities

$$\overline{F}(x) = F(x) = G(x),$$

  and we define $\sigma_x = \mathrm{Id}_{F(x)}$.
- Part of the strong monoidal structure of G gives an isomorphism

$$\overline{F}(\mathbb{1}) = \mathbb{1}^{\mathsf{D}} \xrightarrow[\cong]{\sigma_{\mathbb{1}} = G_0} G(\mathbb{1}) \ .$$

- Recursively, suppose $X, Y \in \mathcal{F}_{\mathsf{im}}(\mathsf{C})$ are objects such that σ_X and σ_Y are already defined. Then $\sigma_{X \otimes Y}$ and σ_{IX} are defined as the composites:

$$\begin{array}{ccc} \overline{F}(X \otimes Y) & \xrightarrow{\sigma_{X\otimes Y}} & G(X \otimes Y) \\ {\scriptstyle \overline{F}_2^{-1}}\downarrow = & & \cong \uparrow {\scriptstyle G_2} \\ \overline{F}(X) \otimes \overline{F}(Y) & \xrightarrow{\sigma_X \otimes \sigma_Y} & G(X) \otimes G(Y) \end{array} \qquad \begin{array}{ccc} \overline{F}(IX) & \xrightarrow{\sigma_{IX}} & G(IX) \\ {\scriptstyle \nu_X^{\overline{F}}}\downarrow = & & \cong \uparrow {\scriptstyle (\nu_X^G)^{-1}} \\ I_{\mathsf{D}}\overline{F}(X) & \xrightarrow{I_{\mathsf{D}}\sigma_X} & I_{\mathsf{D}}G(X) \end{array}$$

The naturality of σ and that it is an involutive monoidal natural transformation are both checked recursively, using the assumption that G is an involutive strong monoidal functor. □

Recall from Definition 4.3.6 that IMCat_0 is the category of small involutive monoidal categories and strict involutive strict monoidal functors.

COROLLARY 5.1.16. *There is an adjunction*

$$\mathsf{Cat} \underset{U}{\overset{\mathcal{F}_{\mathsf{im}}}{\rightleftarrows}} \mathsf{IMCat}_0$$

whose right adjoint U forgets about the involutive structure and the monoidal structure.

PROOF. For a functor $G : \mathsf{C} \longrightarrow \mathsf{D}$ between small categories, we define

$$\mathcal{F}_{\mathsf{im}}(G) : \mathcal{F}_{\mathsf{im}}(\mathsf{C}) \longrightarrow \mathcal{F}_{\mathsf{im}}(\mathsf{D})$$

as the unique strict involutive strict monoidal functor that makes the diagram

$$\begin{array}{ccc} \mathsf{C} & \xrightarrow{G} & \mathsf{D} \\ {\scriptstyle \varepsilon_{\mathsf{C}}}\downarrow & & \downarrow{\scriptstyle \varepsilon_{\mathsf{D}}} \\ \mathcal{F}_{\mathsf{im}}(\mathsf{C}) & \xrightarrow{\mathcal{F}_{\mathsf{im}}(G)} & \mathcal{F}_{\mathsf{im}}(\mathsf{D}) \end{array}$$

in Cat commutative. This defines the functor

$$\mathcal{F}_{\mathsf{im}} : \mathsf{Cat} \longrightarrow \mathsf{IMCat}_0.$$

Theorem 5.1.13 shows that the pair of functors $(\mathcal{F}_{\mathsf{im}}, U)$ satisfies the universal property for an adjunction, as stated in [Mac98] (IV.1 Theorem 2(i)). □

DEFINITION 5.1.17. For a category C, the involutive monoidal category $\mathcal{F}_{\mathsf{im}}(\mathsf{C})$ is called the *free involutive monoidal category generated by* C.

EXAMPLE 5.1.18 (Free Involutive Monoidal Category on One Object). Suppose $\mathbf{1}$ is the category with one object x and only the identity morphism of x. The free involutive monoidal category $\mathcal{F}_{\mathsf{im}}(\mathbf{1})$ generated by one object x has object set $\mathsf{W}_{\mathsf{im}}(\{x\})$, which is the set of involutive monoidal words generated by $\{x\}$. For each involutive monoidal category D and each object $z \in \mathsf{D}$, there is a unique functor

$$F : \mathbf{1} \longrightarrow \mathsf{D} \quad \text{such that} \quad F(x) = z.$$

By Theorem 5.1.13, there exists a unique strict involutive strict monoidal functor

$$\overline{F} : \mathcal{F}_{\mathsf{im}}(\mathbf{1}) \longrightarrow \mathsf{D} \quad \text{such that} \quad \overline{F}(x) = z.$$

For example, we have that

$$\overline{F}\Big(I^4\big(I^2(I\mathbb{1} \otimes I^3x) \otimes I^6x\big)\Big) = I^4_{\mathsf{D}}\big(I^2_{\mathsf{D}}(I_{\mathsf{D}}\mathbb{1}^{\mathsf{D}} \otimes I^3_{\mathsf{D}}z) \otimes I^6_{\mathsf{D}}z\big) \in \mathsf{D}.$$

Since $\mathbf{1}$ is a discrete category with one object, for objects $X, Y \in \mathcal{F}_{\mathsf{im}}(\mathbf{1})$, the morphism set $\mathcal{F}_{\mathsf{im}}(\mathbf{1})(X; Y)$:

- contains a unique isomorphism if $X \equiv Y \pmod 2$;
- is empty otherwise.

It follows that every diagram in $\mathcal{F}_{\mathsf{im}}(\mathbf{1})$ is commutative. As we will see in Section 5.3, this property of the free involutive monoidal category generated by one object implies that every formal diagram in an involutive monoidal category is commutative. ◇

5.2 Free Involutive Strict Monoidal Categories

Recall that a monoidal category is called *strict* if the associativity isomorphism, the left unit isomorphism, and the right unit isomorphism are all identity morphisms. The first purpose of this section is to give an explicit description of the free involutive strict monoidal category generated by a category. This is similar to Section 5.1, but the description is easier. Then we show that the free involutive monoidal category is equivalent to the free involutive strict monoidal category via a strict involutive strict monoidal equivalence.

DEFINITION 5.2.1. Suppose S is a class, and I is a symbol that does not appear in S.

(1) The class $\mathsf{W}_{\mathsf{istm}}(S)$ of *involutive strict monoidal words generated by* S is defined recursively by the following four conditions:

  (i) $\varnothing \in \mathsf{W}_{\mathsf{istm}}(S)$, where $\varnothing$ is the empty sequence.
  (ii) $S \subseteq \mathsf{W}_{\mathsf{istm}}(S)$.
  (iii) If $X \in \mathsf{W}_{\mathsf{istm}}(S)$, then $IX \in \mathsf{W}_{\mathsf{istm}}(S)$, where $I\varnothing = \varnothing$.
  (iv) If $X, Y \in \mathsf{W}_{\mathsf{istm}}(S)$, then so is their concatenation XY. Concatenation is associative by definition.

(2) For $X \in \mathsf{W}_{\mathsf{istm}}(S)$, its *underlying word*

$$\mathsf{word}(X) \in \coprod_{k\geq 0} (\mathbb{Z}_{\geq 0} \times S)^{\times k}$$

is defined as in Definition 5.1.1 with $\mathsf{word}(\varnothing) = \varnothing$.

(3) For $X \in \mathsf{W}_{\mathsf{istm}}(S)$, its *underlying integer sequence*

$$\mathsf{word}_1(X) \in \coprod_{k\geq 0} (\mathbb{Z}_{\geq 0})^{\times k}$$

is obtained from $\mathsf{word}(X)$ by dropping the S-entries.

DEFINITION 5.2.2. Suppose C is a category. Define the category $\mathcal{F}_{\mathsf{istm}}(\mathsf{C})$ with extra structures as in Definition 5.1.3, but with the following modifications:

Objects: $\mathsf{Ob}(\mathcal{F}_{\mathsf{istm}}(\mathsf{C})) = \mathsf{W}_{\mathsf{istm}}(\mathsf{Ob}(\mathsf{C}))$, the class of involutive strict monoidal words generated by $\mathsf{Ob}(\mathsf{C})$.

Monoidal Product: For objects $X, Y \in \mathsf{Ob}(\mathcal{F}_{\mathsf{istm}}(\mathsf{C}))$, their monoidal product $X \otimes Y$ is defined as their concatenation, which is part of the definition of the class $\mathsf{W}_{\mathsf{istm}}(\mathsf{Ob}(\mathsf{C}))$.

Monoidal Unit: The monoidal unit in $\mathcal{F}_{\mathsf{istm}}(\mathsf{C})$ is $\varnothing$.

Monoidal Structure Isomorphisms: Each component of the associativity isomorphism α, the left unit isomorphism λ, and the right unit isomorphism ρ, is an identity morphism.

Involutive Monoidal Structure: $I_0 : \varnothing \longrightarrow \varnothing$ is the identity morphism.

An inspection establishes the following.

LEMMA 5.2.3. *For each category* C*, the tuple*

$$\big((\mathcal{F}_{\mathsf{istm}}(\mathsf{C}), \otimes, \varnothing, \alpha, \lambda, \rho), (I, I_2, I_0), \iota\big)$$

in Definition 5.2.2 is an involutive strict monoidal category.

REMARK 5.2.4. As long as C has at least one object, say x, the involutive strict monoidal category $\mathcal{F}_{\mathsf{istm}}(\mathsf{C})$ is *not* strict involutive because $x \neq I^2 x$. ◇

DEFINITION 5.2.5. For each category C, denote by

$$\mathsf{C} \xrightarrow{\varepsilon_{\mathsf{C}}^{\mathsf{st}}} \mathcal{F}_{\mathsf{istm}}(\mathsf{C})$$

the full and faithful embedding given by

- the inclusion $\mathsf{Ob}(\mathsf{C}) \subseteq \mathsf{W}_{\mathsf{istm}}\big(\mathsf{Ob}(\mathsf{C})\big)$ on objects, and
- the identity function

$$\mathsf{C}(x, y) \xrightarrow{\mathrm{Id}} \mathcal{F}_{\mathsf{istm}}(\mathsf{C})(x; y)$$

on morphism sets for objects $x, y \in \mathsf{C}$.

A simplified version of the proof of Theorem 5.1.13 establishes the following result.

THEOREM 5.2.6. *Suppose* $F : \mathsf{C} \longrightarrow \mathsf{D}$ *is a functor with* C *a category and* D *an involutive strict monoidal category. Then there exists a strict involutive strict monoidal functor*

$$\mathcal{F}_{\mathsf{istm}}(\mathsf{C}) \xrightarrow{\overline{F}} \mathsf{D}$$

such that the following three statements hold.

(1) $\overline{F}$ *extends* F *in the sense that* $\overline{F} \circ \varepsilon_{\mathsf{C}}^{\mathsf{st}} = F$.
(2) As a strict involutive strict monoidal functor extending F, $\overline{F}$ *is unique.*
(3) As an involutive strong monoidal functor extending F, $\overline{F}$ *is unique up to an involutive monoidal natural isomorphism.*

DEFINITION 5.2.7. For a category C, the involutive strict monoidal category $\mathcal{F}_{\mathsf{istm}}(\mathsf{C})$ is called the *free involutive strict monoidal category generated by* C.

Next we observe that the free involutive monoidal category $\mathcal{F}_{\mathsf{im}}(\mathsf{C})$ and the free involutive strict monoidal category $\mathcal{F}_{\mathsf{istm}}(\mathsf{C})$ are equivalent.

THEOREM 5.2.8. *For each category* C*, there is a canonical strict involutive strict monoidal functor*

$$\mathcal{F}_{\mathsf{im}}(\mathsf{C}) \xrightarrow{\ \theta\ } \mathcal{F}_{\mathsf{istm}}(\mathsf{C})$$

that is also an equivalence of categories.

PROOF. The functor θ is the unique strict involutive strict monoidal functor that extends $\varepsilon_{\mathsf{C}}^{\mathsf{st}}$, in the sense that the diagram

$$\begin{array}{ccc} \mathsf{C} & \xrightarrow{\ \varepsilon_{\mathsf{C}}\ } & \mathcal{F}_{\mathsf{im}}(\mathsf{C}) \\ \| & & \downarrow \theta \\ \mathsf{C} & \xrightarrow{\ \varepsilon_{\mathsf{C}}^{\mathsf{st}}\ } & \mathcal{F}_{\mathsf{istm}}(\mathsf{C}) \end{array}$$

is commutative, given by Theorem 5.1.13. In details, the assignment on objects

$$\theta : \mathsf{W}_{\mathsf{im}}(\mathsf{Ob}(\mathsf{C})) \longrightarrow \mathsf{W}_{\mathsf{istm}}(\mathsf{Ob}(\mathsf{C}))$$

is defined recursively as follows:

- $\theta(\mathbb{1}) = \varnothing$.
- $\theta(x) = x$ for $x \in \mathsf{Ob}(\mathsf{C})$.
- If $X, Y \in \mathcal{F}_{\mathsf{im}}(\mathsf{C})$ are objects with $\theta(X), \theta(Y) \in \mathcal{F}_{\mathsf{istm}}(\mathsf{C})$ already defined, then

$$\begin{aligned} \theta(IX) &= I\theta(X), \\ \theta(X \otimes Y) &= \theta(X)\theta(Y). \end{aligned}$$

For example, for objects $x, y \in \mathsf{C}$,

$$\theta\Big(I^4\big(I^2(I\mathbb{1} \otimes I^3x) \otimes I^6y\big)\Big) = I^4\big((I^5x)(I^6y)\big).$$

Since θ does not change the underlying words, there is an equality

$$\mathcal{F}_{\mathsf{im}}(\mathsf{C})(X; Y) = \mathcal{F}_{\mathsf{istm}}(\mathsf{C})(\theta X; \theta Y)$$

of morphism sets, and θ is defined as the identity map on morphism sets. The functor θ is fully faithful on morphism sets, and is surjective on objects. Therefore, θ is an equivalence of categories. □

EXAMPLE 5.2.9 (Free Involutive Strict Monoidal Category on One Object). Suppose $\mathbf{1}$ is the category with one object x and only the identity morphism of x. By Theorem 5.2.6, the free involutive strict monoidal category $\mathcal{F}_{\mathsf{istm}}(\mathbf{1})$ has the following universal property. For each involutive strict monoidal category D and each object $z \in \mathsf{D}$, there exists a unique strict involutive strict monoidal functor

$$F : \mathcal{F}_{\mathsf{istm}}(\mathbf{1}) \longrightarrow \mathsf{D} \quad \text{such that} \quad F(x) = z.$$

Moreover, there is a unique strict involutive strict monoidal equivalence

$$\mathcal{F}_{\mathsf{im}}(\mathbf{1}) \xrightarrow[\sim]{\theta} \mathcal{F}_{\mathsf{istm}}(\mathbf{1}) \quad \text{such that} \quad \theta(x) = x.$$

Notice that θ is surjective but not injective on objects. For example, there are equalities

$$\theta(\mathbb{1}) = \theta(\mathbb{1} \otimes \mathbb{1}) = \varnothing \in \mathsf{W}_{\mathsf{istm}}(\{x\}),$$

so the objects $\mathbb{1} \neq \mathbb{1} \otimes \mathbb{1}$ have the same θ-image. ◇

5.3 Commutativity of Formal Diagrams

One version of Mac Lane's Coherence Theorem [Mac98] (VII.2) says that each formal diagram in each monoidal category commutes. A formal diagram is a diagram in which each morphism involves only the identity morphism, the associativity isomorphism, the left unit isomorphism, the right unit isomorphism, their inverses, monoidal products, and compositions. The purpose of this section is to prove the involutive monoidal version of this coherence result. The proof involves the following category. For functors $F, G : \mathsf{C} \longrightarrow \mathsf{D}$ between the same categories, we denote by $\mathsf{Nat}(F; G)$ the collection of natural transformations $F \longrightarrow G$.

DEFINITION 5.3.1. Suppose C is a small involutive monoidal category as in Definition 4.1.1. Define the category $\mathsf{It}(\mathsf{C})$ with extra structures as follows.

Objects: An object in $\mathsf{It}(\mathsf{C})$ is a pair (m, T) with

- $m \in \mathbb{Z}_{\geq 0}$ and
- $T : \mathsf{C}^{\times m} \longrightarrow \mathsf{C}$ a functor,

where $\mathsf{C}^{\times 0} = \mathbf{1}$ is the discrete category with one object.

Morphisms: For objects $(m, T), (n, U) \in \mathsf{It}(\mathsf{C})$, define the morphism set

$$\mathsf{It}(\mathsf{C})\big((m, T), (n, U)\big) = \begin{cases} \mathsf{Nat}(T; U) & \text{if } m = n, \\ \varnothing & \text{if } m \neq n. \end{cases}$$

Identity morphisms are identity natural transformations. Compositions in $\mathsf{It}(\mathsf{C})$ are vertical compositions of natural transformations.

Monoidal Product: Define the monoidal product

$$(m, T) \otimes (n, U) = (m + n, T \otimes U)$$

with $T \otimes U$ the composite

$$\mathsf{C}^{\times(m+n)} \xrightarrow{\cong} \mathsf{C}^m \times \mathsf{C}^n \xrightarrow{T \times U} \mathsf{C} \times \mathsf{C} \xrightarrow{\otimes} \mathsf{C}\,. \quad (\text{the composite is labeled } T \otimes U)$$

The monoidal product of morphisms is defined componentwise in C. In other words, for morphisms

$$\beta : (m, T) \longrightarrow (m, U) \quad \text{and} \quad \beta' : (k, T') \longrightarrow (k, U'),$$

the monoidal product

$$\beta \otimes \beta' : (m + k, T \otimes T') \longrightarrow (m + k, U \otimes U')$$

is given componentwise by

$$(\beta \otimes \beta')_{(x_1,\ldots,x_m,y_1,\ldots,y_k)} = \beta_{(x_1,\ldots,x_m)} \otimes \beta'_{(y_1,\ldots,y_k)}$$

for objects $x_1, \ldots, x_m, y_1, \ldots, y_k \in \mathsf{C}$.

Monoidal Unit: The monoidal unit in $\mathsf{It}(\mathsf{C})$ is $(0, \mathbb{1})$, where $\mathbb{1} : \mathbf{1} \longrightarrow \mathsf{C}$ denotes the constant functor at the monoidal unit $\mathbb{1}^{\mathsf{C}} \in \mathsf{C}$.

Monoidal Structure Isomorphisms: The associativity isomorphism, the left unit isomorphism, and the right unit isomorphism in $\mathsf{It}(\mathsf{C})$ are defined componentwise in C using the monoidal structure isomorphisms in C.

Involution Functor: Define a functor

$$I : \mathsf{It}(\mathsf{C}) \longrightarrow \mathsf{It}(\mathsf{C})$$

by

- the assignment

$$I(m, T) = \left(m, \ \mathsf{C}^{\times m} \xrightarrow{T} \mathsf{C} \xrightarrow{I_{\mathsf{C}}} \mathsf{C}\right)$$

  on objects, and
- the horizontal composition

$$I\beta = \beta * \mathrm{Id}_{I_{\mathsf{C}}} \quad \text{for} \quad \beta \in \mathsf{It}(\mathsf{C})\big((m, T), (m, U)\big).$$

In other words, the morphism

$$I\beta : I(m, T) \longrightarrow I(m, U)$$

is given componentwise by

$$(I\beta)_{(x_1,\ldots,x_m)} = I_{\mathsf{C}}\big(\beta_{(x_1,\ldots,x_n)}\big).$$

Unit: Define a natural isomorphism

$$\iota : \mathrm{Id}_{\mathsf{It(C)}} \xrightarrow{\cong} I^2$$

componentwise in C using the involutive structure unit $\iota^{\mathsf{C}} : \mathrm{Id}_{\mathsf{C}} \xrightarrow{\cong} I^2_{\mathsf{C}}$ in C. In other words, it is componentwise the horizontal composition

$$\iota_{(m,T)} = \mathrm{Id}_T * \iota^{\mathsf{C}}.$$

Involutive Monoidal Structure: Define the morphism

$$I_0 : (0, \mathbb{1}) \longrightarrow I(0, \mathbb{1})$$

using the morphism $I_0 : \mathbb{1}^{\mathsf{C}} \longrightarrow I_{\mathsf{C}}\mathbb{1}^{\mathsf{C}}$ in C. Similarly, define the natural transformation

$$I(-) \otimes I(-) \xrightarrow{I_2} I(- \otimes -)$$

using the corresponding natural transformation in C.

This finishes the definition of $\mathsf{It(C)}$.

REMARK 5.3.2. In Definition 5.3.1, the smallness assumption on C ensures that for objects (m, T) and (m, U), the collection of morphisms $(m, T) \longrightarrow (m, U)$ is a set. ◇

Using the diagrams in Explanation 4.1.2, an inspection establishes the following.

LEMMA 5.3.3. *$\mathsf{It}(\mathsf{C})$ in Definition 5.3.1 is an involutive monoidal category.*

Recall that $\mathbf{1} = \{x\}$ is a category with one object x and only the identity morphism. The following result is the coherence statement about the commutativity of all formal diagrams in a small involutive monoidal category.

COROLLARY 5.3.4. *Suppose C is a small involutive monoidal category. Then the following statements hold.*

(1) There exists a unique strict involutive strict monoidal functor

$$\mathcal{F}_{\mathsf{im}}(\mathbf{1}) \xrightarrow{\Phi} \mathsf{It}(\mathsf{C}) \quad \textit{such that} \quad \Phi(x) = (1, \mathrm{Id}_{\mathsf{C}}).$$

(2) For objects $X, Y \in \mathcal{F}_{\mathsf{im}}(\mathbf{1})$ with $X \equiv Y \pmod 2$, the unique isomorphism $X \xrightarrow{\cong} Y$ is sent to a unique natural isomorphism

$$\Phi(X) \xrightarrow[\cong]{\phi_{X,Y}} \Phi(Y) \quad : C^{\times|X|} \longrightarrow C$$

that involves only

- *the isomorphisms $\mathrm{Id}_{\mathbb{1}^C}$ and $I_0 : \mathbb{1}^C \longrightarrow I_C \mathbb{1}^C$,*
- *the natural isomorphisms $\mathrm{Id}_{\mathrm{Id}_C}$, ι^C, I_2, α, λ, ρ, and their inverses,*
- *the functors I_C and $\otimes$, and*
- *categorical composites in C.*

(3) The Φ-image of each diagram in $\mathcal{F}_{\mathsf{im}}(\mathbf{1})$ is commutative.

PROOF. Since $\mathsf{It}(\mathsf{C})$ is an involutive monoidal category with $(1, \mathrm{Id}_{\mathsf{C}}) \in \mathsf{It}(\mathsf{C})$ an object, the first assertion is a special case of the Coherence Theorem 5.1.13, as explained in Example 5.1.18. The second assertion is a consequence of the existence of the strict involutive strict monoidal functor Φ. The last assertion is true because every diagram in $\mathcal{F}_{\mathsf{im}}(\mathbf{1})$ is commutative. □

EXAMPLE 5.3.5. Using the construction in the proof of Theorem 5.1.13, let us explain in more details the functor Φ in Corollary 5.3.4.

- The monoidal unit $\mathbb{1} \in \mathcal{F}_{\mathsf{im}}(\mathbf{1})$ is sent to the monoidal unit in $\mathsf{It}(\mathsf{C})$, so

$$\Phi(\mathbb{1}) = (0, \mathbb{1}).$$

- Each object $X \in \mathcal{F}_{\mathsf{im}}(\mathbf{1})$ with $|X| = n$ is sent to a pair

$$\Phi(X) = \left(n, \quad \mathsf{C}^{\times n} \xrightarrow{X_{\mathsf{C}}} \mathsf{C}\right).$$

- With X as above and $k \geq 1$, $I^k X \in \mathcal{F}_{\mathsf{im}}(\mathbf{1})$ is sent to a pair

$$\Phi(I^k X) = \left(n, \quad \mathsf{C}^{\times n} \xrightarrow{X_{\mathsf{C}}} \mathsf{C} \xrightarrow{I^k_{\mathsf{C}}} \mathsf{C}\right).$$

- The object $I^k x \in \mathcal{F}_{\mathsf{im}}(\mathbf{1})$ is sent to

$$\Phi(I^k x) = \left(1, \quad \mathsf{C} \xrightarrow{I^k_{\mathsf{C}}} \mathsf{C}\right).$$

- The x-component of the unit, $\iota_x : x \xrightarrow{\cong} I^2 x \in \mathcal{F}_{\mathsf{im}}(\mathbf{1})$, is sent to the morphism

$$\Phi(\iota_x) = (1, \mathrm{Id}_{\mathsf{C}}) \longrightarrow (1, I^2_{\mathsf{C}}),$$

  which is the unit $\iota^{\mathsf{C}} : \mathrm{Id}_{\mathsf{C}} \xrightarrow{\cong} I^2_{\mathsf{C}}$ in C.
- Similarly, the involutive monoidal structures $\otimes$, I_0, I_2, α, λ, and ρ in $\mathcal{F}_{\mathsf{im}}(\mathbf{1})$ are sent to their corresponding structure morphisms in C.

  For example,

$$\Phi\Big(I^4\big(I^2(I\mathbb{1} \otimes I^3 x) \otimes I^6 x\big)\Big) = \big(2, \mathsf{C}^{\times 2} \longrightarrow \mathsf{C}\big),$$

where $\mathsf{C}^{\times 2} \longrightarrow \mathsf{C}$ is the following composite.

$$\begin{array}{ccccc}
\mathsf{C}\times\mathsf{C} & \longrightarrow & & & \mathsf{C} \\
\cong\big\downarrow & & & & \big\uparrow I^4_{\mathsf{C}} \\
\mathbf{1}\times\mathsf{C}\times\mathsf{C} & & & & \mathsf{C} \\
I_{\mathsf{C}}\mathbb{1}^{\mathsf{C}}\times I^3_{\mathsf{C}}\times I^6_{\mathsf{C}}\big\downarrow & & & & \big\uparrow \otimes \\
\mathsf{C}\times\mathsf{C}\times\mathsf{C} & \xrightarrow{\otimes\times\mathrm{Id}} & \mathsf{C}\times\mathsf{C} & \xrightarrow{I^2_{\mathsf{C}}\times\mathrm{Id}} & \mathsf{C}\times\mathsf{C}
\end{array}$$

Let us define a *formal diagram* in C as a diagram in $\mathsf{It}(\mathsf{C})$ that is the Φ-image of a diagram in $\mathcal{F}_{\mathsf{im}}(\mathbf{1})$. Then Corollary 5.3.4(3) says that every formal diagram in a small involutive monoidal category is commutative. ◇

5.4 Strictification of Involutive Monoidal Categories

Recall that a monoidal category is called *strict* if the associativity isomorphism, the left unit isomorphism, and the right unit isomorphism are all identity morphisms. A monoidal functor (F, F_2, F_0) is called *strong* if the structure morphisms F_2 and F_0 are isomorphisms; it is sometimes called a *tensor functor* in the literature. One version of Mac Lane's Coherence Theorem 1.2.20 for monoidal categories states that each monoidal category is equivalent to a strict monoidal category via strong monoidal functors. The purpose of this section is to prove the involutive monoidal version of this strictification result. Recall that an involutive strict monoidal category is an involutive monoidal category whose underlying monoidal category is strict. Also recall from Definition 2.4.1 the concept of an involutive adjoint equivalence.

THEOREM 5.4.1. *Suppose C is an involutive monoidal category as in Definition 4.1.1. Then there exist canonical choices of*

(i) an involutive strict monoidal category C_{st} *and*
(ii) an involutive adjoint equivalence

$$\mathsf{C} \underset{R}{\overset{L}{\rightleftarrows}} \mathsf{C}_{\mathsf{st}}$$

such that the following five statements hold:

- *The unit* $\epsilon : \mathrm{Id}_{\mathsf{C}} \longrightarrow RL$ *is the identity natural transformation.*
- L *is a strict involutive strong monoidal functor.*
- R *is an involutive strong monoidal functor.*
- *The unit and the counit of the adjunction* $L \dashv R$ *are both monoidal natural transformations and involutive natural transformations.*
- *If the involutive structure on* C *is strict (resp., trivial)–i.e.,* $I^2 = \mathrm{Id}_{\mathsf{C}}$ *(resp.,* $I = \mathrm{Id}_{\mathsf{C}}$*) and* $\iota = \mathrm{Id}_{\mathrm{Id}_{\mathsf{C}}}$*–then the same is true for* C_{st}*.*

PROOF. Suppose C is an involutive monoidal category as in Definition 4.1.1. Since C is a monoidal category, by Mac Lane's Coherence Theorem 1.2.20, there exist a strict monoidal category C_{st} and an adjoint equivalence

$$\mathsf{C} \underset{R}{\overset{L}{\rightleftarrows}} \mathsf{C}_{\mathsf{st}}$$

with (i) both L and R strong monoidal functors, and (ii) $RL = \mathrm{Id}_{\mathsf{C}}$. We will equip C_{st} with the structure of an involutive strict monoidal category. Let us first recall its monoidal category structure and the definition of the adjunction $L \dashv R$.

Objects: Its object set is

$$\mathsf{Ob}(\mathsf{C}_{\mathsf{st}}) = \coprod_{k \geq 0} \mathsf{Ob}(\mathsf{C})^{\times k},$$

so an object in C_{st} is a finite, possibly-empty, sequence of objects in C. For an object $\underline{x} = (x_1, \ldots, x_m) \in \mathsf{C}^{\times m}$, we define the left-normalized monoidal product

$$\underline{x}^{\otimes} = x_1 \otimes \cdots \otimes x_m = \begin{cases} \mathbb{1} & \text{if } m = 0, \\ x_1 & \text{if } m = 1, \\ x_1 \otimes x_2 & \text{if } m = 2, \\ \big(\cdots (x_1 \otimes x_2) \otimes \cdots \big) \otimes x_m & \text{if } m > 2 \end{cases} \tag{5.4.2}$$

in C. When $m > 2$, the left half of each pair of parentheses is at the front.

Morphisms: For objects $\underline{x}$ as above and $\underline{y} = (y_1, \ldots, y_n) \in \mathsf{C}^{\times n}$, C_{st} has the morphism set

$$\mathsf{C}_{\mathsf{st}}\big(\underline{x}; \underline{y}\big) = \mathsf{C}\big(\underline{x}^{\otimes}, \underline{y}^{\otimes}\big),$$

and the identity morphism

$$\mathrm{Id}_{\underline{x}} = \mathrm{Id}_{\underline{x}^{\otimes}} \in \mathsf{C}.$$

Composition in C_{st} is defined in C.

Monoidal Structure: The monoidal product on objects in C_{st} is given by concatenation of sequences, i.e.,

$$\underline{x} \otimes^{\mathsf{st}} \underline{y} = \big(x_1, \ldots, x_m, y_1, \ldots, y_n\big) \in \mathsf{C}^{\times(m+n)}.$$

The monoidal unit is the empty sequence. For morphisms $f \in \mathsf{C}_{\mathsf{st}}\big(\underline{x}; \underline{y}\big)$ and $g \in \mathsf{C}_{\mathsf{st}}\big(\underline{v}; \underline{w}\big)$, their monoidal product $f \otimes^{\mathsf{st}} g$ in C_{st} is the composite

$$(\underline{x}, \underline{v})^{\otimes} \xrightarrow{\cong} \underline{x}^{\otimes} \otimes \underline{v}^{\otimes} \xrightarrow{f \otimes g} \underline{y}^{\otimes} \otimes \underline{w}^{\otimes} \xrightarrow{\cong} (\underline{y}, \underline{w})^{\otimes} \qquad \text{(overall arrow: } f \otimes^{\mathsf{st}} g\text{)} \tag{5.4.3}$$

in C in which:

- The first and the last isomorphisms are the unique isomorphisms guaranteed by Mac Lane's Coherence Theorem [Mac98] (VII.2 Corollary). They involve only identity morphisms, the monoidal structure isomorphisms α, λ, and ρ, their inverses, the monoidal product $\otimes$, and composites.
- $f \otimes g$ is the monoidal product in C.

The associativity isomorphism, the left unit isomorphism, and the right unit isomorphism in C_{st} are identity morphisms.

Left Adjoint: The left adjoint $L : \mathsf{C} \longrightarrow \mathsf{C}_{\mathsf{st}}$ is given by

- $L(x) = (x)$ for objects $x \in \mathsf{C}$, and
- the identity function

$$\mathsf{C}(x, y) = \mathsf{C}_{\mathsf{st}}\big((x); (y)\big)$$

on morphism sets.

The strong monoidal structure on L is given by the isomorphisms

$$\varnothing \xrightarrow{\mathrm{Id}_{\mathbb{1}}} (\mathbb{1}) \quad \text{and} \quad (x)(y) = (x, y) \xrightarrow{\mathrm{Id}_{x \otimes y}} (x \otimes y)$$

for $x, y \in \mathsf{C}$.

Right Adjoint The right adjoint $R : \mathsf{C}_{\mathsf{st}} \longrightarrow \mathsf{C}$ is given by

- $R(\underline{x}) = \underline{x}^{\otimes}$ on objects, and
- the identity function

$$\mathsf{C}_{\mathsf{st}}\big(\underline{x}; \underline{y}\big) = \mathsf{C}(\underline{x}^{\otimes}, \underline{y}^{\otimes})$$

on morphism sets.

The strong monoidal structure on R is given by the isomorphisms

$$\mathbb{1} \xrightarrow{\mathrm{Id}_{\mathbb{1}}} \mathbb{1} = R(\varnothing) \qquad \text{and} \qquad \underline{x}^{\otimes} \otimes \underline{y}^{\otimes} \xrightarrow{\cong} (\underline{x}, \underline{y})^{\otimes}$$

for $\underline{x}, \underline{y} \in \mathsf{C}_{\mathsf{st}}$. The second isomorphism is the unique isomorphism guaranteed by Mac Lane's Coherence Theorem.

Unit and Counit: The unit $\epsilon : \mathrm{Id}_{\mathsf{C}} \longrightarrow RL$ of the adjunction $L \dashv R$ is the identity natural transformation $\mathrm{Id}_{\mathrm{Id}_{\mathsf{C}}}$. The counit $\eta : LR \longrightarrow \mathrm{Id}_{\mathsf{C}_{\mathsf{st}}}$ is componentwise the isomorphism

$$LR(\underline{x}) = L(\underline{x}^{\otimes}) = (\underline{x}^{\otimes}) \xrightarrow{\mathrm{Id}_{\underline{x}^{\otimes}}} \underline{x}$$

for $\underline{x} \in \mathsf{C}_{\mathsf{st}}$.

We equip C_{st} with the involutive structure $(I_{\mathsf{st}}, \iota^{\mathsf{st}})$ as follows.

Involution Functor: On objects

$$I_{\mathsf{st}} : \mathsf{C}_{\mathsf{st}} \longrightarrow \mathsf{C}_{\mathsf{st}}$$

is defined entrywise as

$$I_{\mathsf{st}}(x_1, \ldots, x_m) = \begin{cases} \big(Ix_1, \ldots, Ix_m\big) & \text{if } m > 0,\\ \varnothing & \text{if } m = 0 \end{cases}$$

for $(x_1, \ldots, x_m) \in \mathsf{C}^{\times m}$, where (I, ι) is the involutive structure on C. For a morphism $f : \underline{x} \longrightarrow \underline{y} \in \mathsf{C}_{\mathsf{st}}$, $I_{\mathsf{st}} f$ is defined as the composite

$$\overset{I_{\mathsf{st}}f}{\overbrace{(I_{\mathsf{st}}\underline{x})^{\otimes} \xrightarrow{\cong} I(\underline{x}^{\otimes}) \xrightarrow{If} I(\underline{y}^{\otimes}) \xrightarrow{\cong} (I_{\mathsf{st}}\underline{y})^{\otimes}}} \tag{5.4.4}$$

in C. The first and the last isomorphisms are the unique isomorphisms given by the strong monoidal structure (I_2, I_0) of the involution functor I. The uniqueness of these isomorphisms implies that I_{st} is a functor.

Unit: For an object $\underline{x} \in \mathsf{C}_{\mathsf{st}}$ as above, the $\underline{x}$-component of the unit

$$\iota^{\mathsf{st}} : \mathrm{Id}_{\mathsf{C}_{\mathsf{st}}} \xrightarrow{\cong} I_{\mathsf{st}}^2$$

is defined as the isomorphism

$$\underline{x}^{\otimes} \xrightarrow{\iota_{x_1} \otimes \cdots \otimes \iota_{x_m}} \left(I_{\mathsf{st}}^2 \underline{x}\right)^{\otimes} \in \mathsf{C},$$

which means $\mathrm{Id}_{\mathbb{1}} \in \mathsf{C}$ if $m = 0$. Each $\iota_{x_i} : x_i \xrightarrow{\cong} I^2 x_i$ is the x_i-component of the unit ι in C. The naturality of ι^{st} with respect to a morphism $f : \underline{x} \longrightarrow \underline{y} \in \mathsf{C}_{\mathsf{st}}$ as above is the outer-most diagram below.

$$\begin{array}{ccccc}
\underline{x}^{\otimes} & \xrightarrow{\iota_{x_1} \otimes \cdots \otimes \iota_{x_m}} & (I_{\mathsf{st}}^2 \underline{x})^{\otimes} & \xrightarrow{\cong} & I((I_{\mathsf{st}} \underline{x})^{\otimes}) \\
\| & & & & \downarrow{\scriptstyle I(\cong)} \\
\underline{x}^{\otimes} & & \xrightarrow{\iota_{\underline{x}^{\otimes}}} & & I^2(\underline{x}^{\otimes}) \\
{\scriptstyle f}\downarrow & & & & \downarrow{\scriptstyle I^2 f} \\
\underline{y}^{\otimes} & & \xrightarrow{\iota_{\underline{y}^{\otimes}}} & & I^2(\underline{y}^{\otimes}) \\
\| & & & & \downarrow{\scriptstyle I(\cong)} \\
\underline{y}^{\otimes} & \xrightarrow{\iota_{y_1} \otimes \cdots \otimes \iota_{y_n}} & (I_{\mathsf{st}}^2 \underline{y})^{\otimes} & \xleftarrow{\cong} & I((I_{\mathsf{st}} \underline{y})^{\otimes})
\end{array}$$

The top and the bottom rectangles are commutative by the assumption that $\iota : \mathrm{Id}_{\mathsf{C}} \xrightarrow{\cong} I^2$ is a monoidal natural transformation, i.e., the diagrams (4.1.5) and (4.1.6). The middle rectangle is commutative by the naturality of $\iota : \mathrm{Id}_{\mathsf{C}} \xrightarrow{\cong} I^2$.

The triangle identity (2.1.2) for $(I_{\mathsf{st}}, \iota^{\mathsf{st}})$ follows from the triangle identity for (I, ι) and the naturality of the strong monoidal functor (I, I_2, I_0). So $\left(\mathsf{C}_{\mathsf{st}}, I_{\mathsf{st}}, \iota^{\mathsf{st}}\right)$ is an involutive category. Moreover, if the involutive structure on C is strict (resp., trivial), then the same is true for C_{st}.

Next we equip the involution functor I_{st} with the strict monoidal functor structure $\left((I_{\mathsf{st}})_2, (I_{\mathsf{st}})_0\right)$ as follows.

- The structure morphism

$$\varnothing \xrightarrow{(I_{\mathsf{st}})_0} \varnothing = I_{\mathsf{st}}(\varnothing)$$

is the identity morphism of $\varnothing$ in C_{st}, i.e., $\mathrm{Id}_{\mathbb{1}} \in \mathsf{C}$.

- For objects $\underline{x}, \underline{y} \in \mathsf{C}_{\mathsf{st}}$, the structure morphism

$$I_{\mathsf{st}}\underline{x} \otimes^{\mathsf{st}} I_{\mathsf{st}}\underline{y} \xrightarrow{(I_{\mathsf{st}})_2} I_{\mathsf{st}}\big(\underline{x} \otimes^{\mathsf{st}} \underline{y}\big)$$

is the identity morphism of $\big(Ix_1, \ldots, Ix_m, Iy_1, \ldots, Iy_n\big)$ in C_{st}.

The diagrams (1.2.12), (1.2.13), and (1.2.14) for $\big(I_{\mathsf{st}}, (I_{\mathsf{st}})_2, (I_{\mathsf{st}})_0\big)$ are commutative, so it is a strict monoidal functor.

To see that ι^{st} is a monoidal natural transformation, first observe that the unit diagram (4.1.6) is commutative because every morphism in it is the identity morphism. The diagram (4.1.5) is commutative because

$$\iota^{\mathsf{st}}_{\underline{x} \otimes^{\mathsf{st}} \underline{y}} = \iota^{\mathsf{st}}_{\underline{x}} \otimes^{\mathsf{st}} \iota^{\mathsf{st}}_{\underline{y}},$$

which in turn is true by the naturality of the monoidal structure (α, λ, ρ) in C and the definition (5.4.3) of $\otimes^{\mathsf{st}}$ on morphisms. So C_{st} is an involutive strict monoidal category.

Next we equip the strong monoidal functor $L : \mathsf{C} \longrightarrow \mathsf{C}_{\mathsf{st}}$ with the distributive law

$$\nu^L : LI \longrightarrow I_{\mathsf{st}} L$$

defined componentwise as the identity morphism

$$LI(x) = (Ix) = I_{\mathsf{st}}(x) = I_{\mathsf{st}}L(x), \qquad \nu^L_x : LI(x) \to I_{\mathsf{st}}L(x)$$

for $x \in \mathsf{C}$. This makes (L, ν^L) into a strict involutive functor. Moreover, an inspection shows that the diagrams (4.3.3) and (4.3.4) are commutative, so L is a strict involutive strong monoidal functor.

We now equip the strong monoidal functor $R : \mathsf{C}_{\mathsf{st}} \longrightarrow \mathsf{C}$ with the distributive law

$$\nu^R : RI_{\mathsf{st}} \longrightarrow IR$$

defined componentwise as the unique isomorphism

$$RI_{\mathsf{st}}(\underline{x}) = \big(I_{\mathsf{st}}\underline{x}\big)^{\otimes} \xrightarrow[\cong]{\nu^R_{\underline{x}}} I\big(\underline{x}^{\otimes}\big) = IR(\underline{x}) \tag{5.4.5}$$

given by the strong monoidal functor structure of (I, I_2, I_0).

- The distributive law axiom (2.2.2) for (R, ν^R) holds because $\iota : \mathrm{Id}_{\mathsf{C}} \xrightarrow{\cong} I^2$ is a monoidal natural transformation, i.e., the diagrams (4.1.5) and (4.1.6). So (R, ν^R) is an involutive functor.

- Furthermore, the diagram (4.3.4) is commutative because each path is equal to $I_0 : \mathbb{1} \longrightarrow I\mathbb{1}$.
- The diagram (4.3.3) is commutative because (I, I_2, I_0) is a monoidal functor, i.e., the diagrams (1.2.12), (1.2.13), and (1.2.14).

So R is an involutive strong monoidal functor.

It remains to check that the unit $\epsilon : \mathrm{Id}_{\mathsf{C}} \longrightarrow RL$ and the counit $\eta : LR \longrightarrow \mathrm{Id}_{\mathsf{C}^{\mathsf{st}}}$ are both involutive natural transformations and monoidal natural transformations. The axiom (2.3.2) for the counit η to be an involutive natural transformation is the diagram below.

$$\begin{array}{ccc} LRI_{\mathsf{st}}(\underline{x}) & \xrightarrow{\mathrm{Id}_{(I_{\mathsf{st}}\underline{x})^{\otimes}}} & I_{\mathsf{st}}\underline{x} \\ {\scriptstyle L(\nu^R_{\underline{x}})}\downarrow & & \uparrow{\scriptstyle I_{\mathsf{st}}\eta_{\underline{x}}}\ \cong \\ LIR(\underline{x}) & \overset{\nu^L_{R\underline{x}}}{=\!=\!=} & I_{\mathsf{st}}LR(\underline{x}) \end{array}$$

The left vertical isomorphism is $\nu^R_{\underline{x}}$, while the right vertical isomorphism is $(\nu^R_{\underline{x}})^{-1}$ by the definition of I_{st} applied to a morphism (5.4.4). So the above diagram is commutative. Similarly, that the counit η is a monoidal natural transformation follows from the definition of $\otimes^{\mathsf{st}}$ on morphisms (5.4.3). The unit $\epsilon = \mathrm{Id}_{\mathrm{Id}_{\mathsf{C}}}$ is both involutive and monoidal because it is the identity natural transformation of the identity functor. □

5.5 Free Involutive Monoidal Categories of Involutive Categories

The purpose of this section is to give an explicit description of the free involutive monoidal category generated by an involutive category. Most of the constructions and proofs are adapted from Section 5.1.

DEFINITION 5.5.1. Suppose S is a class equipped with an assignment

$$I_S : S \longrightarrow S,$$

and $\mathbb{1}$, I, and $\otimes$ are symbols that do not appear in S.

(1) The class $\mathsf{W}^{\mathsf{i}}_{\mathsf{im}}(S)$ is defined recursively by the following three conditions:

   (i) $S \subseteq \mathsf{W}^{\mathsf{i}}_{\mathsf{im}}(S)$ and $\mathbb{1} \in \mathsf{W}^{\mathsf{i}}_{\mathsf{im}}(S)$.
   (ii) If $X, Y \in \mathsf{W}^{\mathsf{i}}_{\mathsf{im}}(S)$, then $X \otimes Y \in \mathsf{W}^{\mathsf{i}}_{\mathsf{im}}(S)$. The operation $\otimes$ is not associative by definition.
   (iii) If $X \in \mathsf{W}^{\mathsf{i}}_{\mathsf{im}}(S) \setminus S$, then $IX \in \mathsf{W}^{\mathsf{i}}_{\mathsf{im}}(S)$.

   For an element $x \in S$, we define $Ix = I_S x$.

(2) For $X \in \mathsf{W}^{\mathsf{i}}_{\mathsf{im}}(S)$, its *underlying word*

$$\mathsf{word}(X) \in \coprod_{k \geq 0} S^{\times k}$$

is the possibly-empty finite sequence in S defined recursively by the following four conditions:

(i) $\mathsf{word}(\mathbb{1}) = \varnothing \in S^{\times 0}$, the empty sequence.
(ii) $\mathsf{word}(x) = x$ for $x \in S$.
(iii) If $X, Y \in \mathsf{W}^{\mathsf{i}}_{\mathsf{im}}(S)$ with $\mathsf{word}(X)$ and $\mathsf{word}(Y)$ already defined, then

$$\mathsf{word}(X \otimes Y) = \mathsf{word}(X)\mathsf{word}(Y),$$

the concatenation of $\mathsf{word}(X)$ and $\mathsf{word}(Y)$. Concatenation is associative by definition.
(iv) If $X \in \mathsf{W}^{\mathsf{i}}_{\mathsf{im}}(S)$ with $\mathsf{word}(X) = (x_1, \ldots, x_m) \in S^{\times m}$, then

$$\mathsf{word}(IX) = \begin{cases} \varnothing & \text{if } \mathsf{word}(X) = \varnothing, \\ (Ix_1, \ldots, Ix_m) & \text{if } m > 0. \end{cases}$$

EXAMPLE 5.5.2. If $x, y \in S$, then

$$\mathsf{word}\Big(I^4\big(I^2(I\mathbb{1} \otimes x) \otimes y\big)\Big) = \big(I_S^6 x, I_S^4 y\big),$$

where $I^n = I \cdots I$ with $n > 0$ copies of I. ◇

The next definition is similar to Definition 5.1.3 except for the involutive structure.

DEFINITION 5.5.3. Suppose $(\mathsf{C}, I_{\mathsf{C}}, \iota^{\mathsf{C}})$ is an involutive category. Define the category $\mathcal{F}^{\mathsf{i}}_{\mathsf{im}}(\mathsf{C})$ with extra structures as follows.

Objects: $\mathsf{Ob}\big(\mathcal{F}^{\mathsf{i}}_{\mathsf{im}}(\mathsf{C})\big) = \mathsf{W}^{\mathsf{i}}_{\mathsf{im}}\big(\mathsf{Ob}(\mathsf{C})\big)$, where $\mathsf{Ob}(\mathsf{C})$ is equipped with the assignment $I_{\mathsf{C}} : \mathsf{Ob}(\mathsf{C}) \longrightarrow \mathsf{Ob}(\mathsf{C})$.

Morphisms: For $X, Y \in \mathsf{Ob}\big(\mathcal{F}^{\mathsf{i}}_{\mathsf{im}}(\mathsf{C})\big)$ with

$$\mathsf{word}(X) = (x_1, \ldots, x_j) \in \mathsf{Ob}(\mathsf{C})^{\times j} \quad \text{and} \quad \mathsf{word}(Y) = (y_1, \ldots, y_k) \in \mathsf{Ob}(\mathsf{C})^{\times k},$$

it has the morphism set

$$\mathcal{F}^{\mathsf{i}}_{\mathsf{im}}(\mathsf{C})(X; Y) = \begin{cases} \prod_{i=1}^{j} \mathsf{C}(x_i, y_i) & \text{if } j = k, \\ \varnothing & \text{if } j \neq k, \end{cases}$$

where an empty product, for the case $j = 0$, is taken as a one-point set. Identity morphisms and composition are defined in C.

Monoidal Product: For objects $X, Y \in \mathsf{Ob}(\mathcal{F}^{\mathsf{i}}_{\mathsf{im}}(\mathsf{C}))$, their monoidal product $X \otimes Y$ is defined as part of the definition of the class $\mathsf{W}^{\mathsf{i}}_{\mathsf{im}}(\mathsf{Ob}(\mathsf{C}))$. The monoidal product on morphisms in $\mathcal{F}^{\mathsf{i}}_{\mathsf{im}}(\mathsf{C})$ is defined as concatenation of finite sequences of morphisms in C.

Monoidal Unit: The monoidal unit in $\mathcal{F}^{\mathsf{i}}_{\mathsf{im}}(\mathsf{C})$ is $\mathbb{1} \in \mathsf{W}^{\mathsf{i}}_{\mathsf{im}}(\mathsf{Ob}(\mathsf{C}))$.

Associativity Isomorphisms: For X, Y as above and $Z \in \mathsf{Ob}(\mathcal{F}^{\mathsf{i}}_{\mathsf{im}}(\mathsf{C}))$ with

$$\mathsf{word}(Z) = (z_1, \ldots, z_l) \in \mathsf{Ob}(\mathsf{C})^{\times l},$$

the associativity isomorphism

$$(X \otimes Y) \otimes Z \xrightarrow[\cong]{\alpha_{X,Y,Z}} X \otimes (Y \otimes Z) \quad \in \mathcal{F}^{\mathsf{i}}_{\mathsf{im}}(\mathsf{C})$$

is defined as

$$\left\{ \{\mathrm{Id}_{x_i}\}_{i=1}^{j}, \{\mathrm{Id}_{y_i}\}_{i=1}^{k}, \{\mathrm{Id}_{z_i}\}_{i=1}^{l} \right\} \in \prod_{i=1}^{j} \mathsf{C}(x_i, x_i) \times \prod_{i=1}^{k} \mathsf{C}(y_i, y_i) \times \prod_{i=1}^{l} \mathsf{C}(z_i, z_i).$$

Left and Right Units: For X as above, the left and the right unit isomorphisms

$$\mathbb{1} \otimes X \xrightarrow[\cong]{\lambda_X} X \xleftarrow[\cong]{\rho_X} X \otimes \mathbb{1} \quad \in \mathcal{F}^{\mathsf{i}}_{\mathsf{im}}(\mathsf{C})$$

are both given by $\{\mathrm{Id}_{x_i}\}_{i=1}^{j}$.

Involution Functor: Define a functor

$$I : \mathcal{F}^{\mathsf{i}}_{\mathsf{im}}(\mathsf{C}) \longrightarrow \mathcal{F}^{\mathsf{i}}_{\mathsf{im}}(\mathsf{C})$$

by the assignment $X \longmapsto IX$ on objects, and the function

$$\begin{array}{ccc} \mathcal{F}^{\mathsf{i}}_{\mathsf{im}}(X; Y) = \prod_{i=1}^{j} \mathsf{C}(x_i, y_i) & \ni & \{f_i\}_{i=1}^{j} \\ \big\downarrow I & & \big\downarrow \\ \mathcal{F}^{\mathsf{i}}_{\mathsf{im}}(IX; IY) = \prod_{i=1}^{j} \mathsf{C}(I_{\mathsf{C}} x_i, I_{\mathsf{C}} y_i) & \ni & \{I_{\mathsf{C}} f_i\}_{i=1}^{j} \end{array}$$

on morphisms. If $j = 0$, then this function is the unique function between two one-point sets.

Unit: Define a natural isomorphism

$$\iota : \mathrm{Id}_{\mathcal{F}^{\mathsf{i}}_{\mathsf{im}}(\mathsf{C})} \xrightarrow{\cong} I^2$$

by

$$\iota_X = \big\{\iota^{\mathsf{C}}_{x_i}\big\}_{i=1}^{j} : X \xrightarrow{\cong} I^2 X \in \mathcal{F}^{\mathsf{i}}_{\mathsf{im}}(\mathsf{C})(X; I^2 X) = \prod_{i=1}^{j} \mathsf{C}\big(x_i, I^2_{\mathsf{C}} x_i\big)$$

for each object $X \in \mathcal{F}^{\mathsf{i}}_{\mathsf{im}}(\mathsf{C})$ as above.

Involutive Monoidal Structure: Define the morphism $I_0 : \mathbb{1} \longrightarrow I\mathbb{1}$ as the unique isomorphism in

$$\mathcal{F}^{\mathsf{i}}_{\mathsf{im}}(\mathsf{C})(\mathbb{1}; I\mathbb{1}) = \{*\}.$$

Finally, for objects $X, Y \in \mathcal{F}^{\mathsf{i}}_{\mathsf{im}}(\mathsf{C})$ as above, define the isomorphism

$$IX \otimes IY \xrightarrow[\cong]{(I_2)_{X,Y}} I(X \otimes Y) \in \mathcal{F}^{\mathsf{i}}_{\mathsf{im}}(\mathsf{C})\big(IX \otimes IY, I(X \otimes Y)\big)$$

as

$$\Big\{\{\mathrm{Id}_{I_{\mathsf{C}} x_i}\}_{i=1}^{j}, \{\mathrm{Id}_{I_{\mathsf{C}} y_i}\}_{i=1}^{k}\Big\} \in \prod_{i=1}^{j} \mathsf{C}\big(I_{\mathsf{C}} x_i, I_{\mathsf{C}} x_i\big) \times \prod_{i=1}^{k} \mathsf{C}\big(I_{\mathsf{C}} y_i, I_{\mathsf{C}} y_i\big).$$

This finishes the definition of $\mathcal{F}^{\mathsf{i}}_{\mathsf{im}}(\mathsf{C})$.

An inspection of the diagrams in Explanation 4.1.2 establishes the following.

LEMMA 5.5.4. *For each involutive category $(\mathsf{C}, I_{\mathsf{C}}, \iota^{\mathsf{C}})$, the tuple*

$$\big((\mathcal{F}^{\mathsf{i}}_{\mathsf{im}}(\mathsf{C}), \otimes, \mathbb{1}, \alpha, \lambda, \rho), (I, I_2, I_0), \iota\big)$$

in Definition 5.5.3 is an involutive monoidal category.

DEFINITION 5.5.5. For each involutive category C, denote by

$$\mathsf{C} \xrightarrow{\varepsilon_{\mathsf{C}}} \mathcal{F}^{\mathsf{i}}_{\mathsf{im}}(\mathsf{C})$$

the full and faithful embedding given by

- the inclusion $\mathsf{Ob}(\mathsf{C}) \subseteq \mathsf{W}^{\mathsf{i}}_{\mathsf{im}}\big(\mathsf{Ob}(\mathsf{C})\big)$ on objects, and

- the identity function

$$\mathsf{C}(x, y) \xrightarrow{\text{Id}} \mathcal{F}^{\mathsf{i}}_{\mathsf{im}}(\mathsf{C})(x; y)$$

on morphism sets for objects $x, y \in \mathsf{C}$.

The following observation says that $\mathcal{F}^{\mathsf{i}}_{\mathsf{im}}(\mathsf{C})$ is the free involutive monoidal category generated by an involutive category C. The proof is adapted from that of Theorem 5.1.13.

THEOREM 5.5.6. *Suppose $(F, \nu^F) : \mathsf{C} \longrightarrow \mathsf{D}$ is an involutive functor with C an involutive category and D an involutive monoidal category. Then there exists an involutive strict monoidal functor*

$$\mathcal{F}^{\mathsf{i}}_{\mathsf{im}}(\mathsf{C}) \xrightarrow{\overline{F}} \mathsf{D}$$

such that the following three statements hold.

(1) $\overline{F}$ extends F in the sense that $\overline{F} \circ \varepsilon_{\mathsf{C}} = F$.
(2) If F is a strict involutive functor, then so is $\overline{F}$, which is furthermore unique.
(3) As an involutive strong monoidal functor extending F, $\overline{F}$ is unique up to an involutive monoidal natural isomorphism.

PROOF. On the objects in $\mathcal{F}^{\mathsf{i}}_{\mathsf{im}}(\mathsf{C})$, $\overline{F}$ is defined recursively as follows:

- $\overline{F}(x) = F(x)$ for $x \in \mathsf{Ob}(\mathsf{C}) \subseteq \mathsf{Ob}\big(\mathcal{F}^{\mathsf{i}}_{\mathsf{im}}(\mathsf{C})\big)$.
- $\overline{F}(\mathbb{1}) = \mathbb{1}^{\mathsf{D}}$, the monoidal unit in D.
- For $X, Y \in \mathsf{Ob}\big(\mathcal{F}^{\mathsf{i}}_{\mathsf{im}}(\mathsf{C})\big)$ with $\overline{F}(X), \overline{F}(Y) \in \mathsf{D}$ already defined, we set

$$\overline{F}(X \otimes Y) = \overline{F}(X) \otimes \overline{F}(Y),$$

$$\overline{F}(IX) = \begin{cases} F(IX) & \text{if } X \in \mathsf{Ob}(\mathsf{C}), \\ I_{\mathsf{D}}\overline{F}(X) & \text{if } X \in \mathsf{Ob}\big(\mathcal{F}^{\mathsf{i}}_{\mathsf{im}}(\mathsf{C})\big) \setminus \mathsf{Ob}(\mathsf{C}). \end{cases}$$

To define $\overline{F}$ on morphisms, we call an object in $\mathcal{F}^{\mathsf{i}}_{\mathsf{im}}(\mathsf{C})$ *reduced* if either

(i) it is $\mathbb{1}$, or
(ii) it has the form

$$\big(\cdots(x_1 \otimes x_2) \otimes \cdots\big) \otimes x_k$$

for some $k > 0$ and objects $x_1, \ldots, x_k \in \mathsf{C}$, where the left half of each pair of parentheses is at the front.

A *basic isomorphism* in $\mathcal{F}^{\mathsf{i}}_{\mathsf{im}}(\mathsf{C})$ is defined as in Definition 5.1.7 but with ι omitted. For each object $X \in \mathcal{F}^{\mathsf{i}}_{\mathsf{im}}(\mathsf{C})$, there exists a canonical choice of a basic isomorphism

$$X \xrightarrow[\cong]{\delta_X} X' \in \mathcal{F}^{\mathsf{i}}_{\mathsf{im}}(\mathsf{C})$$

with X' reduced. The proof is the same as for Lemma 5.1.10, except that in step (2), the morphism

$$\delta_2^i : I_{\mathsf{C}}^{m_i} x_i \longrightarrow I_{\mathsf{C}}^{m_i} x_i$$

is now the identity morphism $\mathrm{Id}_{I_{\mathsf{C}}^{m_i} x_i}$ if $x_i \in \mathsf{C}$, which implies $I_{\mathsf{C}}^{m_i} x_i \in \mathsf{C}$. For instance, if x and y are objects in C, then an example of a basic isomorphism δ_X with a reduced codomain is the following composite.

$$\begin{array}{ccc}
X = I^4(I^2(I\mathbb{1} \otimes x) \otimes y) & \xrightarrow{\delta_X} & I_{\mathsf{C}}^6 x \otimes I_{\mathsf{C}}^4 y = X' \\
\downarrow{\scriptstyle (I_2^{-1})^{\circ 4}} & & \\
I^6(I\mathbb{1} \otimes x) \otimes I_{\mathsf{C}}^4 y & & \uparrow{\scriptstyle \lambda \otimes \mathrm{Id}} \\
\downarrow{\scriptstyle (I_2^{-1})^{\circ 6} \otimes \mathrm{Id}} & & \\
(I^7\mathbb{1} \otimes I_{\mathsf{C}}^6 x) \otimes I_{\mathsf{C}}^4 y & \xrightarrow{[(I_0^{-1})^{\circ 7} \otimes \mathrm{Id}] \otimes \mathrm{Id}} & (\mathbb{1} \otimes I_{\mathsf{C}}^6 x) \otimes I_{\mathsf{C}}^4 y
\end{array}$$

Following Lemma 5.1.11, there is an equality

$$\mathsf{word}(X) = \mathsf{word}(X')$$

because none of the constituent morphisms of $\delta_X = \delta_4\delta_3\delta_2\delta_1$ changes the underlying words. The equality of underlying words implies the equality of morphism sets

$$\mathcal{F}^{\mathsf{i}}_{\mathsf{im}}(\mathsf{C})(X; Y) = \mathcal{F}^{\mathsf{i}}_{\mathsf{im}}(\mathsf{C})(X'; Y')$$

with X' and Y' reduced objects. As in (5.1.15), the last equality allows us to define the morphism

$$\overline{F}(f) = \overline{F}(\delta_Y)^{-1} \circ \overline{F}(f') \circ \overline{F}(\delta_X) \tag{5.5.7}$$

for $f \in \mathcal{F}^{\mathsf{i}}_{\mathsf{im}}(\mathsf{C})(X; Y)$. This makes

$$\overline{F} : \mathcal{F}^{\mathsf{i}}_{\mathsf{im}}(\mathsf{C}) \longrightarrow \mathsf{D}$$

into a functor extending F.

The functor $\overline{F}$ is made into an involutive functor with the distributive law

$$\nu^{\overline{F}} : \overline{F}I \longrightarrow I_{\mathsf{D}}\overline{F}$$

defined componentwise as

$$\Big(\overline{F}I(X) \xrightarrow{\nu^{\overline{F}}_X} I_{\mathsf{D}}\overline{F}(X) \Big) = \begin{cases} \nu^F_X & \text{if } X \in \mathsf{Ob}(\mathsf{C}), \\ \mathrm{Id}_{I_{\mathsf{D}}\overline{F}(X)} & \text{if } X \in \mathsf{Ob}\big(\mathcal{F}^{\mathsf{i}}_{\mathsf{im}}(\mathsf{C})\big) \setminus \mathsf{Ob}(\mathsf{C}). \end{cases}$$

The functor $\overline{F}$ is assigned the strict monoidal functor structure. This makes $\overline{F}$ into an involutive strict monoidal functor. If F is a strict involutive functor–i.e., if ν^F is the identity natural transformation–then so is $\overline{F}$. The uniqueness statements of $\overline{F}$ are proved exactly as in the last two paragraphs of the proof of Theorem 5.1.13. □

Recall from Definition 3.1.1 that ICat_0 is the category of small involutive categories and strict involutive functors. Also recall from Definition 4.3.6 that IMCat_0 is the category of small involutive monoidal categories and strict involutive strict monoidal functors.

COROLLARY 5.5.8. *There is an adjunction*

$$\mathsf{ICat}_0 \underset{U}{\overset{\mathcal{F}^{\mathsf{i}}_{\mathsf{im}}}{\rightleftarrows}} \mathsf{IMCat}_0$$

whose right adjoint U forgets about the monoidal structure.

PROOF. For each involutive category C, the functor $\varepsilon_{\mathsf{C}} : \mathsf{C} \longrightarrow \mathcal{F}^{\mathsf{i}}_{\mathsf{im}}(\mathsf{C})$ is a strict involutive functor. For a strict involutive functor $G : \mathsf{C} \longrightarrow \mathsf{D}$ between small involutive categories, we define

$$\mathcal{F}^{\mathsf{i}}_{\mathsf{im}}(G) : \mathcal{F}^{\mathsf{i}}_{\mathsf{im}}(\mathsf{C}) \longrightarrow \mathcal{F}^{\mathsf{i}}_{\mathsf{im}}(\mathsf{D})$$

as the unique strict involutive strict monoidal functor that makes the diagram

$$\begin{array}{ccc} \mathsf{C} & \xrightarrow{G} & \mathsf{D} \\ {\scriptstyle \varepsilon_{\mathsf{C}}}\downarrow & & \downarrow{\scriptstyle \varepsilon_{\mathsf{D}}} \\ \mathcal{F}^{\mathsf{i}}_{\mathsf{im}}(\mathsf{C}) & \xrightarrow{\mathcal{F}^{\mathsf{i}}_{\mathsf{im}}(G)} & \mathcal{F}^{\mathsf{i}}_{\mathsf{im}}(\mathsf{D}) \end{array}$$

in ICat_0 commutative. This defines the functor

$$\mathcal{F}^{\mathsf{i}}_{\mathsf{im}} : \mathsf{ICat}_0 \longrightarrow \mathsf{IMCat}_0.$$

Theorem 5.5.6 shows that $(\mathcal{F}^{\mathsf{i}}_{\mathsf{im}}, U)$ satisfies the universal property for an adjunction, as stated in [Mac98] (IV.1 Theorem 2(i)). □

5.6 Free Involutive Strict Monoidal Categories of Involutive Categories

The purpose of this section is to give an explicit description of the free involutive strict monoidal category generated by an involutive category. This is a simplification of Section 5.5, as Section 5.2 is a simplification of Section 5.1. We also observe that, for an involutive category, the free involutive monoidal category is equivalent to the free involutive strict monoidal category via a strict involutive strict monoidal equivalence.

DEFINITION 5.6.1. Suppose S is a class equipped with an assignment

$$I_S : S \longrightarrow S,$$

and I is a symbol that does not appear in S.

(1) The class $\mathsf{W}^{\mathsf{i}}_{\mathsf{istm}}(S)$ is defined recursively by thc following conditions:

   (i) $\varnothing \in \mathsf{W}^{\mathsf{i}}_{\mathsf{istm}}(S)$, where $\varnothing$ is the empty sequence.
   (ii) $S \subseteq \mathsf{W}^{\mathsf{i}}_{\mathsf{istm}}(S)$.
   (iii) If $X \in \mathsf{W}^{\mathsf{i}}_{\mathsf{istm}}(S) \setminus S$, then $IX \in \mathsf{W}^{\mathsf{i}}_{\mathsf{istm}}(S)$, where $I\varnothing = \varnothing$.
   (iv) If $X, Y \in \mathsf{W}^{\mathsf{i}}_{\mathsf{istm}}(S)$, then so is their concatenation XY. Concatenation is associative by definition.

   For an element $x \in S$, we define $Ix = I_S x$.

(2) For $X \in \mathsf{W}^{\mathsf{i}}_{\mathsf{istm}}(S)$, its *underlying word*

$$\mathsf{word}(X) \in \coprod_{k \geq 0} S^{\times k}$$

   is defined as in Definition 5.5.1 with $\mathsf{word}(\varnothing) = \varnothing$.

DEFINITION 5.6.2. Suppose $(\mathsf{C}, I_{\mathsf{C}}, \iota^{\mathsf{C}})$ is an involutive category. Define the category $\mathcal{F}^{\mathsf{i}}_{\mathsf{istm}}(\mathsf{C})$ with extra structures as in Definition 5.5.3, but with the following modifications:

Objects: $\mathsf{Ob}(\mathcal{F}^{\mathsf{i}}_{\mathsf{istm}}(\mathsf{C})) = \mathsf{W}^{\mathsf{i}}_{\mathsf{istm}}(\mathsf{Ob}(\mathsf{C}))$, where $\mathsf{Ob}(\mathsf{C})$ is equipped with the assignment $I_{\mathsf{C}} : \mathsf{Ob}(\mathsf{C}) \longrightarrow \mathsf{Ob}(\mathsf{C})$.

Monoidal Product: For objects $X, Y \in \mathsf{Ob}(\mathcal{F}^{\mathsf{i}}_{\mathsf{istm}}(\mathsf{C}))$, their monoidal product $X \otimes Y$ is defined as their concatenation, which is part of the definition of the class $\mathsf{W}^{\mathsf{i}}_{\mathsf{istm}}(\mathsf{Ob}(\mathsf{C}))$.

Monoidal Unit: The monoidal unit in $\mathcal{F}^{\mathsf{i}}_{\mathsf{istm}}(\mathsf{C})$ is $\varnothing$.

Monoidal Structure Isomorphisms: Each component of the associativity isomorphism α, the left unit isomorphism λ, and the right unit isomorphism ρ, is an identity morphism.

Involutive Monoidal Structure: $I_0 : \varnothing \longrightarrow \varnothing$ is the identity morphism.

An inspection establishes the following.

LEMMA 5.6.3. *For each involutive category* C*, the tuple*

$$\big((\mathcal{F}^{\mathsf{i}}_{\mathsf{istm}}(\mathsf{C}), \otimes, \varnothing, \alpha, \lambda, \rho), (I, I_2, I_0), \iota\big)$$

in Definition 5.6.2 is an involutive strict monoidal category.

DEFINITION 5.6.4. For each involutive category C, denote by

$$\mathsf{C} \xrightarrow{\varepsilon^{\mathsf{st}}_{\mathsf{C}}} \mathcal{F}^{\mathsf{i}}_{\mathsf{istm}}(\mathsf{C})$$

the full and faithful embedding given by

- the inclusion $\mathsf{Ob}(\mathsf{C}) \subseteq \mathsf{W}^{\mathsf{i}}_{\mathsf{istm}}\big(\mathsf{Ob}(\mathsf{C})\big)$ on objects, and
- the identity function

$$\mathsf{C}(x, y) \xrightarrow{\mathrm{Id}} \mathcal{F}^{\mathsf{i}}_{\mathsf{istm}}(\mathsf{C})(x; y)$$

on morphism sets for objects $x, y \in \mathsf{C}$.

A simplified version of the proof of Theorem 5.5.6 establishes the following result.

THEOREM 5.6.5. *Suppose* $(F, \nu^F) : \mathsf{C} \longrightarrow \mathsf{D}$ *is a strict involutive functor with* C *an involutive category and* D *an involutive strict monoidal category. Then there exists a unique strict involutive strict monoidal functor*

$$\mathcal{F}^{\mathsf{i}}_{\mathsf{istm}}(\mathsf{C}) \xrightarrow{\overline{F}} \mathsf{D}$$

such that $\overline{F} \circ \varepsilon^{\mathsf{st}}_{\mathsf{C}} = F$.

DEFINITION 5.6.6. For an involutive category C, the involutive strict monoidal category $\mathcal{F}^{\mathsf{i}}_{\mathsf{istm}}(\mathsf{C})$ is called the *free involutive strict monoidal category generated by* C.

Next we observe that the free involutive monoidal category $\mathcal{F}^{\mathsf{i}}_{\mathsf{im}}(\mathsf{C})$ and the free involutive strict monoidal category $\mathcal{F}^{\mathsf{i}}_{\mathsf{istm}}(\mathsf{C})$ are equivalent.

COROLLARY 5.6.7. *For each involutive category* C*, there is a canonical strict involutive strict monoidal functor*

$$\mathcal{F}^{\mathsf{i}}_{\mathsf{im}}(\mathsf{C}) \xrightarrow{\ \theta\ } \mathcal{F}^{\mathsf{i}}_{\mathsf{istm}}(\mathsf{C})$$

that is also an equivalence of categories.

PROOF. The functor θ is the unique strict involutive strict monoidal functor that extends $\varepsilon^{\mathsf{st}}_{\mathsf{C}}$, in the sense that the diagram

$$\begin{array}{ccc} \mathsf{C} & \xrightarrow{\ \varepsilon_{\mathsf{C}}\ } & \mathcal{F}^{\mathsf{i}}_{\mathsf{im}}(\mathsf{C}) \\ \| & & \downarrow \theta \\ \mathsf{C} & \xrightarrow{\ \varepsilon^{\mathsf{st}}_{\mathsf{C}}\ } & \mathcal{F}^{\mathsf{i}}_{\mathsf{istm}}(\mathsf{C}) \end{array}$$

is commutative, given by Theorem 5.5.6. It exists because $\varepsilon^{\mathsf{st}}_{\mathsf{C}}$ is a strict involutive functor. As in the proof of Theorem 5.2.8, the functor θ is the identity function on morphism sets, and is surjective on objects. Therefore, θ is an equivalence of categories. □

5.7 Exercises and Notes

(1) In the last paragraph of the proof of Theorem 5.1.13, check that $\sigma : \overline{F} \longrightarrow \mathsf{G}$ is an involutive monoidal natural isomorphism.
(2) Write down a detailed proof of Theorem 5.2.6.
(3) Prove Lemma 5.3.3.
(4) In the proof of Theorem 5.4.1, check:

- The triangle identity (2.1.2) for $(I_{\mathsf{st}}, \iota^{\mathsf{st}})$.
- That the counit η is a monoidal natural transformation.

(5) Write down a detailed proof of Theorem 5.6.5.
(6) Prove that there is a monad in Cat whose algebras are exactly small involutive monoidal categories. Do the same with monoidal replaced by strict monoidal.
(7) Repeat the previous exercise with Cat replaced by ICat_0.
(8) Prove the strict involutive variants of all of the main results in this chapter. In other words:

- Give explicit constructions of
  - the free strict involutive monoidal category and
  - the free strict involutive strict monoidal category

  of a category.

- Prove that they are equivalent via a canonical strict involutive strict monoidal functor, and so forth.

Notes. Theorem 5.2.8 is the involutive analogue of Theorem 1.2 in [JS93], which shows the existence of a strict monoidal equivalence from the free monoidal category of a category to its free strict monoidal category. The monoidal category $\mathsf{It}(\mathsf{C})$ in Definition 5.3.1 is the same one defined by Mac Lane [Mac98] (VII.2 Corollary). In the proof of Theorem 5.4.1, in the definitions of the morphism $I_{\mathsf{st}} f$ in (5.4.4) and of the distributive law $\nu^R_{\underline{x}}$ in (5.4.5), the uniqueness of the isomorphisms is due to Epstein's Coherence Theorem [Eps66] restricted to monoidal categories.

Chapter 6
Coherence of Involutive Symmetric Monoidal Categories

The purpose of this chapter is to study coherence of involutive symmetric monoidal categories. Most of the constructions are adapted from Chapter 5, where we studied coherence of involutive monoidal categories. In Section 6.1 and Section 6.2, we give explicit constructions of the free involutive symmetric monoidal category and of the free involutive strict symmetric monoidal category generated by a category. We observe that they are equivalent via a strict involutive strict symmetric monoidal functor. In Section 6.3 we show that in a small involutive symmetric monoidal category, every suitably defined formal diagram is commutative. In Section 6.4 we show that every involutive symmetric monoidal category can be strictified to an involutive strict symmetric monoidal category via an involutive adjoint equivalence involving involutive strong symmetric monoidal functors. The remaining two sections contain explicit constructions of the free involutive (strict) symmetric monoidal category generated by an involutive category.

Our strictification results–namely, Theorem 5.4.1 in the non-symmetric case and Theorem 6.4.1 in the symmetric case–are important in applications. When involutive (symmetric) monoidal categories are used in practice, for example in [BBS19, BSW19b], one often ignores parentheses in multiple monoidal products and uses expressions such as $x \otimes y \otimes z$. What this really means is that, if the ambient involutive (symmetric) monoidal category C is not strict monoidal, such as $\mathsf{Vect}(\mathbb{C})$ and $\mathsf{Chain}(\mathbb{C})$, then one first replaces C by an involutively equivalent involutive strict (symmetric) monoidal category C_{st}, in which multiple monoidal products are unambiguous even without parentheses. In the literature, this first step of replacing C by an involutively equivalent C_{st} is usually performed without explicit comments. Our strictification results, for both involutive monoidal categories and involutive symmetric monoidal categories, ensure that this strictification exists and is canonical.

© The Author(s), under exclusive license to Springer Nature Switzerland AG 2020

D. Yau, *Involutive Category Theory*, Lecture Notes in Mathematics 2279,
https://doi.org/10.1007/978-3-030-61203-0_6

6.1 Free Involutive Symmetric Monoidal Categories

The purpose of this section is to give an explicit construction of the free involutive symmetric monoidal category generated by a category. First we need some notations.

The symmetric group on k letters is denoted by Σ_k. Its unit is denoted by id_k or simply id. For a sequence of variables $(y_1, \ldots, y_k)$ and a permutation $\sigma \in \Sigma_k$, we define the σ-action

$$(y_1, \ldots, y_k)\sigma = \big(y_{\sigma(1)}, \ldots, y_{\sigma(k)}\big).$$

We also need the following notations for block permutations and block sums.

DEFINITION 6.1.1. Suppose $n \geq 1$, $k_i \geq 0$ for $1 \leq i \leq n$, and $k = k_1 + \cdots + k_n$.

(1) For $\sigma \in \Sigma_n$, the *block permutation*

$$\sigma\langle k_1, \ldots, k_n\rangle \in \Sigma_k$$

is defined by

$$\big(\sigma\langle k_1, \ldots, k_n\rangle\big)(k_1 + \cdots + k_{i-1} + j) = \underbrace{k_{\sigma^{-1}(1)} + \cdots + k_{\sigma^{-1}(\sigma(i)-1)}}_{0 \text{ if } \sigma(i)=1} + j$$

for $1 \leq j \leq k_i$ and $1 \leq i \leq n$.

(2) For $\tau_i \in \Sigma_{k_i}$ for $1 \leq i \leq n$, the *block sum*

$$\tau_1 \times \cdots \times \tau_n \in \Sigma_k$$

is defined by

$$\big(\tau_1 \times \cdots \times \tau_n\big)(k_1 + \cdots + k_{i-1} + j) = \underbrace{k_1 + \cdots + k_{i-1}}_{0 \text{ if } i=1} + \tau_i(j)$$

for $1 \leq j \leq k_i$ and $1 \leq i \leq n$.

EXPLANATION 6.1.2. The block permutation $\sigma\langle k_1, \ldots, k_n\rangle$ permutes n consecutive blocks of lengths $k_1, \ldots, k_n$ as σ permutes $\{1, \ldots, n\}$, leaving the relative order within each block unchanged. The block sum $\tau_1 \times \cdots \times \tau_n$ leaves the order of the n consecutive blocks unchanged and only permutes within the ith block using τ_i for $1 \leq i \leq n$. ◇

The following definition is the symmetric analogue of Definition 5.1.3.

DEFINITION 6.1.3. Suppose C is a category. Define the category $\mathcal{F}_{\mathsf{isym}}(\mathsf{C})$ with extra structures as follows.

Objects: $\mathsf{Ob}(\mathcal{F}_{\mathsf{isym}}(\mathsf{C})) = \mathsf{W}_{\mathsf{im}}(\mathsf{Ob}(\mathsf{C}))$, the class of involutive monoidal words generated by $\mathsf{Ob}(\mathsf{C})$ as in Definition 5.1.1.

Morphisms: For $X, Y \in \mathsf{Ob}(\mathcal{F}_{\mathsf{isym}}(\mathsf{C}))$ with underlying words

$$\begin{aligned} \mathsf{word}(X) &= (m_1, x_1) \cdots (m_j, x_j) \in \big(\mathbb{Z}_{\geq 0} \times \mathsf{Ob}(\mathsf{C})\big)^{\times j}, \\ \mathsf{word}(Y) &= (n_1, y_1) \cdots (n_k, y_k) \in \big(\mathbb{Z}_{\geq 0} \times \mathsf{Ob}(\mathsf{C})\big)^{\times k} \end{aligned} \tag{6.1.4}$$

as in (5.1.4), the morphism set $\mathcal{F}_{\mathsf{isym}}(\mathsf{C})(X; Y)$ is the set of pairs

$$\big(\sigma; \{f_i\}_{i=1}^{j}\big) \in \Sigma_j \times \prod_{i=1}^{j} \mathsf{C}(x_i, y_{\sigma(i)})$$

such that

$$(m_1, \ldots, m_j) \equiv (n_1, \ldots, n_j)\sigma \pmod 2$$

if $j = k$, and is empty if $j \neq k$. An empty product, for the case $j = 0$, is taken as a one-point set. We call σ the *underlying permutation* of the morphism $(\sigma; \{f_i\})$.

Identities: Identity morphisms are defined using identity permutations and identity morphisms in C.

Composition: Categorical composition is defined using multiplication in the symmetric groups and composition in C, i.e.,

$$\big(\tau; \{g_i\}\big) \circ \big(\sigma; \{f_i\}\big) = \big(\tau\sigma; \{g_{\sigma(i)} f_i\}\big). \tag{6.1.5}$$

Monoidal Product: For $X, Y \in \mathsf{Ob}(\mathcal{F}_{\mathsf{isym}}(\mathsf{C}))$, their monoidal product $X \otimes Y$ is defined as part of the definition of the class $\mathsf{W}_{\mathsf{im}}(\mathsf{Ob}(\mathsf{C}))$. The monoidal product on morphisms in $\mathcal{F}_{\mathsf{isym}}(\mathsf{C})$ is defined using the block sum inclusion

$$\Sigma_l \times \Sigma_j \longrightarrow \Sigma_{l+j}, \qquad (\tau; \sigma) \longmapsto \tau \times \sigma$$

and concatenation of finite sequences of morphisms in C. In other words, it is defined as

$$\big(\tau; \{g_i\}_{i=1}^{l}\big) \otimes \big(\sigma; \{f_i\}_{i=1}^{j}\big) = \big(\tau \times \sigma; \{g_i\}_{i=1}^{l}, \{f_i\}_{i=1}^{j}\big). \tag{6.1.6}$$

Monoidal Unit: The monoidal unit in $\mathcal{F}_{\mathsf{isym}}(\mathsf{C})$ is $\mathbb{1} \in \mathsf{W}_{\mathsf{im}}(\mathsf{Ob}(\mathsf{C}))$.

Monoidal Structure Isomorphisms: The associativity isomorphism α, the left unit isomorphism λ, and the right unit isomorphism ρ are defined using identity permutations and identity morphisms in C.

Symmetry: Define a natural isomorphism with components

$$X \otimes Y \xrightarrow{\xi_{X,Y}} Y \otimes X \quad \text{for} \quad X, Y \in \mathsf{Ob}(\mathcal{F}_{\mathsf{isym}}(\mathsf{C})) \tag{6.1.7}$$

given by

$$\xi_{X,Y} = \big((1,2)\langle j,k\rangle; \{\mathrm{Id}_{x_i}\}_{i=1}^{j}, \{\mathrm{Id}_{y_i}\}_{i=1}^{k}\big).$$

Here $(1,2)\langle j,k\rangle \in \Sigma_{j+k}$ is the block permutation induced by the transposition $(1,2) \in \Sigma_2$ that swaps the first j entries as a single interval with the last k entries as a single interval.

Involution Functor: Define a functor

$$I : \mathcal{F}_{\mathsf{isym}}(\mathsf{C}) \longrightarrow \mathcal{F}_{\mathsf{isym}}(\mathsf{C})$$

by the assignment $X \longmapsto IX$ on objects, and the identity function

$$\mathcal{F}_{\mathsf{isym}}(X; Y) \xrightarrow{\mathrm{Id}} \mathcal{F}_{\mathsf{isym}}(IX; IY)$$

on morphism sets.

Unit: Define a natural isomorphism

$$\iota : \mathrm{Id}_{\mathcal{F}_{\mathsf{isym}}(\mathsf{C})} \xrightarrow{\cong} I^2$$

using identity permutations and identity morphisms in C.

Involutive Monoidal Structure: Define the morphism $I_0 : \mathbb{1} \longrightarrow I\mathbb{1}$ as the unique isomorphism in

$$\mathcal{F}_{\mathsf{isym}}(\mathsf{C})(\mathbb{1}; I\mathbb{1}) = \{*\}.$$

Finally, for objects $X, Y \in \mathcal{F}_{\mathsf{isym}}(\mathsf{C})$, the isomorphism

$$IX \otimes IY \xrightarrow[\cong]{(I_2)_{X,Y}} I(X \otimes Y) \in \mathcal{F}_{\mathsf{im}}(\mathsf{C})\big(IX \otimes IY, I(X \otimes Y)\big)$$

is defined using identity permutations and identity morphisms in C.

LEMMA 6.1.8. *For each category* C*, the tuple*

$$\big((\mathcal{F}_{\mathsf{isym}}(C), \otimes, \mathbb{1}, \alpha, \lambda, \rho, \xi), (I, I_2, I_0), \iota\big)$$

in Definition 6.1.3 is an involutive symmetric monoidal category.

PROOF. An inspection shows that $\mathcal{F}_{\mathsf{isym}}(\mathsf{C})$ is a symmetric monoidal category, that $\big(\mathcal{F}_{\mathsf{isym}}(\mathsf{C}), I, \iota\big)$ is an involutive category, and that the diagrams in Explanation 4.1.2 are commutative. □

DEFINITION 6.1.9. For each category C, denote by

$$\mathsf{C} \xrightarrow{\varepsilon_{\mathsf{C}}} \mathcal{F}_{\mathsf{isym}}(\mathsf{C})$$

the full and faithful embedding given by

- the inclusion $\mathsf{Ob}(\mathsf{C}) \subseteq \mathsf{W}_{\mathsf{im}}\big(\mathsf{Ob}(\mathsf{C})\big)$ on objects, and
- the identity function

$$\mathsf{C}(x, y) \xrightarrow{\mathrm{Id}} \mathcal{F}_{\mathsf{isym}}(\mathsf{C})(x; y)$$

on morphism sets for objects $x, y \in \mathsf{C}$.

The following observation says that $\mathcal{F}_{\mathsf{isym}}(\mathsf{C})$ is the free involutive symmetric monoidal category generated by C.

THEOREM 6.1.10. *Suppose $F : C \longrightarrow D$ is a functor with C a category and D an involutive symmetric monoidal category. Then there exists a unique strict involutive strict symmetric monoidal functor*

$$\mathcal{F}_{\mathsf{isym}}(C) \xrightarrow{\overline{F}} D$$

such that $\overline{F} \circ \varepsilon_{C} = F$.

PROOF. This proof is adapted from that of Theorem 5.1.13, so we only point out the modifications, which involve what $\overline{F}$ does to morphisms. Using the notations in Definition 6.1.3, consider a morphism $f \in \mathcal{F}_{\mathsf{isym}}(\mathsf{C})(X; Y)$ of the form

$$\big(\sigma; \{f_i\}_{i=1}^{j}\big) \in \Sigma_j \times \prod_{i=1}^{j} \mathsf{C}(x_i, y_{\sigma(i)})$$

such that

$$(m_1, \ldots, m_j) \equiv (n_1, \ldots, n_j)\sigma \pmod 2.$$

Then the morphism $\overline{F}(f) : \overline{F}(X) \longrightarrow \overline{F}(Y)$ in D is defined as the composite

$$\begin{array}{ccccc} \overline{F}(X) & & \xrightarrow{\quad\overline{F}(f)\quad} & & \overline{F}(Y) \\ {\scriptstyle \overline{F}(\delta_X)}\big\downarrow & & & & \big\uparrow{\scriptstyle \overline{F}(\delta_Y)^{-1}} \\ \overline{F}(X') & \xrightarrow{\overline{F}(f')} & \mathsf{cod}\big(\overline{F}(f')\big) & \xrightarrow[\cong]{\sigma} & \overline{F}(Y') \end{array} \tag{6.1.11}$$

in (5.1.15), but with an extra isomorphism

$$\begin{array}{c} \mathsf{cod}\big(\overline{F}(f')\big) = I_{\mathsf{D}}^{\epsilon_1} F(y_{\sigma(1)}) \otimes \cdots \otimes I_{\mathsf{D}}^{\epsilon_j} F(y_{\sigma(j)}) \\ \cong \big\downarrow \sigma \\ \overline{F}(Y') = \mathsf{dom}\big(\overline{F}(\delta_Y)^{-1}\big) = I_{\mathsf{D}}^{\epsilon_{\sigma^{-1}(1)}} F(y_1) \otimes \cdots \otimes I_{\mathsf{D}}^{\epsilon_{\sigma^{-1}(j)}} F(y_j) \end{array}$$

between $\overline{F}(f')$ and $\overline{F}(\delta_Y)^{-1}$. Both the domain and the codomain of σ have the left-normalized parenthesization as in (5.1.8). The isomorphism σ is the unique natural isomorphism guaranteed by Mac Lane's Coherence Theorem [Mac98] (XI.1 Theorem 1) for the symmetric monoidal category D, corresponding to the permutation $\sigma \in \Sigma_j$. It involves only identity morphisms, the associativity isomorphism, the symmetry isomorphism, their inverses, monoidal products, and composites in D.

The assignment $\overline{F}$ still preserves identity morphisms. To check that $\overline{F}$ preserves composition of morphisms, suppose $g \in \mathcal{F}_{\mathsf{isym}}(\mathsf{C})(Y; Z)$ is a morphism of the form

$$\big(\tau; \{g_i\}_{i=1}^{j}\big) \in \Sigma_j \times \prod_{i=1}^{j} \mathsf{C}(y_i, z_{\tau(i)})$$

such that

$$(n_1, \ldots, n_j) \equiv (p_1, \ldots, p_j)\tau \pmod 2$$

with

$$\mathsf{word}(Z) = (p_1, z_1) \cdots (p_j, z_j) \in \big(\mathbb{Z}_{\geq 0} \times \mathsf{Ob}(\mathsf{C})\big)^{\times j}.$$

To show that

$$\overline{F}(g) \circ \overline{F}(f) = \overline{F}(g \circ f),$$

we paste the diagrams defining $\overline{F}(f)$ and $\overline{F}(g)$ together along the edges labeled by $\overline{F}(\delta_Y)^{\pm 1}$. The resulting pasted diagram is equal to the one defining $\overline{F}(g \circ f)$ because the diagram

$$\begin{array}{ccc}
I_{\mathsf{D}}^{\epsilon_1}F(y_{\sigma(1)}) \otimes \cdots \otimes I_{\mathsf{D}}^{\epsilon_j}F(y_{\sigma(j)}) & \xrightarrow[\cong]{\sigma} & I_{\mathsf{D}}^{\epsilon_{\sigma^{-1}(1)}}F(y_1) \otimes \cdots \otimes I_{\mathsf{D}}^{\epsilon_{\sigma^{-1}(j)}}F(y_j) \\
{\scriptstyle \overline{F}(\{g_{\sigma(i)}\}')}\Big\downarrow & & \Big\downarrow{\scriptstyle \overline{F}(g')} \\
I_{\mathsf{D}}^{\epsilon_1}F(z_{\tau\sigma(1)}) \otimes \cdots \otimes I_{\mathsf{D}}^{\epsilon_j}F(z_{\tau\sigma(j)}) & \xrightarrow[\cong]{\sigma} & I_{\mathsf{D}}^{\epsilon_{\sigma^{-1}(1)}}F(z_{\tau(1)}) \otimes \cdots \otimes I_{\mathsf{D}}^{\epsilon_{\sigma^{-1}(j)}}F(z_{\tau(j)})
\end{array}$$

in D, in which all four objects have the left-normalized parenthesization, is commutative by the naturality of the natural isomorphism σ. So $\overline{F}$ is a functor.

Equipped with the identity distributive law and the strict monoidal functor structure, $\overline{F}$ becomes a strict involutive strict monoidal functor. It is a symmetric monoidal functor because of the σ factor in the definition of $\overline{F}(f)$ in (6.1.11).

To check that $\overline{F}$ is unique, observe that the original morphism $f = (\sigma; \{f_i\})$ factors as

(6.1.12)
$$\begin{array}{ccccc}
\mathsf{X} & & \xrightarrow{\quad f \quad} & & \mathsf{Y} \\
{\scriptstyle \delta_{\mathsf{X}}}\Big\downarrow & & & & \Big\uparrow{\scriptstyle \delta_{\mathsf{Y}}^{-1}} \\
\mathsf{X}' = I_{\mathsf{D}}^{\epsilon_1}x_1 \otimes \cdots \otimes I_{\mathsf{D}}^{\epsilon_j}x_j & & & & I_{\mathsf{D}}^{\epsilon_{\sigma^{-1}(1)}}y_1 \otimes \cdots \otimes I_{\mathsf{D}}^{\epsilon_{\sigma^{-1}(j)}}y_j = \mathsf{Y}' \\
& \searrow^{(\mathrm{id}_j;\{f_i\})} & & \nearrow_{(\sigma;\{\mathrm{Id}_{y_{\sigma(i)}}\})} & \\
& & I_{\mathsf{D}}^{\epsilon_1}y_{\sigma(1)} \otimes \cdots \otimes I_{\mathsf{D}}^{\epsilon_j}y_{\sigma(j)} & &
\end{array}$$

in $\mathcal{F}_{\mathsf{isym}}(\mathsf{C})$, in which the three objects in the lower half have the left-normalized parenthesization as in (5.1.8). The strict involutive strict symmetric monoidal functor requirement then forces $\overline{F}(f)$ to be equal to the composite in (6.1.11), proving the uniqueness of $\overline{F}$. □

Recall from Definition 4.3.6 that $\mathsf{ISymCat}_0$ is the category whose objects are small involutive symmetric monoidal categories and whose morphisms are strict involutive strict symmetric monoidal functors.

COROLLARY 6.1.13. *There is an adjunction*

$$\mathsf{Cat} \underset{U}{\overset{\mathcal{F}_{\mathsf{isym}}}{\rightleftarrows}} \mathsf{ISymCat}_0$$

whose right adjoint U forgets about the involutive structure and the symmetric monoidal structure.

PROOF. For a functor $G : \mathsf{C} \longrightarrow \mathsf{D}$ between small categories, we define

$$\mathcal{F}_{\mathsf{isym}}(G) : \mathcal{F}_{\mathsf{isym}}(\mathsf{C}) \longrightarrow \mathcal{F}_{\mathsf{isym}}(\mathsf{D})$$

as the unique strict involutive strict symmetric monoidal functor that makes the diagram

$$\begin{array}{ccc} \mathsf{C} & \xrightarrow{G} & \mathsf{D} \\ {\scriptstyle \varepsilon_{\mathsf{C}}}\downarrow & & \downarrow{\scriptstyle \varepsilon_{\mathsf{D}}} \\ \mathcal{F}_{\mathsf{isym}}(\mathsf{C}) & \xrightarrow{\mathcal{F}_{\mathsf{isym}}(G)} & \mathcal{F}_{\mathsf{isym}}(\mathsf{D}) \end{array}$$

in Cat commutative. This defines the functor

$$\mathcal{F}_{\mathsf{isym}} : \mathsf{Cat} \longrightarrow \mathsf{ISymCat}_0.$$

Theorem 6.1.10 shows that the pair of functors $(\mathcal{F}_{\mathsf{isym}}, U)$ satisfies the universal property for an adjunction, as stated in [Mac98] (IV.1 Theorem 2(i)). □

DEFINITION 6.1.14. For a category C, the involutive symmetric monoidal category $\mathcal{F}_{\mathsf{isym}}(\mathsf{C})$ is called the *free involutive symmetric monoidal category generated by* C.

EXAMPLE 6.1.15 (Free Involutive Symmetric Monoidal Category on One Object). Suppose $\mathbf{1}$ is the category with one object x and only the identity morphism of x. The free involutive symmetric monoidal category $\mathcal{F}_{\mathsf{isym}}(\mathbf{1})$ generated by one object x has object set $\mathsf{W}_{\mathsf{im}}(\{x\})$, which is the set of involutive monoidal words generated by $\{x\}$. For each involutive symmetric monoidal category D and each object $z \in \mathsf{D}$, there is a unique functor

$$F : \mathbf{1} \longrightarrow \mathsf{D} \quad \text{such that} \quad F(x) = z.$$

By Theorem 6.1.10, there exists a unique strict involutive strict symmetric monoidal functor

$$\overline{F} : \mathcal{F}_{\mathsf{isym}}(\mathbf{1}) \longrightarrow \mathsf{D} \quad \text{such that} \quad \overline{F}(x) = z.$$

Since $\mathbf{1}$ is a discrete category with one object, a typical morphism $f : X \longrightarrow Y$ in $\mathcal{F}_{\mathsf{isym}}(\mathbf{1})$ has the form

$$f = \left(\sigma; \{\mathrm{Id}_x\}_{i=1}^{j}\right) \in \Sigma_j \times \prod_{i=1}^{j} \mathbf{1}(x, x) \cong \Sigma_j$$

such that, using the notations in (6.1.4),

$$(m_1, \ldots, m_j) \equiv (n_1, \ldots, n_j)\sigma \pmod 2. \tag{6.1.16}$$

It is uniquely determined by its underlying permutation σ. It follows that two morphisms in $\mathcal{F}_{\mathsf{isym}}(\mathbf{1})$ with the same domain and the same codomain are equal if and only if their underlying permutations are equal. ◇

More generally, we have the following result.

COROLLARY 6.1.17. *For each discrete category* C, *two morphisms in* $\mathcal{F}_{\mathsf{isym}}(\mathsf{C})$ *with the same domain and the same codomain are equal if and only if their underlying permutations are equal.*

PROOF. The argument is the same as in the case $\mathbf{1}$. Indeed, a typical morphism $f : X \longrightarrow Y$ in $\mathcal{F}_{\mathsf{isym}}(\mathsf{C})$ has the form

$$f = \left(\sigma; \{\mathrm{Id}_{x_i}\}_{i=1}^{j}\right) \in \Sigma_j \times \prod_{i=1}^{j} \mathsf{C}(x_i, x_i) \cong \Sigma_j$$

such that (6.1.16) holds, because C only has identity morphisms. So it is uniquely determined by its underlying permutation. □

6.2 Free Involutive Strict Symmetric Monoidal Categories

The first purpose of this section is to give an explicit description of the free involutive strict symmetric monoidal category generated by a category. Then we show that the free involutive symmetric monoidal category is equivalent to the free involutive strict symmetric monoidal category via a strict involutive strict symmetric monoidal equivalence.

DEFINITION 6.2.1. Suppose C is a category. Define the category $\mathcal{F}_{\mathsf{iss}}(\mathsf{C})$ with extra structures as in Definition 6.1.3, but with the following modifications:

Objects: $\mathsf{Ob}(\mathcal{F}_{\mathsf{iss}}(\mathsf{C})) = \mathsf{W}_{\mathsf{istm}}(\mathsf{Ob}(\mathsf{C}))$ as in Definition 5.2.1.

Monoidal Product: For objects $X, Y \in \mathsf{Ob}(\mathcal{F}_{\mathsf{iss}}(\mathsf{C}))$, their monoidal product $X \otimes Y$ is defined as their concatenation, which is part of the definition of the class $\mathsf{W}_{\mathsf{istm}}(\mathsf{Ob}(\mathsf{C}))$.

Monoidal Unit: The monoidal unit in $\mathcal{F}_{\mathsf{iss}}(\mathsf{C})$ is $\varnothing$.

Monoidal Structure Isomorphisms: Each component of the associativity isomorphism α, the left unit isomorphism λ, and the right unit isomorphism ρ, is an identity morphism.

Involutive Monoidal Structure: $I_0 : \varnothing \longrightarrow \varnothing$ is the identity morphism.

An inspection establishes the following.

LEMMA 6.2.2. *For each category* C, *the tuple*

$$\left((\mathcal{F}_{\mathsf{iss}}(\mathsf{C}), \otimes, \varnothing, \alpha, \lambda, \rho, \xi), (I, I_2, I_0), \iota\right)$$

in Definition 6.2.1 is an involutive strict symmetric monoidal category.

DEFINITION 6.2.3. For each category C, denote by

$$\mathsf{C} \xrightarrow{\varepsilon^{\mathsf{st}}_{\mathsf{C}}} \mathcal{F}_{\mathsf{iss}}(\mathsf{C})$$

the full and faithful embedding given by

- the inclusion $\mathsf{Ob}(\mathsf{C}) \subseteq \mathsf{W}_{\mathsf{istm}}\big(\mathsf{Ob}(\mathsf{C})\big)$ on objects, and
- the identity function

$$\mathsf{C}(x, y) \xrightarrow{\mathrm{Id}} \mathcal{F}_{\mathsf{iss}}(\mathsf{C})(x; y)$$

on morphism sets for objects $x, y \in \mathsf{C}$.

An adaptation of the proof of Theorem 6.1.10 establishes the following result.

THEOREM 6.2.4. *Suppose $F : \mathsf{C} \longrightarrow \mathsf{D}$ is a functor with C a category and D an involutive strict symmetric monoidal category. Then there exists a unique strict involutive strict symmetric monoidal functor*

$$\mathcal{F}_{\mathsf{iss}}(\mathsf{C}) \xrightarrow{\overline{F}} \mathsf{D}$$

such that $\overline{F} \circ \varepsilon^{\mathsf{st}}_{\mathsf{C}} = F$.

DEFINITION 6.2.5. For a category C, the involutive strict symmetric monoidal category $\mathcal{F}_{\mathsf{iss}}(\mathsf{C})$ is called the *free involutive strict symmetric monoidal category generated by* C.

Next we observe that the free involutive symmetric monoidal category and the free involutive strict symmetric monoidal category are equivalent.

THEOREM 6.2.6. *For each category C, there is a canonical strict involutive strict symmetric monoidal functor*

$$\mathcal{F}_{\mathsf{isym}}(\mathsf{C}) \xrightarrow{\theta} \mathcal{F}_{\mathsf{iss}}(\mathsf{C})$$

that is also an equivalence of categories.

PROOF. The functor θ is the unique strict involutive strict symmetric monoidal functor that makes the diagram

$$\begin{array}{ccc} \mathsf{C} & \xrightarrow{\varepsilon_{\mathsf{C}}} & \mathcal{F}_{\mathsf{isym}}(\mathsf{C}) \\ \| & & \downarrow \theta \\ \mathsf{C} & \xrightarrow{\varepsilon^{\mathsf{st}}_{\mathsf{C}}} & \mathcal{F}_{\mathsf{iss}}(\mathsf{C}) \end{array}$$

commutative, given by Theorem 6.1.10. The same argument as in the proof of Theorem 5.2.8 shows that θ is an equivalence of categories. □

6.3 Commutativity of Formal Diagrams

The purpose of this section is to prove that all formal diagrams, suitably defined, in an involutive symmetric monoidal category commute. Recall from Definition 5.3.1 that for a small involutive monoidal category C, $\mathsf{It}(\mathsf{C})$ is the involutive monoidal category whose objects are pairs (m, T) with $m \geq 0$ an integer and $T : \mathsf{C}^{\times m} \longrightarrow \mathsf{C}$ a functor, and whose morphisms are natural transformations.

DEFINITION 6.3.1. Suppose C is a small involutive symmetric monoidal category with symmetry isomorphism ξ. We equip the involutive monoidal category $\mathsf{It}(\mathsf{C})$ with the natural isomorphism

$$(T \otimes U)(\underline{x}, \underline{y}) = T(\underline{x}) \otimes U(\underline{y}) \xrightarrow{\xi_{T(\underline{x}),U(\underline{y})}} U(\underline{y}) \otimes T(\underline{x}) = (U \otimes T)(\underline{y}, \underline{x})$$

for $(m, T), (n, U) \in \mathsf{It}(\mathsf{C})$, $\underline{x} \in \mathsf{C}^{\times m}$, and $\underline{y} \in \mathsf{C}^{\times n}$.

Using the diagrams in Explanation 4.1.2, an inspection establishes the following.

LEMMA 6.3.2. *$(\mathsf{It}(\mathsf{C}), \xi)$ in Definition 6.3.1 is an involutive symmetric monoidal category.*

With $\mathbf{1} = \{x\}$ denoting a discrete category with one object x, the following result is the coherence statement about the commutativity of all formal diagrams in a small involutive symmetric monoidal category.

COROLLARY 6.3.3. *Suppose C is a small involutive symmetric monoidal category. Then the following statements hold.*

(1) There exists a unique strict involutive strict symmetric monoidal functor

$$\mathcal{F}_{\mathsf{isym}}(\mathbf{1}) \xrightarrow{\Phi} \mathsf{It}(\mathsf{C}) \quad \textit{such that} \quad \Phi(x) = (1, \mathrm{Id}_{\mathsf{C}}).$$

(2) Each morphism $f : X \longrightarrow Y$ in $\mathcal{F}_{\mathsf{isym}}(\mathbf{1})$ of the form

$$f = \left(\sigma; \{\mathrm{Id}_x\}_{i=1}^{j}\right) \in \Sigma_j \times \prod_{i=1}^{j} \mathbf{1}(x,x) \cong \Sigma_j$$

satisfying (6.1.16) *is sent by* Φ *to a natural isomorphism*

$$\Phi(X) \xrightarrow[\cong]{\phi_{X,Y}} \Phi(Y) \ : \mathsf{C}^{\times j} \longrightarrow \mathsf{C}$$

that involves only

- *the isomorphisms* $\mathrm{Id}_{\mathbb{1}^{\mathsf{C}}}$ *and* $I_0 : \mathbb{1}^{\mathsf{C}} \longrightarrow I_{\mathsf{C}}\mathbb{1}^{\mathsf{C}}$,
- *the natural isomorphisms* $\mathrm{Id}_{\mathrm{Id}_{\mathsf{C}}}$, ι^{C}, I_2, α, λ, ρ, ξ, *and their inverses,*
- *the functors* I_{C} *and* $\otimes$, *and*
- *categorical composites in* C.

PROOF. Since $\mathsf{It}(\mathsf{C})$ is an involutive symmetric monoidal category with an object $(1, \mathrm{Id}_{\mathsf{C}}) \in \mathsf{It}(\mathsf{C})$, the first assertion follows from the Coherence Theorem 6.1.10. The second assertion follows from the factorization (6.1.12) of morphisms in $\mathcal{F}_{\mathsf{isym}}(\mathbf{1})$. □

Let us define a *formal diagram* in a small involutive symmetric monoidal category C as a diagram in $\mathsf{It}(\mathsf{C})$ that is the Φ-image of a commutative diagram in $\mathcal{F}_{\mathsf{isym}}(\mathbf{1})$. Then Corollary 6.3.3 implies that every formal diagram in a small involutive symmetric monoidal category is commutative.

6.4 Strictification of Involutive Symmetric Monoidal Categories

The purpose of this section is to show that each involutive symmetric monoidal category can be strictified to an involutive strict symmetric monoidal category.

THEOREM 6.4.1. *Suppose* C *is an involutive symmetric monoidal category as in Definition 4.1.1. Then there exist canonical choices of*

(i) an involutive strict symmetric monoidal category C_{st} *and*
(ii) an involutive adjoint equivalence

$$\mathsf{C} \underset{R}{\overset{L}{\rightleftarrows}} \mathsf{C}_{\mathsf{st}}$$

such that the following five statements hold:

- *The unit* $\epsilon : \mathrm{Id}_{\mathsf{C}} \longrightarrow RL$ *is the identity natural transformation.*
- L *is a strict involutive strong symmetric monoidal functor.*
- R *is an involutive strong symmetric monoidal functor.*
- *The unit and the counit of the adjunction* $L \dashv R$ *are both monoidal natural transformations and involutive natural transformations.*
- *If the involutive structure on* C *is strict (resp., trivial)–i.e.,* $I^2 = \mathrm{Id}_{\mathsf{C}}$ *(resp.,* $I = \mathrm{Id}_{\mathsf{C}}$*) and* $\iota = \mathrm{Id}_{\mathrm{Id}_{\mathsf{C}}}$*–then the same is true for* C_{st}.

PROOF. Since C is an involutive symmetric monoidal category, it suffices to equip the involutive adjoint equivalence in Theorem 5.4.1 with a compatible symmetry structure. First we equip the involutive strict monoidal category C_{st} with the natural isomorphism

$$\underline{x} \otimes^{\mathsf{st}} \underline{y} \xrightarrow{\xi_{\underline{x},\underline{y}}} \underline{y} \otimes^{\mathsf{st}} \underline{x} \in \mathsf{C}_{\mathsf{st}}$$

for objects $\underline{x}, \underline{y} \in \mathsf{C}_{\mathsf{st}}$, given by the composite

$$\begin{array}{ccc} (\underline{x},\underline{y})^{\otimes} & \xrightarrow{\xi_{\underline{x},\underline{y}}} & (\underline{y},\underline{x})^{\otimes} \\ \cong\downarrow & & \uparrow\cong \\ \underline{x}^{\otimes} \otimes \underline{y}^{\otimes} & \xrightarrow[\cong]{\xi_{\underline{x}^{\otimes},\underline{y}^{\otimes}}} & \underline{y}^{\otimes} \otimes \underline{x}^{\otimes} \end{array}$$

in C. In this composite:

- The first and the last isomorphisms are the unique natural isomorphisms guaranteed by Mac Lane's Coherence Theorem for monoidal categories [Mac98] (VII.2 Corollary). Each one involves only identity morphisms, the associativity isomorphism, the left/right unit isomorphisms, their inverses, the monoidal product, and composites.
- The bottom isomorphism $\xi_{\underline{x}^{\otimes},\underline{y}^{\otimes}}$ is the symmetry isomorphism in C.

An inspection shows that the diagram (4.1.7) is commutative, so $(\mathsf{C}_{\mathsf{st}}, \xi)$ is an involutive strict symmetric monoidal category. Another inspection shows that the adjoint functors $L : \mathsf{C} \longrightarrow \mathsf{C}_{\mathsf{st}}$ and $R : \mathsf{C}_{\mathsf{st}} \longrightarrow \mathsf{C}$ are both symmetric monoidal functors. □

6.5 Free Involutive Symmetric Monoidal Categories of Involutive Categories

The purpose of this section is to give an explicit description of the free involutive symmetric monoidal category generated by an involutive category.

DEFINITION 6.5.1. Suppose $(\mathsf{C}, I_{\mathsf{C}}, \iota^{\mathsf{C}})$ is an involutive category. Define the category $\mathcal{F}^{\mathsf{i}}_{\mathsf{isym}}(\mathsf{C})$ with extra structures as follows.

Objects: $\mathsf{Ob}(\mathcal{F}^{\mathsf{i}}_{\mathsf{isym}}(\mathsf{C})) = \mathsf{W}^{\mathsf{i}}_{\mathsf{im}}(\mathsf{Ob}(\mathsf{C}))$ as in Definition 5.5.1, where $\mathsf{Ob}(\mathsf{C})$ is equipped with the assignment $I_{\mathsf{C}} : \mathsf{Ob}(\mathsf{C}) \longrightarrow \mathsf{Ob}(\mathsf{C})$.

Morphisms: For $X, Y \in \mathsf{Ob}(\mathcal{F}^{\mathsf{i}}_{\mathsf{isym}}(\mathsf{C}))$ with

$$\mathsf{word}(X) = (x_1, \ldots, x_j) \in \mathsf{Ob}(\mathsf{C})^{\times j} \quad \text{and} \quad \mathsf{word}(Y) = (y_1, \ldots, y_k) \in \mathsf{Ob}(\mathsf{C})^{\times k},$$

a morphism $f : X \longrightarrow Y$ in $\mathcal{F}^{\mathsf{i}}_{\mathsf{isym}}(\mathsf{C})$ is a pair

$$\left(\sigma; \{f_i\}_{i=1}^{j}\right) \in \Sigma_j \times \prod_{i=1}^{j} \mathsf{C}\left(x_i, y_{\sigma(i)}\right)$$

if $j = k$. No such morphisms exist if $j \neq k$. We call σ the *underlying permutation* of the morphism $(\sigma; \{f_i\})$. The identity morphism of X is $\left(\mathrm{id}_j; \{\mathrm{Id}_{x_i}\}_{i=1}^{j}\right)$. Composition is defined using products in the symmetric groups and composition in C as in (6.1.5).

Monoidal Product: For objects $X, Y \in \mathsf{Ob}(\mathcal{F}^{\mathsf{i}}_{\mathsf{isym}}(\mathsf{C}))$, the monoidal product $X \otimes Y$ is defined as part of the definition of the class $\mathsf{W}^{\mathsf{i}}_{\mathsf{im}}(\mathsf{Ob}(\mathsf{C}))$. The monoidal product on morphisms in $\mathcal{F}^{\mathsf{i}}_{\mathsf{isym}}(\mathsf{C})$ is defined using block sum inclusion and concatenation of finite sequences of morphisms in C, as in (6.1.6).

Monoidal Unit: The monoidal unit in $\mathcal{F}^{\mathsf{i}}_{\mathsf{im}}(\mathsf{C})$ is $\mathbb{1} \in \mathsf{W}^{\mathsf{i}}_{\mathsf{im}}(\mathsf{Ob}(\mathsf{C}))$.

Monoidal Structure Isomorphisms: The associativity isomorphism, the left unit isomorphism, and the right unit isomorphism are defined using identity permutations and identity morphisms in C.

Symmetry: Define a natural isomorphism

$$X \otimes Y \xrightarrow{\xi_{X,Y}} Y \otimes X \quad \text{for} \quad X, Y \in \mathsf{Ob}(\mathcal{F}^{\mathsf{i}}_{\mathsf{isym}}(\mathsf{C}))$$

as in (6.1.7) using block permutations and identity morphisms in C.

Involution Functor: Define a functor

$$I : \mathcal{F}^{\mathsf{i}}_{\mathsf{isym}}(\mathsf{C}) \longrightarrow \mathcal{F}^{\mathsf{i}}_{\mathsf{isym}}(\mathsf{C})$$

by the assignment $X \longmapsto IX$ on objects, and the function

$$\begin{array}{ccc} \mathcal{F}^{\mathsf{i}}_{\mathsf{isym}}(X;Y) & \ni & \Big(\sigma \in \Sigma_j; \big\{f_i \in \mathsf{C}(x_i, y_{\sigma(i)})\big\}_{i=1}^{j}\Big) \\ \Big\downarrow I & & \Big\downarrow \\ \mathcal{F}^{\mathsf{i}}_{\mathsf{isym}}(IX;IY) & \ni & \Big(\sigma; \big\{I_{\mathsf{C}} f_i \in \mathsf{C}(I_{\mathsf{C}} x_i, I_{\mathsf{C}} y_{\sigma(i)})\big\}_{i=1}^{j}\Big) \end{array}$$

on morphisms. If $j = 0$, then this function is the unique function between two one-point sets.

Unit: Define a natural isomorphism

$$\iota : \mathrm{Id}_{\mathcal{F}^{\mathsf{i}}_{\mathsf{isym}}(\mathsf{C})} \xrightarrow{\cong} I^2$$

by

$$\iota_X = \Big(\mathrm{id}_j; \big\{\iota^{\mathsf{C}}_{x_i}\big\}_{i=1}^{j}\Big) : X \xrightarrow{\cong} I^2 X \in \mathcal{F}^{\mathsf{i}}_{\mathsf{isym}}(\mathsf{C})(X; I^2 X)$$

for each object $X \in \mathcal{F}^{\mathsf{i}}_{\mathsf{isym}}(\mathsf{C})$ as above, where $\iota^{\mathsf{C}}_{x_i} : x_i \xrightarrow{\cong} I^2_{\mathsf{C}} x_i$ is the unit in C applied to x_i.

Involutive Monoidal Structure: Define the morphism $I_0 : \mathbb{1} \longrightarrow I\mathbb{1}$ as the unique isomorphism in

$$\mathcal{F}^{\mathsf{i}}_{\mathsf{isym}}(\mathsf{C})(\mathbb{1}; I\mathbb{1}) = \{*\}.$$

Finally, for objects $X, Y \in \mathcal{F}^{\mathsf{i}}_{\mathsf{isym}}(\mathsf{C})$ as above, the isomorphism

$$IX \otimes IY \xrightarrow[\cong]{(I_2)_{X,Y}} I(X \otimes Y) \quad \in \mathcal{F}^{\mathsf{i}}_{\mathsf{isym}}(\mathsf{C})\big(IX \otimes IY, I(X \otimes Y)\big)$$

is defined using identity permutation and identity morphisms in C.

An inspection of the diagrams in Explanation 4.1.2 establishes the following.

LEMMA 6.5.2. *For each involutive category $(\mathsf{C}, I_{\mathsf{C}}, \iota^{\mathsf{C}})$, the tuple*

$$\Big(\big(\mathcal{F}^{\mathsf{i}}_{\mathsf{isym}}(\mathsf{C}), \otimes, \mathbb{1}, \alpha, \lambda, \rho, \xi\big), (I, I_2, I_0), \iota\Big)$$

in Definition 6.5.1 is an involutive symmetric monoidal category.

DEFINITION 6.5.3. For each involutive category C, denote by

$$\mathsf{C} \xrightarrow{\varepsilon_{\mathsf{C}}} \mathcal{F}^{\mathsf{i}}_{\mathsf{isym}}(\mathsf{C})$$

the full and faithful embedding given by

- the inclusion $\mathsf{Ob}(\mathsf{C}) \subseteq \mathsf{W}^{\mathsf{i}}_{\mathsf{im}}(\mathsf{Ob}(\mathsf{C}))$ on objects, and
- the identity function

$$\mathsf{C}(x, y) \xrightarrow{\text{Id}} \mathcal{F}^{\mathsf{i}}_{\mathsf{isym}}(\mathsf{C})(x; y)$$

on morphism sets for objects $x, y \in \mathsf{C}$.

The following observation says that $\mathcal{F}^{\mathsf{i}}_{\mathsf{isym}}(\mathsf{C})$ is the free involutive symmetric monoidal category generated by an involutive category C.

THEOREM 6.5.4. *Suppose $(F, \nu^F) : C \longrightarrow D$ is a strict involutive functor with C an involutive category and D an involutive symmetric monoidal category. Then there exists a unique strict involutive strict symmetric monoidal functor*

$$\mathcal{F}^{\mathsf{i}}_{\mathsf{isym}}(C) \xrightarrow{\overline{F}} D$$

such that $\overline{F} \circ \varepsilon_C = F$.

PROOF. The proof is adapted from that of Theorem 5.5.6, so we only point out the modifications, which involve what $\overline{F}$ does to morphisms. In the definition of a basic isomorphism, the underlying permutation is the identity permutation. For a morphism $f \in \mathcal{F}^{\mathsf{i}}_{\mathsf{isym}}(X; Y)$, instead of the composite in (5.5.7), we define $\overline{F}(f)$ as the composite in (6.1.11). The proof of Theorem 6.1.10 shows that $\overline{F}$ is a functor. When equipped with the identity distributive law and the strict monoidal functor structure, $\overline{F}$ is a strict involutive strict symmetric monoidal functor. Its uniqueness is proved exactly as in the last paragraph of the proof of Theorem 6.1.10. □

6.6 Free Involutive Strict Symmetric Monoidal Categories of Involutive Categories

The purpose of this section is to give an explicit description of the free involutive strict symmetric monoidal category generated by an involutive category. We also observe that, for an involutive category, the free involutive symmetric monoidal category is equivalent to the free involutive strict symmetric monoidal category via a strict involutive strict symmetric monoidal equivalence.

DEFINITION 6.6.1. Suppose $(\mathsf{C}, I_{\mathsf{C}}, \iota^{\mathsf{C}})$ is an involutive category. Define the category $\mathcal{F}^{\mathsf{i}}_{\mathsf{iss}}(\mathsf{C})$ with extra structures as in Definition 6.5.1, but with the following modifications:

Objects: $\mathsf{Ob}(\mathcal{F}^{\mathsf{i}}_{\mathsf{iss}}(\mathsf{C})) = \mathsf{W}^{\mathsf{i}}_{\mathsf{istm}}(\mathsf{Ob}(\mathsf{C}))$ as in Definition 5.6.1, where $\mathsf{Ob}(\mathsf{C})$ is equipped with the assignment $I_{\mathsf{C}} : \mathsf{Ob}(\mathsf{C}) \longrightarrow \mathsf{Ob}(\mathsf{C})$.

Monoidal Product: For objects $X, Y \in \mathsf{Ob}(\mathcal{F}^{\mathsf{i}}_{\mathsf{iss}}(\mathsf{C}))$, their monoidal product $X \otimes Y$ is defined as their concatenation, which is part of the definition of the class $\mathsf{W}^{\mathsf{i}}_{\mathsf{istm}}(\mathsf{Ob}(\mathsf{C}))$.

Monoidal Unit: The monoidal unit in $\mathcal{F}^{\mathsf{i}}_{\mathsf{iss}}(\mathsf{C})$ is $\varnothing$.

Monoidal Structure Isomorphisms: Each component of the associativity isomorphism α, the left unit isomorphism λ, and the right unit isomorphism ρ, is an identity morphism.

Involutive Monoidal Structure: $I_0 : \varnothing \longrightarrow \varnothing$ is the identity morphism.

An inspection establishes the following.

LEMMA 6.6.2. *For each involutive category* C*, the tuple*

$$\big((\mathcal{F}^{\mathsf{i}}_{\mathsf{iss}}(\mathsf{C}), \otimes, \varnothing, \alpha, \lambda, \rho, \xi), (I, I_2, I_0), \iota\big)$$

in Definition 6.6.1 is an involutive strict symmetric monoidal category.

DEFINITION 6.6.3. For each involutive category C, denote by

$$\mathsf{C} \xrightarrow{\varepsilon^{\mathsf{st}}_{\mathsf{C}}} \mathcal{F}^{\mathsf{i}}_{\mathsf{iss}}(\mathsf{C})$$

the full and faithful embedding given by

- the inclusion $\mathsf{Ob}(\mathsf{C}) \subseteq \mathsf{W}^{\mathsf{i}}_{\mathsf{istm}}(\mathsf{Ob}(\mathsf{C}))$ on objects, and
- the identity function

$$\mathsf{C}(x, y) \xrightarrow{\mathrm{Id}} \mathcal{F}^{\mathsf{i}}_{\mathsf{iss}}(\mathsf{C})(x; y)$$

on morphism sets for objects $x, y \in \mathsf{C}$.

Essentially the same proof as that of Theorem 6.5.4 establishes the following result.

THEOREM 6.6.4. *Suppose* $(F, \nu^F) : \mathsf{C} \longrightarrow \mathsf{D}$ *is a strict involutive functor with* C *an involutive category and* D *an involutive strict symmetric monoidal category. Then there exists a unique strict involutive strict symmetric monoidal functor*

$$\mathcal{F}^{\mathsf{i}}_{\mathsf{iss}}(\mathsf{C}) \xrightarrow{\overline{F}} \mathsf{D}$$

such that $\overline{F} \circ \varepsilon^{\mathsf{st}}_{\mathsf{C}} = F$.

DEFINITION 6.6.5. For an involutive category C, the involutive strict symmetric monoidal category $\mathcal{F}^{\mathsf{i}}_{\mathsf{iss}}(\mathsf{C})$ is called the *free involutive strict symmetric monoidal category generated by* C.

Next we observe that, as in the non-symmetric case, the free involutive symmetric monoidal category and the free involutive strict symmetric monoidal category are equivalent.

COROLLARY 6.6.6. *For each involutive category* C*, there is a canonical strict involutive strict symmetric monoidal functor*

$$\mathcal{F}^{\mathsf{i}}_{\mathsf{isym}}(\mathsf{C}) \xrightarrow{\ \theta\ } \mathcal{F}^{\mathsf{i}}_{\mathsf{iss}}(\mathsf{C})$$

that is also an equivalence of categories.

PROOF. The functor θ is the unique strict involutive strict symmetric monoidal functor that extends $\varepsilon^{\mathsf{st}}_{\mathsf{C}}$, in the sense that the diagram

$$\begin{array}{ccc} \mathsf{C} & \xrightarrow{\ \varepsilon_{\mathsf{C}}\ } & \mathcal{F}^{\mathsf{i}}_{\mathsf{isym}}(\mathsf{C}) \\ \| & & \downarrow \theta \\ \mathsf{C} & \xrightarrow{\ \varepsilon^{\mathsf{st}}_{\mathsf{C}}\ } & \mathcal{F}^{\mathsf{i}}_{\mathsf{iss}}(\mathsf{C}) \end{array}$$

is commutative, given by Theorem 6.5.4. It exists because $\varepsilon^{\mathsf{st}}_{\mathsf{C}}$ is a strict involutive functor. As in the proof of Theorem 5.2.8, the functor θ is the identity function on morphism sets, and is surjective on objects. Therefore, θ is an equivalence of categories. □

6.7 Exercises

(1) For integers $n \geq 1$ and $k_1, \ldots, k_n \geq 0$ with $k = k_1 + \cdots + k_n$, prove the following statements about block permutations and block sums.

- The equality

$$(\sigma\tau)\langle k_1, \ldots, k_n\rangle = \sigma\langle k_{\tau^{-1}(1)}, \ldots, k_{\tau^{-1}(n)}\rangle \cdot \tau\langle k_1, \ldots, k_n\rangle \in \Sigma_k$$

  holds for $\sigma, \tau \in \Sigma_n$.
- The equality

$$(\sigma_1\tau_1) \times \cdots \times (\sigma_n\tau_n) = \big(\sigma_1 \times \cdots \times \sigma_n\big) \cdot \big(\tau_1 \times \cdots \times \tau_n\big) \in \Sigma_k$$

holds for $\sigma_i, \tau_i \in \Sigma_{k_i}$ for $1 \leq i \leq n$.

- The equality

$$\sigma\langle k_1, \ldots, k_n\rangle \cdot \big(\tau_1 \times \cdots \times \tau_n\big) = \big(\tau_{\sigma^{-1}(1)} \times \cdots \times \tau_{\sigma^{-1}(n)}\big) \cdot \sigma\langle k_1, \ldots, k_n\rangle$$

holds for $\sigma \in \Sigma_n$ and $\tau_i \in \Sigma_{k_i}$ for $1 \leq i \leq n$.

(2) Prove Lemma 6.1.8.

(3) In the proof of Theorem 6.1.10, check that:

- (6.1.12) is actually a decomposition of the morphism $f = \big(\sigma; \{f_i\}\big)$.
- The decomposition (6.1.12) forces $\overline{F}(f)$ to be equal to the composite in (6.1.11).

(4) Give a detailed proof of Theorem 6.2.4.

(5) For each of the involutive symmetric monoidal categories in Examples 4.1.12 and 4.1.13, describe explicitly:

- the involutive strict symmetric monoidal category C_{st};
- the involutive adjoint equivalence (L, R) in Theorem 6.4.1.

(6) Prove that there is a monad in Cat whose algebras are exactly small involutive symmetric monoidal categories. Do the same with symmetric monoidal replaced by strict symmetric monoidal.

(7) Repeat the previous exercise with Cat replaced by ICat_0.

(8) Give a detailed proof of Theorem 6.6.4.

(9) Prove the strict involutive variants of all of the main results in this chapter. In other words:

- Give explicit constructions of
  - the free strict involutive symmetric monoidal category and
  - the free strict involutive strict symmetric monoidal category

  of a category.
- Prove that they are equivalent via a canonical strict involutive strict symmetric monoidal functor, and so forth.

Chapter 7
Categorical Gelfand–Naimark–Segal Construction

This chapter contains a categorical analogue of the Gelfand–Naimark–Segal (GNS) construction, which is originally a functional analytic statement. For a C^*-algebra A, the GNS construction gives a correspondence between states on A and cyclic $*$-representations of A. The categorical GNS correspondence, given in Theorem 7.2.7 below, is an explicit isomorphism between the category of quantum probability objects and the category of monoid inner product objects. A *quantum probability object* is a categorical abstraction of a complex unital $*$-algebra equipped with a state over $\mathbb{C}$. A *monoid inner product object* is a categorical abstraction of a complex unital $*$-algebra equipped with a Hermitian form for which the equality

$$\langle ba, c\rangle = \langle a, b^*c\rangle$$

holds.

In Section 7.1 we define the category of monoid inner product objects. The main result of this section is an explicit description of the free-forgetful adjunction between the category of reversing involutive monoids and the category of monoid inner product objects. Quantum probability objects and the categorical GNS construction are discussed in Section 7.2. The proofs of several statements in Sections 7.1 and 7.2 require checking that some large diagrams are commutative. Those proofs are given in Section 7.3.

Throughout the rest of this chapter,

$$\big((\mathsf{C}, \otimes, \mathbb{1}, \alpha, \lambda, \rho, \xi), (I, I_2, I_0), \iota\big)$$

denotes an involutive symmetric monoidal category as in Definition 4.1.1. It will serve as our base category.

© The Author(s), under exclusive license to Springer Nature Switzerland AG 2020
D. Yau, *Involutive Category Theory*, Lecture Notes in Mathematics 2279,
https://doi.org/10.1007/978-3-030-61203-0_7

7.1 Monoid Inner Product Objects

The purposes of this section are (i) to define monoid inner product objects and (ii) to establish a free-forgetful adjunction between the categories of reversing involutive monoids and of monoid inner product objects. We first define a categorical version of an object equipped with a Hermitian form.

Recall from Definition 2.5.1 that $\mathsf{Inv}(\mathsf{C})$ is the category of involutive objects in C. An involutive object (X, j) in C consists of an object $X \in \mathsf{C}$ and an involutive structure map $j : X \longrightarrow IX$ such that

$$Ij \circ j = \iota_X : X \xrightarrow{\cong} I^2 X.$$

By Proposition 2.5.4(2), using the adjunction $I \dashv I$, an involutive object can equivalently be described as a pair

$$\big(X, j' : IX \longrightarrow X\big)$$

such that

$$j' \circ Ij' = \iota_X^{-1} : I^2 X \xrightarrow{\cong} X.$$

We will use these two equivalent descriptions of an involutive object interchangeably.

To define inner product objects, we first need the following Lemma, whose proof is given in Section 7.3.

LEMMA 7.1.1. *For each object $A \in \mathsf{C}$, when equipped with the composite*

(7.1.2)
$$\begin{array}{ccccc}
I(IA \otimes A) & & \xrightarrow{\quad j^{IA \otimes A} \quad} & & IA \otimes A \\
\Big\downarrow{\scriptstyle I(IA \otimes \iota_A)} & & & & \Big\uparrow{\scriptstyle \xi_{A,IA}} \\
I(IA \otimes I^2 A) & \xrightarrow{I(I_2)} & I^2(A \otimes IA) & \xrightarrow{\iota_{A \otimes IA}^{-1}} & A \otimes IA
\end{array}$$

in C, the pair

$$\big(IA \otimes A, j^{IA \otimes A}\big)$$

is an involutive object in C.

Lemma 7.1.1 is used in the next definition.

DEFINITION 7.1.3. Define the category $\mathsf{IP}(\mathsf{C})$ as follows.

Objects: An object in $\mathsf{IP}(\mathsf{C})$ is a tuple

$$\big(A, p, (X, j^X)\big),$$

called an *inner product object* in C, in which:

- A is an object in C, called the *space object.*
- (X, j^X) is an involutive object in C, called the *value object.*
- p is a morphism

$$\big(IA \otimes A, j^{IA\otimes A}\big) \xrightarrow{\ p\ } (X, j^X)$$

of involutive objects in C, called the *inner product.*

We sometimes abbreviate such an inner product object to p.

Morphisms: A morphism

$$\big(A, p, (X, j^X)\big) \xrightarrow{(f,g)} \big(B, q, (Y, j^Y)\big)$$

in $\mathsf{IP}(\mathsf{C})$ is a pair (f, g) in which:

- $f : A \longrightarrow B$ is a morphism in C.
- $g : (X, j^X) \longrightarrow (Y, j^Y)$ is a morphism of involutive objects in C.
- The diagram

(7.1.4)

$$\begin{array}{ccc} IA \otimes A & \xrightarrow{If \otimes f} & IB \otimes B \\ {\scriptstyle p}\downarrow & & \downarrow{\scriptstyle q} \\ X & \xrightarrow{\ \ g\ \ } & Y \end{array}$$

in C is commutative.

We sometimes abbreviate such a morphism to $(f, g) : p \longrightarrow q$.

Identities: The identity morphism of an object $\big(A, p, (X, j^X)\big)$ is $(\mathrm{Id}_A, \mathrm{Id}_X)$.

Composition: Composition in $\mathsf{IP}(\mathsf{C})$ is defined using composition in C and in $\mathsf{Inv}(\mathsf{C})$.

EXAMPLE 7.1.5 (Hermitian Form as Inner Product Object). In Definition 7.1.3, that the inner product $p : IA \otimes A \longrightarrow X$ is a morphism of involutive objects in C means that the diagram

(7.1.6)

$$\begin{array}{ccccc}
I(IA \otimes A) & & \xrightarrow{\quad Ip \quad} & & IX \\
\Big\downarrow{\scriptstyle I(IA\otimes\iota_A)} & & & & \Big\downarrow{\scriptstyle j^X} \\
I(IA \otimes I^2A) & & & & X \\
\Big\downarrow{\scriptstyle I(I_2)} & & & & \Big\uparrow{\scriptstyle p} \\
I^2(A \otimes IA) & \xrightarrow{\iota^{-1}_{A\otimes IA}} & A \otimes IA & \xrightarrow{\xi_{A,IA}} & IA \otimes A
\end{array}$$

is commutative.

For example, if $\mathsf{C} = \mathsf{Vect}(\mathbb{C})$ and $X = \mathbb{C}$ with complex conjugation as its involutive structure, then an inner product object $(A, p, \mathbb{C})$ is exactly a *Hermitian form* on the complex vector space A. Indeed, the $\mathbb{C}$-linearity of

$$\overline{A} \otimes A \xrightarrow{\quad p \quad} \mathbb{C}$$

means that it is linear in the second variable and conjugate linear in the first variable, i.e., sesquilinear. Furthermore, the commutative diagram (7.1.6) means that p is conjugate symmetric, i.e., satisfies the equality

$$p(y, x) = \overline{p(x, y)}$$

for $x, y \in A$. Sesquilinearity and conjugate symmetry are of course two of the requirements for a complex inner product space. ◇

Next we incorporate a categorical version of a unital $*$-algebra structure. Recall from Definition 4.7.2 that $\mathsf{RIMon}(\mathsf{C})$ is the category of reversing involutive monoids in C. By Proposition 4.7.7, a reversing involutive monoid is exactly a tuple $\big(X, \mu, 1, j : IX \longrightarrow X\big)$ such that:

- $(X, \mu, 1)$ is a monoid.
- (X, j) is an involutive object.
- The diagrams (4.7.8) and (4.7.9) are commutative.

Examples of reversing involutive monoids include complex unital $*$-algebras and their differential graded analogues.

DEFINITION 7.1.7. Define the category $\mathsf{MIP}(\mathsf{C})$ as follows.

Objects: An object in $\mathsf{MIP}(\mathsf{C})$ is a tuple

$$\big((A, \mu, 1, j), p\big),$$

called a *monoid inner product object* in C, in which:

- $\big(A, \mu, 1, j : IA \longrightarrow A\big)$ is a reversing involutive monoid in C.
- $p : IA \otimes A \longrightarrow X$ is an inner product object in C.
- The diagram

(7.1.8)

$$\begin{array}{ccccc}
(IA \otimes IA) \otimes A & \xrightarrow{I_2 \otimes A} & I(A \otimes A) \otimes A & \xrightarrow{I\mu \otimes A} & IA \otimes A \\
{\scriptstyle \zeta \otimes A}\uparrow & & & & \downarrow{\scriptstyle p} \\
(IA \otimes IA) \otimes A & & & & X \\
{\scriptstyle \alpha}\downarrow & & & & \uparrow{\scriptstyle p} \\
IA \otimes (IA \otimes A) & \xrightarrow{IA \otimes (j \otimes A)} & IA \otimes (A \otimes A) & \xrightarrow{IA \otimes \mu} & IA \otimes A
\end{array}$$

is commutative.

We sometimes abbreviate such an object to p. We call $(A, \mu, 1, j)$ the *underlying reversing involutive monoid* of the monoid inner product object.

Morphisms: A morphism

$$\big((A, \mu^A, 1^A, j^A), p\big) \xrightarrow{(f,g)} \big((B, \mu^B, 1^B, j^B), q\big)$$

in $\mathsf{MIP}(\mathsf{C})$ consists of a morphism

$$(f, g) : p \longrightarrow q \in \mathsf{IP}(\mathsf{C})$$

such that $f : A \longrightarrow B$ is also a morphism of reversing involutive monoids.

Identities: The identity morphism of $\big((A, \mu, 1, j), p\big)$ as above is $(\mathrm{Id}_A, \mathrm{Id}_X)$.

Composition: Composition in $\mathsf{MIP}(\mathsf{C})$ is defined using composition in $\mathsf{RIMon}(\mathsf{C})$ and in $\mathsf{Inv}(\mathsf{C})$.

EXAMPLE 7.1.9. Continuing Examples 4.7.6 and 7.1.5, a monoid inner product object in $\mathsf{Vect}(\mathbb{C})$ of the form

$$\big(A, p : \overline{A} \otimes A \longrightarrow \mathbb{C}\big)$$

consists of

- a complex unital $*$-algebra A, and
- a Hermitian form $p = \langle , \rangle$ on A

such that the diagram (7.1.8) is commutative. In this case, (7.1.8) is the equality

$$\big\langle a, b^* \cdot_A c\big\rangle = \big\langle a \cdot_{\overline{A}} b, c\big\rangle$$

for $a, b, c \in A$, in which $b^* \cdot_A c$ and $a \cdot_{\overline{A}} b$ are the multiplications in A and $\overline{A}$, respectively. ◇

There is a forgetful functor from the category of monoid inner product objects to the category of reversing involutive monoids. The rest of this section is about its left adjoint, for which we need the following Lemma, whose proof is given in Section 7.3.

LEMMA 7.1.10. *Suppose* $\big(A, \mu, \mathbb{1}, j : IA \longrightarrow A\big)$ *is a reversing involutive monoid in* C. *Then the composite*

$$\begin{tikzcd} IA \otimes A \arrow[r, "j \otimes A"] \arrow[rr, bend left, "p^A"] & A \otimes A \arrow[r, "\mu"] & A \end{tikzcd}$$

in C *is a morphism of involutive objects in* C, *where* $IA \otimes A$ *is equipped with the involutive structure in Lemma 7.1.1.*

LEMMA 7.1.11. *The pair* (A, p^A) *in Lemma 7.1.10 is a monoid inner product object in* C.

PROOF. In view of Lemma 7.1.10, it remains to prove the axiom (7.1.8) for p^A, which is the outer-most diagram below.

$$\begin{tikzcd} (IA \otimes IA) \otimes A \arrow[r, "I_2 \otimes A"] & I(A \otimes A) \otimes A \arrow[r, "I\mu \otimes A"] & IA \otimes A \arrow[d, "j \otimes A"] \\ (IA \otimes IA) \otimes A \arrow[u, "\xi \otimes A"] \arrow[r, "(j \otimes j) \otimes A"] \arrow[d, "\alpha"'] & (A \otimes A) \otimes A \arrow[r, "\mu \otimes A"] \arrow[d, "\alpha"] & A \otimes A \arrow[d, "\mu"] \\ IA \otimes (IA \otimes A) \arrow[r, "j \otimes (j \otimes A)"] \arrow[d, "IA \otimes (j \otimes A)"'] & A \otimes (A \otimes A) \arrow[dr, "A \otimes \mu"] & A \\ IA \otimes (A \otimes A) \arrow[r, "IA \otimes \mu"] & IA \otimes A \arrow[r, "j \otimes A"] & A \otimes A \arrow[u, "\mu"'] \end{tikzcd}$$

The top rectangle is commutative by (4.7.9). The middle left rectangle is commutative by the naturality of the associativity isomorphism α. The bottom trapezoid is commutative by the functoriality of $\otimes$. The middle right trapezoid is commutative by the associativity axiom of the monoid A in Definition 1.2.8. □

The next observation will play a crucial role in the proof of Theorem 7.2.7 below.

PROPOSITION 7.1.12. *There is an adjunction*

$$\begin{tikzcd} \mathsf{RIMon}(\mathsf{C}) \arrow[r, shift left, "\mathcal{F}"] & \mathsf{MIP}(\mathsf{C}) \arrow[l, shift left, "U"] \end{tikzcd}$$

in which:

- *The right adjoint* U *is defined by* $U(A, p) = A$, *which forgets about the inner product.*
- *The left adjoint sends* $A \in \mathsf{RIMon}(\mathsf{C})$ *to the monoid inner product object*

$$\mathcal{F}(A) = (A, p^A)$$

in Lemma 7.1.11.

- *The unit ε : Id $\longrightarrow$ $U\mathcal{F}$ of the adjunction is the identity.*

PROOF. The image of a morphism $f : A \longrightarrow B \in \mathsf{RIMon(C)}$ under $\mathcal{F}$ is the pair (f, f). To see that this is a morphism of monoid inner product objects, it is enough to show that it is a morphism of inner product objects. The diagram (7.1.4) is the outer-most diagram below.

$$
\begin{array}{ccccc}
 & & p^A & & \\
IA \otimes A & \xrightarrow{\;j^A \otimes A\;} & A \otimes A & \xrightarrow{\;\mu^A\;} & A \\
{\scriptstyle If \otimes f}\downarrow & & {\scriptstyle f \otimes f}\downarrow & & \downarrow{\scriptstyle f} \\
IB \otimes B & \xrightarrow{\;j^B \otimes B\;} & B \otimes B & \xrightarrow{\;\mu^B\;} & B \\
 & & p^B & &
\end{array}
$$

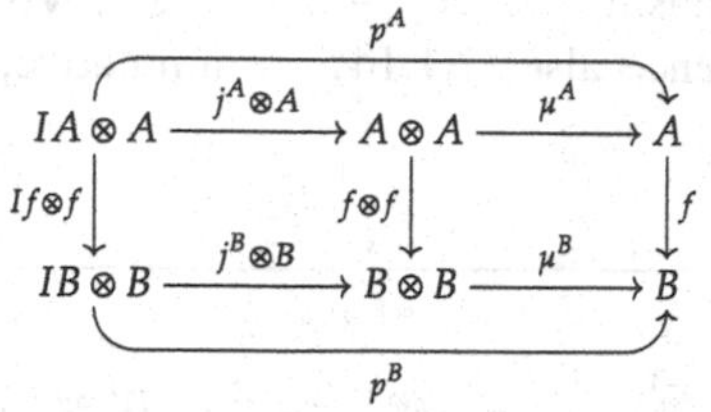

This is commutative because f is compatible with both the involutive structures and the multiplications. This defines the functor $\mathcal{F}$.

To check that $(\mathcal{F}, U)$ forms an adjunction, it suffices to show that, for each morphism $f : A \longrightarrow B$ in $\mathsf{RIMon(C)}$ with $\big(B, q : IB \otimes B \longrightarrow Y\big)$ a monoid inner product object in C, there exists a unique morphism $g : A \longrightarrow Y$ of involutive objects in C such that the diagram (7.1.4), which in this case is the diagram

(7.1.13)
$$
\begin{array}{ccccc}
 & & p^A & & \\
IA \otimes A & \xrightarrow{\;j^A \otimes A\;} & A \otimes A & \xrightarrow{\;\mu^A\;} & A \\
{\scriptstyle If \otimes f}\downarrow & & & & \downarrow{\scriptstyle g} \\
IB \otimes B & & \xrightarrow{\;q\;} & & Y,
\end{array}
$$

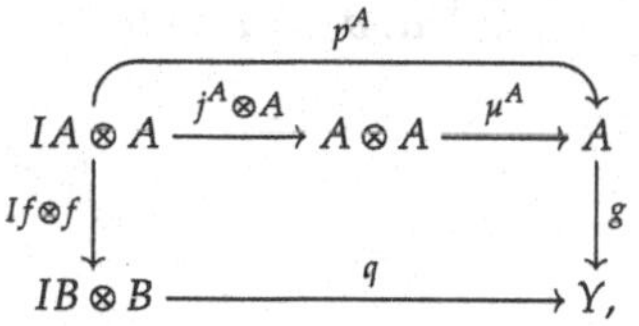

is commutative.

Suppose $g : A \longrightarrow Y$ is any morphism in C. Consider the following diagram in C.

(7.1.14)
$$
\begin{array}{ccccccccccc}
 & & & & & 1^A \otimes A & & & & & \\
A & \xrightarrow{\lambda_A^{-1}} & 1 \otimes A & \xrightarrow{I_0 \otimes A} & I\mathbb{1} \otimes A & \xrightarrow{I1^A \otimes A} & IA \otimes A & \xrightarrow{j^A \otimes A} & A \otimes A \\
{\scriptstyle f}\downarrow & & {\scriptstyle 1 \otimes f}\downarrow & & {\scriptstyle I\mathbb{1} \otimes f}\downarrow & & {\scriptstyle If \otimes f}\downarrow & & \downarrow{\scriptstyle \mu^A} \\
 & & & & & & & & A \\
 & & & & & & & & \downarrow{\scriptstyle g} \\
B & \xrightarrow{\lambda_B^{-1}} & 1 \otimes B & \xrightarrow{I_0 \otimes B} & I\mathbb{1} \otimes B & \xrightarrow{I1^B \otimes B} & IB \otimes B & \xrightarrow{q} & Y
\end{array}
$$

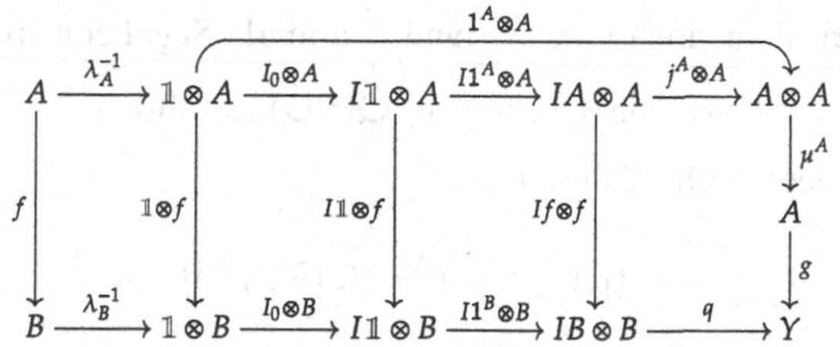

The left-most rectangle is commutative by the naturality of λ. The two middle rectangles are commutative by definition and the fact that f preserves units. The right-most rectangle is the diagram (7.1.13). There are equalities

(7.1.15) $$j^A \circ I1^A \circ I_0 = 1^A : \mathbb{1} \longrightarrow A$$

by (4.7.8) and

$$\mu^A \circ (1^A \otimes A) \circ \lambda_A^{-1} = \mathrm{Id}_A \tag{7.1.16}$$

by the unity axiom of a monoid in Definition 1.2.8. So the top-right composite in (7.1.14) is g.

Therefore, *if* a morphism $g : A \longrightarrow Y \in \mathsf{Inv(C)}$ exists that makes the diagram (7.1.13), and hence also (7.1.14), commutative, then it is unique and is given by the composite

$$\begin{array}{ccccccc} A & & & \xrightarrow{\quad g \quad} & & & Y \\ {\scriptstyle f}\downarrow & & & & & & \uparrow{\scriptstyle q} \\ B & \xrightarrow{\lambda_B^{-1}} & \mathbb{1} \otimes B & \xrightarrow{I_0 \otimes B} & I\mathbb{1} \otimes B & \xrightarrow{I1^B \otimes B} & IB \otimes B \end{array} \tag{7.1.17}$$

in C, which is the left-bottom composite in (7.1.14). In Lemma 7.3.4 below, we will prove that this composite is indeed a morphism of involutive objects in C that makes the diagram (7.1.13) commutative. □

EXAMPLE 7.1.18. With $\mathsf{C} = \mathsf{Vect}(\mathbb{C})$, a reversing involutive monoid A in C is exactly a complex unital $*$-algebra. The composite $p^A : IA \otimes A \longrightarrow A$ in Lemma 7.1.11 is given by the multiplication

$$p^A(a, b) = a^* \cdot b$$

for $a, b \in A$. ◇

7.2 Correspondence Between States and Inner Products

The purpose of this section is to show that there is a correspondence between quantum probability objects, to be defined below, and monoid inner product objects. This is a categorical version of the Gelfand–Naimark–Segal construction.

DEFINITION 7.2.1. Define the category QP(C) as follows.

Objects: An object in QP(C) is a tuple

$$\big((A, \mu, 1, j^A), f, (X, j^X)\big),$$

called a *quantum probability object* in C, in which:

- $(A, \mu, 1, j^A)$ is a reversing involutive monoid in C, called the *object of observables*.
- (X, j^X) is an involutive object in C, called the *value object*.
- $f : A \longrightarrow X$ is a morphism of involutive objects in C, called a *state on A over X*.

We sometimes abbreviate such an object to f.

Morphisms: A morphism

$$\big((A,\mu^A,\mathbb{1}^A,j^A),f,(X,j^X)\big) \xrightarrow{(h,h')} \big((B,\mu^B,\mathbb{1}^B,j^B),g,(Y,j^Y)\big)$$

in $\mathsf{QP(C)}$ is a pair (h,h') such that:

- $h : A \longrightarrow B$ is a morphism of reversing involutive monoids in C.
- $h' : X \longrightarrow Y$ is a morphism of involutive objects in C.
- The diagram

(7.2.2)
$$\begin{array}{ccc} A & \xrightarrow{h} & B \\ {\scriptstyle f}\downarrow & & \downarrow{\scriptstyle g} \\ X & \xrightarrow{h'} & Y \end{array}$$

in $\mathsf{Inv(C)}$ is commutative.

Identities: The identity morphism of an object $\big(A,f,(X,j^X)\big)$ is $(\mathrm{Id}_A,\mathrm{Id}_X)$.

Composition: Composition in $\mathsf{QP(C)}$ is defined using composition in $\mathsf{RIMon(C)}$ and in $\mathsf{Inv(C)}$.

REMARK 7.2.3. Suppose $U : \mathsf{RIMon(C)} \longrightarrow \mathsf{Inv(C)}$ is the forgetful functor defined by

$$U(A,\mu,\mathbb{1},j) = (A,j).$$

In other words, it forgets about the monoid structure. Then

$$\mathsf{QP(C)} = U \downarrow \mathsf{Inv(C)},$$

which is the category under U. ◇

EXAMPLE 7.2.4. Suppose $\mathsf{C} = \mathsf{Vect}(\mathbb{C})$ and $X = \mathbb{C}$ with complex conjugation as the involutive structure. Then a quantum probability object of the form $f : A \longrightarrow \mathbb{C}$ consists of

- a complex unital $*$-algebra A and
- a state f on A over $\mathbb{C}$.

This data is sometimes called a *quantum probability space*, which inspired our terminology in Definition 7.2.1. ◇

Next is the first part of a categorical version of the Gelfand–Naimark–Segal construction. It associates to each state an inner product.

LEMMA 7.2.5. *There is a functor*

$$\mathsf{QP}(C) \xrightarrow{\Phi} \mathsf{MIP}(C)$$

defined by

$$\Phi\Big(A \xrightarrow{f} X \Big) = \Big(IA \otimes A \xrightarrow{p^A} A \xrightarrow{f} X \Big)$$

on objects, with $p^A = \mu^A \circ (j^A \otimes A)$ *as in Lemma 7.1.10.*

PROOF. Since (A, p^A) is a monoid inner product object in C by Lemma 7.1.11, so is (A, fp^A) by an inspection. By definition Φ sends a morphism $(h, h') : f \longrightarrow g$ in $\mathsf{QP}(\mathsf{C})$ as in Definition 7.2.1 to (h, h'). To see that

$$\Phi(h, h') : \Phi(f) \longrightarrow \Phi(g)$$

is a well-defined morphism in $\mathsf{MIP}(\mathsf{C})$, note that the required axiom (7.1.4) in this case is the outer-most diagram below.

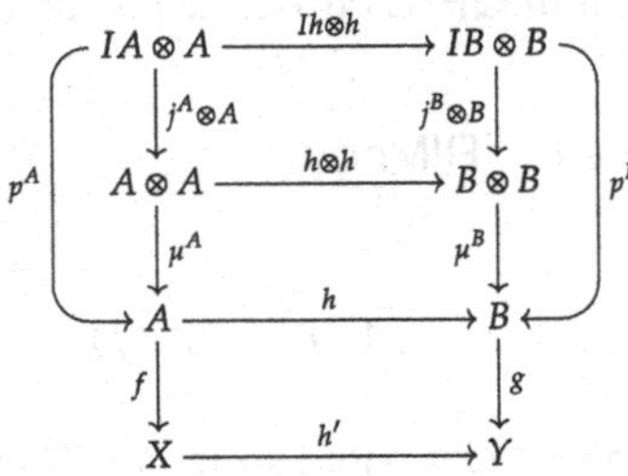

The top two rectangles are commutative because h is compatible with both the multiplications and the involutive structures. The bottom rectangle is the commutative diagram (7.2.2). □

Next we go backward and associate to each inner product a state. The proof of the following Lemma is given in Section 7.3.

LEMMA 7.2.6. *There is a functor*

$$\mathsf{MIP}(C) \xrightarrow{\Psi} \mathsf{QP}(C)$$

defined by

$$\Psi\Big(IA \otimes A \xrightarrow{p} X \Big) = \Big(A \xrightarrow{\lambda_A^{-1}} \mathbb{1} \otimes A \xrightarrow{I_0 \otimes A} I\mathbb{1} \otimes A \xrightarrow{I1^A \otimes A} IA \otimes A \xrightarrow{p} X \Big)$$

on objects.

The following observation is a categorical version of the Gelfand–Naimark–Segal construction.

THEOREM 7.2.7. *The functors*

$$QP(C) \underset{\Psi}{\overset{\Phi}{\rightleftarrows}} MIP(C)$$

in Lemmas 7.2.5 and 7.2.6 are inverse isomorphisms.

PROOF. Given a quantum probability object $f : A \longrightarrow X$ in C, the object $\Psi\Phi(f)$ is the following composite.

$$
\begin{array}{c}
A \xrightarrow{\hspace{4cm} \Psi\Phi(f) \hspace{4cm}} X \\
\lambda_A^{-1} \downarrow \hspace{9cm} \uparrow f \\
1 \otimes A \xrightarrow{I_0 \otimes A} I1 \otimes A \xrightarrow{I1^A \otimes A} IA \otimes A \xrightarrow{j^A \otimes A} A \otimes A \xrightarrow{\mu^A} A
\end{array}
$$

This composite is equal to f by (7.1.15) and (7.1.16).

Conversely, suppose $p : IA \otimes A \longrightarrow X$ is a monoid inner product object in C. By (7.1.17), the counit of the adjunction $\mathcal{F} \dashv U$ in Proposition 7.1.12 applied to p is

$$\big(\mathrm{Id}_A, \Psi(p)\big) : \mathcal{F}U(p) \longrightarrow p.$$

Expressed in the form (7.1.4), this morphism in MIP(C) is the commutative diagram

$$
\begin{array}{c}
IA \otimes A \xrightarrow{\mathrm{Id}_{IA} \otimes \mathrm{Id}_A} IA \otimes A \\
{\scriptstyle j^A \otimes A} \downarrow \hspace{4cm} \\
p^A A \otimes A \hspace{3cm} p \\
{\scriptstyle \mu^A} \downarrow \hspace{4cm} \\
A \xrightarrow{\hspace{1cm} \Psi(p) \hspace{1cm}} X
\end{array}
$$

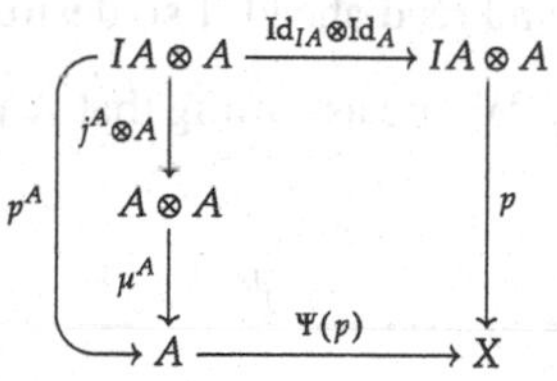

in C. The commutativity of this diagram means that

$$p = \Psi(p) \circ p^A = \Phi\Psi(p),$$

proving the converse. □

EXAMPLE 7.2.8. Continuing Example 7.2.4, given a quantum probability space $f : A \longrightarrow \mathbb{C}$ on A over $\mathbb{C}$, the monoid inner product object

$$\Phi(f) = \langle,\rangle : \overline{A} \otimes A \longrightarrow \mathbb{C}$$

in $\mathsf{Vect}(\mathbb{C})$ is given by

$$\langle a, b\rangle = f\big(a^* \cdot b\big) \quad \text{for} \quad a, b \in A.$$

Conversely, given a monoid inner product object in $\mathsf{Vect}(\mathbb{C})$ of the form

$$p = \langle , \rangle : \overline{A} \otimes A \longrightarrow \mathbb{C},$$

the quantum probability space

$$\Psi(p) : A \longrightarrow \mathbb{C}$$

is given by

$$\Psi(p)(a) = \langle 1^A, a\rangle \quad \text{for} \quad a \in A.$$

Restricting to $\mathsf{C} = \mathsf{Vect}(\mathbb{C})$ and codomain $\mathbb{C}$, Theorem 7.2.7 says that these constructions are inverses of each other. ◇

Before reading the next section, the reader is asked to consider Exercise 2 in Section 7.4.

7.3 Proofs of Technical Lemmas

The purpose of this section is to prove some of the technical assertions used in Sections 7.1 and 7.2. Nothing in this section is used in other chapters, so the reader may safely skip this section and read ahead if so desired.

PROOF OF LEMMA 7.1.1. We are assuming that $A \in \mathsf{C}$ is an object, and we want to show that the composite

$$\begin{array}{ccccc}
I(IA \otimes A) & & \xrightarrow{\quad j^{IA\otimes A} \quad} & & IA \otimes A \\
{\scriptstyle I(IA\otimes\iota_A)}\downarrow & & & & \uparrow{\scriptstyle \xi_{A,IA}} \\
I(IA \otimes I^2A) & \xrightarrow{I(I_2)} & I^2(A \otimes IA) & \xrightarrow{\iota^{-1}_{A\otimes IA}} & A \otimes IA
\end{array}$$

in (7.1.2) makes $(IA \otimes A, j^{IA\otimes A})$ into an involutive object in C. In other words, we must show that the composite

$$I^2(IA \otimes A) \xrightarrow{I(j^{IA\otimes A})} I(IA \otimes A) \xrightarrow{j^{IA\otimes A}} IA \otimes A$$

in C is equal to $\iota^{-1}_{IA\otimes A}$.

To prove this, consider the following diagram.

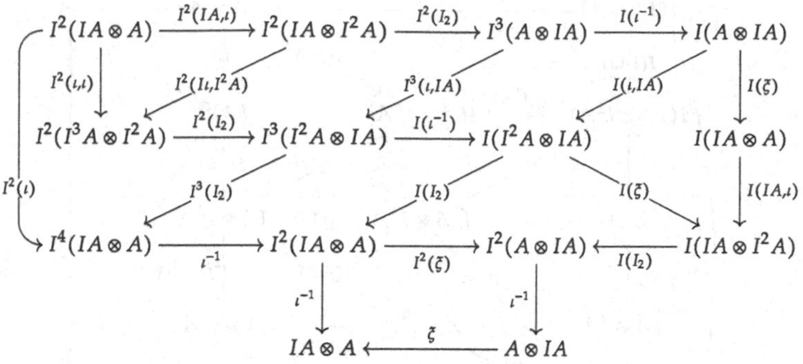

The composite along the top, right, and bottom is $j^{IA\otimes A} \circ I(j^{IA\otimes A})$, while the composite along the left is $\iota^{-1}_{IA\otimes A}$ because $I^2\iota = \iota_{I^2}$ by the triangle identity (2.1.2). So it suffices to observe the commutativity of its eight sub-diagrams.

- The left-most sub-diagram is commutative by (4.1.5), which holds because ι is a monoidal natural transformation.
- The top left triangle is commutative by the triangle identity $I\iota = \iota_I$ in (2.1.2).
- The three sub-diagrams to its right are commutative by the naturality of I_2, ι^{-1}, and ξ, respectively.
- The middle parallelogram is commutative by the naturality of ι^{-1} and the triangle identity in the form $I\iota^{-1} = \iota_I^{-1}$.
- The triangle to its right is commutative by (4.1.7), which holds because I is a symmetric monoidal functor.
- The bottom rectangle is commutative by the naturality of ι^{-1}.

Therefore, we have proved the equality

$$j^{IA\otimes A} \circ I(j^{IA\otimes A}) = \iota^{-1}_{IA\otimes A}.$$

So $(IA \otimes A, j^{IA\otimes A})$ is an involutive object in C. □

Proof of Lemma 7.1.10. We are assuming that $(A, \mu, 1, j) \in \mathsf{RIMon}(\mathsf{C})$, and we want to show that $p^A = \mu \circ (j \otimes A)$ is a morphism of involutive objects in C. In other words, we need to show that p^A is compatible with the involutive structures in $IA \otimes A$ and in A, which is the outer-most diagram below.

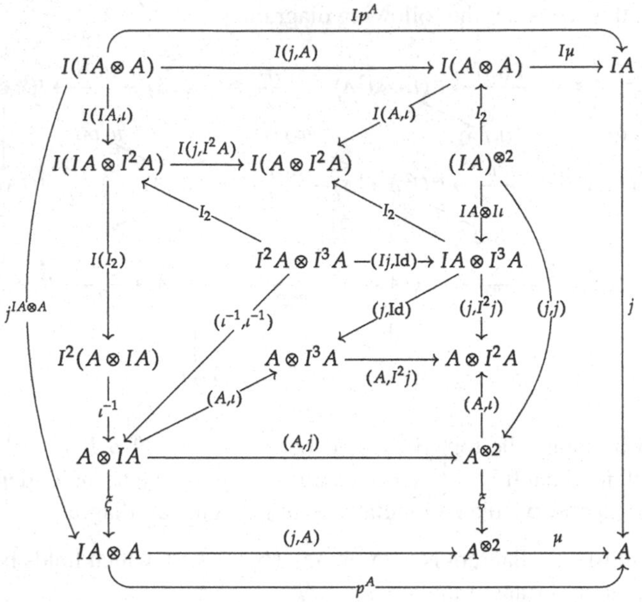

In the above diagram:

- The top, left, and bottom thin strips are commutative by the definitions of p^A and of $j^{IA\otimes A}$ in (7.1.2).
- The right-most sub-diagram is commutative by (4.7.9), the naturality of ξ, and the symmetry axiom $\xi \circ \xi = \mathrm{Id}$ in (1.2.23).
- The other nine sub-diagrams are commutative by the functoriality of $\otimes$, the naturality of I_2, ι, and ξ, the triangle identity $I\iota = \iota_I$ (2.1.2), $j \circ Ij = \iota^{-1}$, and that ι is a monoidal natural transformation (4.1.5).
- ι and I_2 are isomorphisms by definition and Corollary 4.1.10, respectively.

Therefore, the commutativity of the above diagram implies that p^A is a morphism of involutive objects in C. □

Our next objective is to prove Lemma 7.3.4. This is about the morphism g in (7.1.17) defined as the composite

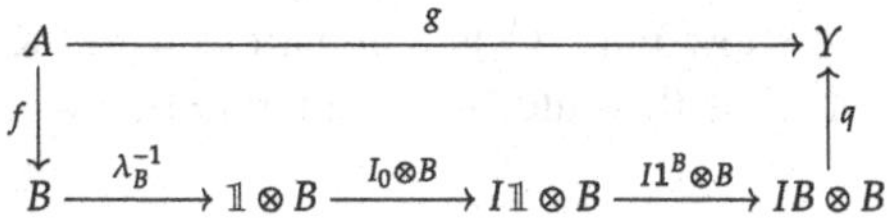

in C, with $f : A \longrightarrow B$ a morphism in $\mathsf{RIMon(C)}$ and $q : IB \otimes B \longrightarrow Y$ a monoid inner product object in C.

LEMMA 7.3.1. *The morphism* $g : A \longrightarrow Y$ *in (7.1.17) is a morphism of involutive objects in* C.

PROOF. We will paste together several diagrams. The desired condition that g be a morphism of involutive objects is the outer-most diagram below.

(7.3.2)

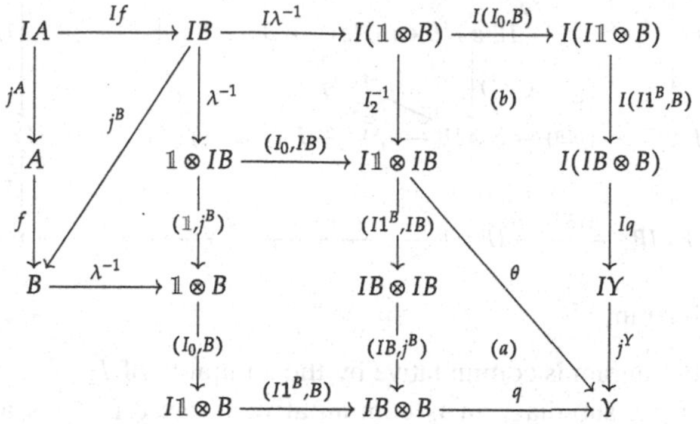

In the above diagram:

- The upper left triangle is commutative because f is a morphism of involutive objects. The triangle under it is commutative by the naturality of λ. The upper middle square is commutative by the left unity axiom (1.2.13) of the monoidal functor (I, I_2, I_0). The lower middle rectangle is commutative by the functoriality of $\otimes$.
- The morphism θ is by definition the composite

(7.3.3)

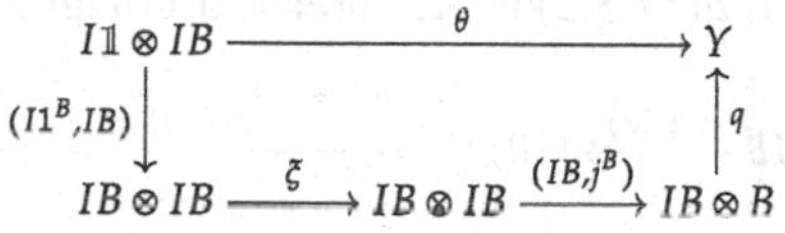

$$
\begin{array}{ccc}
I\mathbb{1} \otimes IB & \xrightarrow{\quad \theta \quad} & Y \\
{\scriptstyle (I\mathbb{1}^B,IB)} \downarrow & & \uparrow {\scriptstyle q} \\
IB \otimes IB & \xrightarrow{\xi} IB \otimes IB \xrightarrow{(IB,j^B)} & IB \otimes B
\end{array}
$$

in C.

Set theoretically, if we write $\langle , \rangle$ for q and $(-)^*$ for j^B and j^Y, then the sub-diagrams (a) and (b) represent the equalities

$$\langle 1^B, b^* \rangle = \langle b, (1^B)^* \rangle = \langle 1^B, b \rangle^* \quad \text{for} \quad b \in B.$$

The sub-diagram (b) in (7.3.2) is the outer-most diagram below, in which $1 = 1^B$, and both j^B and j^Y are denoted by j.

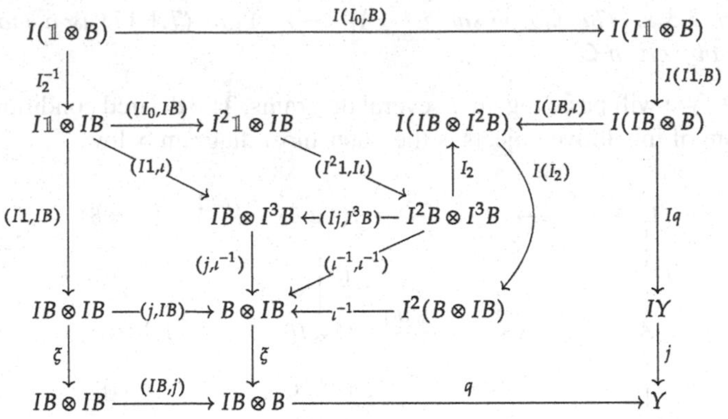

In the above diagram:

- The top sub-diagram is commutative by the naturality of I_2.
- The lower right sub-diagram is commutative because (i) q is a morphism of involutive objects, and (ii) $IB \otimes B$ has the involutive structure in (7.1.2).
- The parallelogram is commutative by (i) the triangle identity $I\iota = \iota_I$ in (2.1.2), and (ii) that $j : IB \longrightarrow B$ preserves units as in (4.7.8).
- The left trapezoid is commutative by definition.
- The bottom left square is commutative by the naturality of ξ.
- The middle triangle is commutative by $j \circ Ij = \iota^{-1}$ in Proposition 2.5.4(2).
- The remaining sub-diagram in the middle is commutative by (4.1.5).

This proves the commutativity of the sub-diagram (b) in (7.3.2).

The sub-diagram (a) in (7.3.2) is the outer-most diagram below.

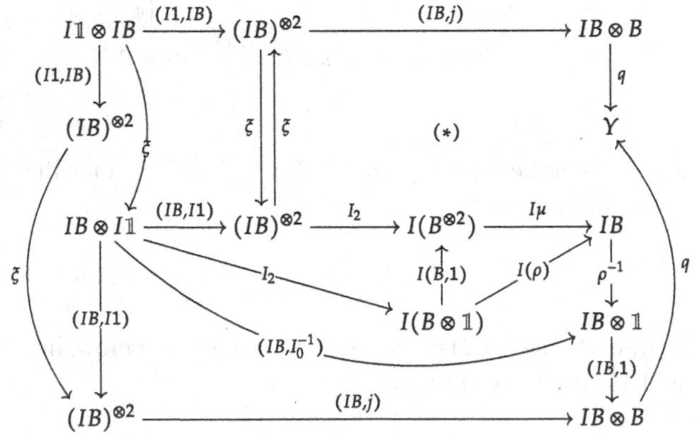

In the above diagram, the sub-diagrams other than $(*)$ are commutative by the naturality of ξ and I_2, the symmetry axiom $\xi \circ \xi = \mathrm{Id}$ in (1.2.23), the right unity axiom (1.2.14) of the monoidal functor (I, I_2, I_0), the right unity axiom of the monoid B in Definition 1.2.8, and (4.7.8).

Set theoretically, the sub-diagram $(*)$ is the equality

$$\langle ab, 1\rangle = \langle a, b^*\rangle \quad \text{for} \quad a, b \in B.$$

More precisely, $(*)$ is the outer-most diagram below.

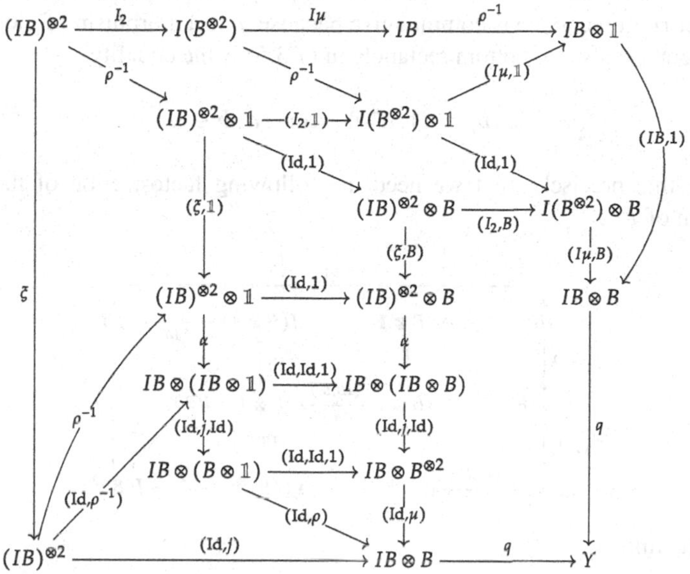

In the above diagram:

- The lower right rectangle is commutative by the axiom (7.1.8) for the monoid inner product object q.
- The other sub-diagrams are commutative by the naturality of ρ and α, the functoriality of $\otimes$, the right diagram in (1.2.6), and the right unity of the monoid B.

This proves that $(*)$ is commutative, so (a) in (7.3.2) is also commutative. □

LEMMA 7.3.4. *The composite morphism* $g : A \longrightarrow Y \in \mathsf{C}$ *in* (7.1.17) *is a morphism of involutive objects that makes the diagram* (7.1.13) *commutative.*

PROOF. Because of Lemma 7.3.1, it remains to prove that the diagram (7.1.13), which is the outer-most diagram below, is commutative.

$(7.3.5)$

$$
\begin{array}{ccccccc}
IA \otimes A & \xrightarrow{(j,A)} & A^{\otimes 2} & \xrightarrow{\mu} & A & \xrightarrow{f} & B \\
{\scriptstyle (If,f)}\downarrow & & \downarrow{\scriptstyle f^{\otimes 2}} & & & & \| \\
IB \otimes B & \xrightarrow{(j,B)} & B^{\otimes 2} & & \xrightarrow{\mu} & & B \\
{\scriptstyle q}\downarrow & & & & & & \downarrow{\scriptstyle \lambda^{-1}} \\
Y & \xleftarrow{\ q\ } IB \otimes B & \xleftarrow{(I1,B)} & I1 \otimes B & \xleftarrow{(I_0,B)} & 1 \otimes B
\end{array}
$$

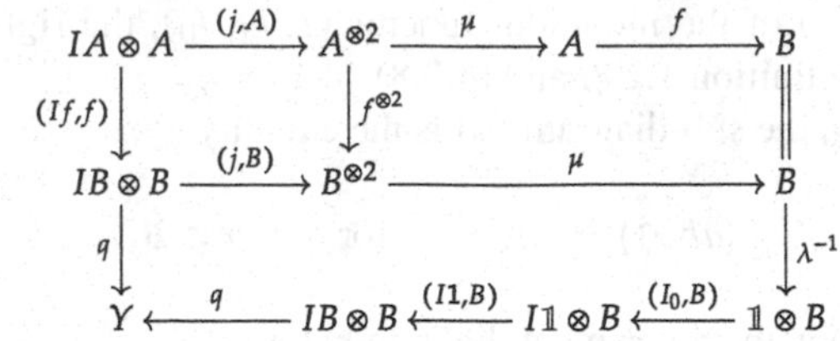

The upper left square is commutative because f is a morphism of involutive objects. The upper right rectangle is commutative because f is a morphism of monoids.

Set theoretically, the bottom rectangle in (7.3.5) is the equality

$$\langle a, b \rangle = \langle 1, a^* b \rangle \quad \text{for} \quad a, b \in B.$$

To prove this precisely, first we need the following factorization of the identity morphism of IB.

$(7.3.6)$

$$
\begin{array}{c}
\xrightarrow{\hspace{4cm}\text{Id}\hspace{4cm}} \\
\begin{array}{ccccccc}
IB & \xrightarrow{\rho^{-1}} & IB \otimes 1 & & I(B \otimes 1) & \xrightarrow{I\rho} & IB \\
{\scriptstyle \lambda^{-1}}\downarrow & & \| & & {\scriptstyle I_2}\uparrow & & \uparrow \\
1 \otimes IB & \xrightarrow{\xi} & IB \otimes 1 & \xrightarrow{(IB,I_0)} & IB \otimes I1 \quad I(B,1) & & {\scriptstyle I\mu} \\
{\scriptstyle (I_0,IB)}\downarrow & & & & \downarrow{\scriptstyle (IB,I1)} & & \\
I1 \otimes IB & \xrightarrow{(I1,IB)} & (IB)^{\otimes 2} & \xrightarrow{\xi} & (IB)^{\otimes 2} & \xrightarrow{I_2} & I(B^{\otimes 2})
\end{array}
\end{array}
$$

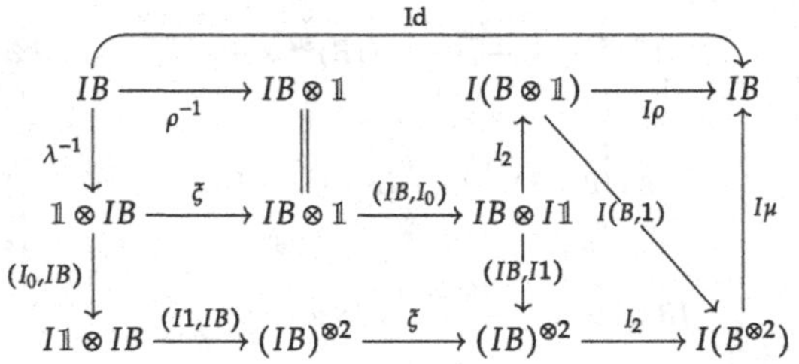

In this diagram:

- The top middle sub-diagram is commutative by the right unity axiom (4.1.4) of the monoidal functor (I, I_2, I_0).
- The top left square is commutative by the unit axiom (1.2.23) in a symmetric monoidal category.
- The bottom rectangle and the triangle to its right are commutative by the naturality of ξ and I_2, respectively.
- The right-most triangle is commutative by the right unity axiom of the monoid B.

So the diagram (7.3.6) is commutative.

Using the factorization of Id_{IB} in (7.3.6), the bottom rectangle in (7.3.5) is the outer-most diagram below.

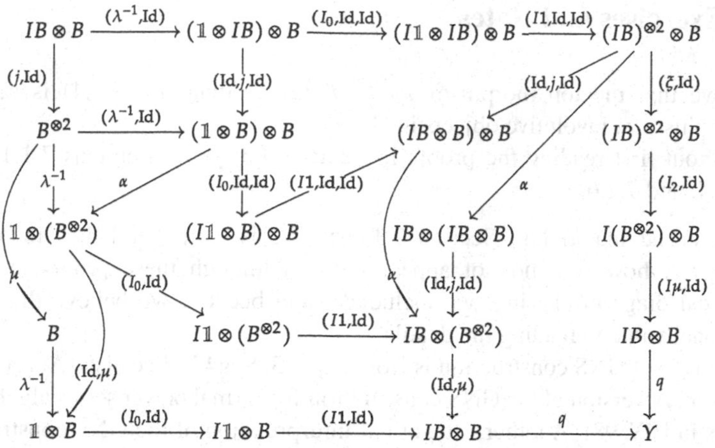

The right-most trapezoid is commutative by the monoid inner product axiom (7.1.8). The other seven sub-diagrams are commutative by the functoriality of $\otimes$, the naturality of λ and α, and the monoidal category property (1.2.6). The commutativity of this diagram implies that of (7.3.5). $\square$

PROOF OF LEMMA 7.2.6. First we apply Lemma 7.3.1 with $f = \mathrm{Id}_A$ and $(B, q) = (A, p)$ to infer that the composite

$$\Psi\left(IA \otimes A \xrightarrow{\;p\;} X \right) = \left(A \xrightarrow{\;\lambda_A^{-1}\;} 1 \otimes A \xrightarrow{\;I_0 \otimes A\;} I1 \otimes A \xrightarrow{\;I1^A \otimes A\;} IA \otimes A \xrightarrow{\;p\;} X \right)$$

is actually a morphism of involutive objects in C. So Ψ is well-defined on objects.

For a morphism $(f, g) : p \longrightarrow q$ in MIP(C) as in Definition 7.1.7, the morphism

$$\Psi(f, g) : \Psi(p) \longrightarrow \Psi(q) \in \mathsf{QP(C)}$$

is defined as the pair (f, g). To see that it is a well-defined morphism in QP(C), note that the required axiom (7.2.2) is the outer-most diagram below.

$$
\begin{array}{ccccccccc}
& & & & \Psi(p) & & & & \\
A & \xrightarrow{\lambda_A^{-1}} & 1 \otimes A & \xrightarrow{I_0 \otimes A} & I1 \otimes A & \xrightarrow{I1^A \otimes A} & IA \otimes A & \xrightarrow{p} & X \\
{\scriptstyle f}\downarrow & & \downarrow{\scriptstyle 1 \otimes f} & & \downarrow{\scriptstyle I1 \otimes f} & & \downarrow{\scriptstyle If \otimes f} & & \downarrow{\scriptstyle g} \\
B & \xrightarrow[\lambda_B^{-1}]{} & 1 \otimes B & \xrightarrow[I_0 \otimes B]{} & I1 \otimes B & \xrightarrow[I1^B \otimes B]{} & IB \otimes B & \xrightarrow[q]{} & Y \\
& & & & \Psi(q) & & & &
\end{array}
$$

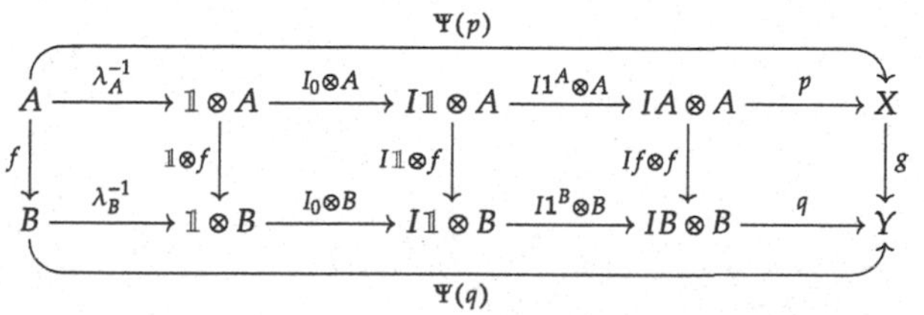

From left to right, the left-most square is commutative by the naturality of λ. The next square is commutative by definition. The next square is commutative because f preserves units. The right-most square is the commutative diagram (7.1.4). $\square$

7.4 Exercises and Notes

(1) Prove that the top morphism $If \otimes f$ in the diagram (7.1.4) is actually a morphism of involutive objects in C.
(2) Without first reading the proofs in Section 7.3, prove Lemmas 7.1.1, 7.1.10, 7.3.4, and 7.2.6.

Notes. The material in Section 7.1 and 7.2 is from [Jac12]. The proofs in Section 7.3, however, did not appear there. Although these proofs are mostly categorical diagram chasing, we included them because we believe that there is educational value in reading the details.

The original GNS construction is from [GN43, Seg47]. See also [Arv76, Sch90] for a proof. A version of the GNS construction for formal power series algebras over $\mathbb{C}[[\hbar]]$ is in [BW98]. Another categorical interpretation of the GNS construction is in [Par18]. Discussion of the GNS construction in quantum field theories can be found in [Haa96, KM15, Mor18].

Chapter 8
Involutive Operads

Operad theory is a versatile language for studying algebraic structures with multiple inputs and one output. An operad is a bookkeeping device for keeping track of operations with multiple inputs and one output, their composition, and their equivariance. In the literature an operad is also called a multicategory or a symmetric multicategory, depending on whether there is a symmetric group action or not.

In this chapter we extend operads to their involutive analogues. Since we do not assume any prior knowledge of operads, in Section 8.1 we recall some basic aspects of operads. There are several equivalent definitions of an operad. One such definition says that an operad is a monoid in a monoidal category of symmetric sequences. We recall this definition of an operad in Section 8.1.

An operad, like a monoid, can act on objects. In Section 8.2 we recall the monad associated to an operad, and define an algebra over an operad as an algebra of the associated monad. After unpacking this definition to an explicit list of operations and axioms, we provide a series of examples of algebras over an operad, including associative monoids, commutative monoids, and their diagrams.

Involutive operads are defined in Section 8.3 as the involutive monoids in an involutive monoidal category of symmetric sequences. This involves first equipping the category of symmetric sequences with an involutive structure, making it an involutive monoidal category. It follows that the category of operads inherits an involutive structure. Involutive operads are the involutive objects in this involutive category of operads. Unpacking this definition we observe that an involutive operad is equipped with an operad structure and an involutive structure in symmetric sequences such that two axioms about the operad units and composition are satisfied. The rest of this section contains examples, including the involutive associative operad, a diagram version of it, the non-reversing involutive associative operad, and the involutive commutative operad.

Just as each operad has an associated monad, in Section 8.4 we show that each involutive operad yields an involutive monad. The involutive algebras of this involutive monad are defined in Section 8.5 as those for the involutive operad.

© The Author(s), under exclusive license to Springer Nature Switzerland AG 2020
D. Yau, *Involutive Category Theory*, Lecture Notes in Mathematics 2279,
https://doi.org/10.1007/978-3-030-61203-0_8

Unpacking this definition we observe that an involutive algebra over an involutive operad has an underlying algebra over the underlying operad and an involutive structure, together with a compatibility axiom between the two structures. The rest of this section contains examples of involutive algebras over an involutive operad, including involutive monoids, involutive commutative monoids, reversing involutive monoids, and their diagrams.

In Section 8.6 we observe that each involutive symmetric monoidal functor induces an involutive functor between the categories of involutive operads. This allows one to change the base involutive symmetric monoidal category when considering involutive operads. Examples of involutive operads arising from this induced involutive functor include those in Section 8.5.

Algebraic quantum field theory is a mathematical framework for quantum field theory. An algebraic quantum field theory is a functor whose domain category is regarded as a category of suitable spacetime regions, and whose codomain category is some category of algebraic objects, such as modules or chain complexes. There is an operadic formulation of algebraic quantum field theory using the AQFT involutive operads. In Section 8.7 we observe that the AQFT involutive operads are all obtained from the induced involutive functor in Section 8.6.

Conventions. Unless otherwise specified, in Sections 8.1 and 8.2, our base category

$$\big(\mathsf{C}, \otimes, \mathbb{1}, \alpha, \lambda, \rho, \xi\big)$$

is a *strict* symmetric monoidal category. In particular, multiple monoidal products in C do not require parentheses. For a symmetric monoidal category that is not strict, we first use Mac Lane's Coherence Theorem [Mac98] (XI.3) to replace it by an equivalent strict symmetric monoidal category via strong symmetric monoidal functors.

From Section 8.3, our base category

$$\big((\mathsf{C}, \otimes, \mathbb{1}, \alpha, \lambda, \rho, \xi), (I, I_2, I_0), \iota\big)$$

is an involutive *strict* symmetric monoidal category as in Definition 4.1.1. For an involutive symmetric monoidal category that is not strict symmetric monoidal (e.g., $\mathsf{Vect}(\mathbb{C})$ and $\mathsf{Chain}(\mathbb{C})$), we first use the Strictification Theorem 6.4.1 to replace it by an involutively equivalent involutive strict symmetric monoidal category.

Furthermore, we always assume that C is cocomplete, and that its monoidal product commutes with colimits on both sides. We write $\varnothing$ for an initial object in C.

8.1 Operads

The purpose of this section is to recall some basic definitions and facts about operads in a symmetric monoidal category C. For a more thorough introduction to operads, the reader is referred to [Yau16, Yau∞]. An operad is a monoid in a monoidal category of symmetric sequences, which we define first. The monoidal product in the category of symmetric sequences involves a coend. In Proposition 8.1.13, we provide a more explicit and elementary description of an operad that does not involve any coends.

The symmetric group on k letters is denoted by Σ_k. Its unit is denoted by id_k.

DEFINITION 8.1.1. For the rest of this chapter, $\mathfrak{C}$ denotes a set, whose elements are called *colors*.

(1) The set of $\mathfrak{C}$*-profiles* is defined as

$$\mathsf{Prof}(\mathfrak{C}) = \coprod_{n\geq 0} \mathfrak{C}^{\times n},$$

so a $\mathfrak{C}$-profile is a finite, possibly empty, sequence of elements in $\mathfrak{C}$. We usually write a $\mathfrak{C}$-profile as

$$\underline{a} = (a_1, \ldots, a_m) \in \mathfrak{C}^{\times m},$$

in which case $|\underline{a}| = m$. The *empty* $\mathfrak{C}$*-profile* is denoted by $\varnothing$, which is not to be confused with an initial object in C.

(2) For $\mathfrak{C}$-profiles $\underline{a} = (a_1, \ldots, a_m)$ and $\underline{b}$, a *left permutation* $\sigma : \underline{a} \longrightarrow \underline{b}$ is a permutation $\sigma \in \Sigma_{|\underline{a}|}$ such that

$$\sigma\underline{a} = (a_{\sigma^{-1}(1)}, \ldots, a_{\sigma^{-1}(m)}) = \underline{b}$$

(3) The *groupoid of* $\mathfrak{C}$*-profiles*, with $\mathsf{Prof}(\mathfrak{C})$ as object set and with left permutations as the isomorphisms, is denoted by $\Sigma_{\mathfrak{C}}$. The opposite groupoid $\Sigma_{\mathfrak{C}}^{\mathsf{op}}$ is regarded as the groupoid of $\mathfrak{C}$-profiles with *right permutations*

$$\underline{a}\sigma = (a_{\sigma(1)}, \ldots, a_{\sigma(m)})$$

as isomorphisms.

(4) The objects of the diagram category

$$\mathsf{SSeq}_{\mathsf{C}}^{\mathfrak{C}} = \mathsf{C}^{\Sigma_{\mathfrak{C}}^{\mathsf{op}} \times \mathfrak{C}}$$

are called $\mathfrak{C}$*-colored symmetric sequences* in C.

(5) An object in $\Sigma_{\mathfrak{C}}^{\mathsf{op}} \times \mathfrak{C}$ is written vertically as $\binom{d}{\underline{c}}$ with $d \in \mathfrak{C}$ and $\underline{c} \in \Sigma_{\mathfrak{C}}^{\mathsf{op}}$. A typical entry of a $\mathfrak{C}$-colored symmetric sequence X in C is denoted by $X\binom{d}{\underline{c}}$.

(6) For a set S, an object in the product category

$$\mathsf{C}^S = \prod_{s \in S} \mathsf{C}$$

is called an *S-colored object*, denoted by $X = \{X_s\}_{s \in S}$.

Our next objective is to equip the category $\mathsf{SSeq}_{\mathsf{C}}^{\mathfrak{C}}$ of $\mathfrak{C}$-colored symmetric sequences with a monoidal structure, for which we need the categorical concept of a coend. For an in-depth discussion of coends, see [Mac98] (IX.6) or [Lor∞].

DEFINITION 8.1.2. Suppose given a functor $F : \mathsf{D}^{\mathsf{op}} \times \mathsf{D} \longrightarrow \mathsf{B}$.

(1) A *wedge* of F is a pair (X, ζ) consisting of

- an object $X \in \mathsf{B}$ and
- a morphism $\zeta_d : F(d, d) \longrightarrow X$ for each object $d \in \mathsf{D}$

such that the diagram

$$\begin{array}{ccc} F(d',d) & \xrightarrow{F(d',g)} & F(d',d') \\ {\scriptstyle F(g,d)}\downarrow & & \downarrow{\scriptstyle \zeta_{d'}} \\ F(d,d) & \xrightarrow{\zeta_d} & X \end{array}$$

is commutative for each morphism $g : d \longrightarrow d' \in \mathsf{D}$.

(2) A morphism $f : (X, \zeta) \longrightarrow (Y, \phi)$ of wedges of F is a morphism $f : X \longrightarrow Y$ in B such that the diagram

$$\begin{array}{ccc} F(d,d) & \xrightarrow{\zeta_d} & X \\ & {\scriptstyle \phi_d}\searrow & \downarrow{\scriptstyle f} \\ & & Y \end{array}$$

is commutative for each object $d \in \mathsf{D}$.

(3) A *coend* of F, if it exists, is an initial wedge $\left(\int^{d \in \mathsf{D}} F(d, d), \omega\right)$.

We usually denote such a coend by either $\int^{d \in \mathsf{D}} F(d, d)$ or $\int^{d} F(d, d)$, suppressing ω and sometimes D from the notation.

EXPLANATION 8.1.3. A coend of $F : \mathsf{D}^{\mathsf{op}} \times \mathsf{D} \longrightarrow \mathsf{B}$ is a wedge $\int^{d \in \mathsf{D}} F(d, d)$ of F such that given any wedge (X, ζ) of F, there exists a unique morphism

$$\int^{d \in \mathsf{D}} F(d, d) \xrightarrow{\;h\;} X \in \mathsf{B}$$

such that the diagram

$$\begin{array}{ccc} F(d,d) & \xrightarrow{\omega_d} & \int^{d\in \mathsf{D}} F(d,d) \\ & \searrow^{\zeta_d} & \downarrow h \\ & & X \end{array}$$

is commutative for each object $d \in \mathsf{D}$. One can think of a coend $\int^d F(d,d)$ as a parametrized average of the objects $F(d,d)$ for $d \in \mathsf{D}$, with relations parametrized by the indexing category D. ◇

EXAMPLE 8.1.4. Suppose $F : \mathsf{D}^{\mathsf{op}} \times \mathsf{D} \longrightarrow \mathsf{B}$ is a functor with D a small discrete category, so it has no non-identity morphisms. If B has coproducts, then

$$\int^{d\in \mathsf{D}} F(d,d) \cong \coprod_{d\in \mathsf{D}} F(d,d),$$

a coproduct of the objects $F(d,d)$ for $d \in \mathsf{D}$. ◇

NOTATION 8.1.5. For a set S and an object $X \in \mathsf{C}$, we sometimes use the dot notation

$$S \cdot X = \coprod_{s\in S} X$$

for coproducts.

We now define a monoidal structure on the category $\mathsf{SSeq}_{\mathsf{C}}^{\mathfrak{C}}$ of $\mathfrak{C}$-colored symmetric sequences.

DEFINITION 8.1.6. Suppose $X, Y \in \mathsf{SSeq}_{\mathsf{C}}^{\mathfrak{C}}$.

(1) For each $\underline{c} = (c_1, \ldots, c_m) \in \Sigma_{\mathfrak{C}}$, define the object $Y^{\underline{c}} \in \mathsf{C}^{\Sigma_{\mathfrak{C}}^{\mathsf{op}}}$ entrywise as the coend

$$Y^{\underline{c}}(\underline{b}) = \int^{\{\underline{a}_j\}\in \prod_{j=1}^{m} \Sigma_{\mathfrak{C}}^{\mathsf{op}}} \Sigma_{\mathfrak{C}}^{\mathsf{op}}(\underline{a}_1, \ldots, \underline{a}_m; \underline{b}) \cdot \Big[\bigotimes_{j=1}^{m} Y\tbinom{c_j}{\underline{a}_j}\Big] \in \mathsf{C} \tag{8.1.7}$$

for $\underline{b} \in \Sigma_{\mathfrak{C}}^{\mathsf{op}}$, in which $(\underline{a}_1, \ldots, \underline{a}_m) \in \Sigma_{\mathfrak{C}}^{\mathsf{op}}$ is the concatenation. The object $Y^{\underline{c}}$ is natural in $\underline{c} \in \Sigma_{\mathfrak{C}}$ via left permutations of the tensor factors in $\bigotimes_{j=1}^{m} Y\tbinom{c_j}{\underline{a}_j}$.

(2) The $\mathfrak{C}$-*colored symmetric circle product*

$$X \circ Y \in \mathsf{SSeq}_{\mathsf{C}}^{\mathfrak{C}}$$

is defined entrywise as the coend

$$(X \circ Y)\binom{d}{\underline{b}} = \int^{\underline{c} \in \Sigma_{\mathfrak{C}}} X\binom{d}{\underline{c}} \otimes Y^{\underline{c}}(\underline{b}) \in \mathsf{C} \tag{8.1.8}$$

for $\binom{d}{\underline{b}} \in \Sigma_{\mathfrak{C}}^{\mathsf{op}} \times \mathfrak{C}$.

(3) Define an object $\mathsf{I} \in \mathsf{SSeq}_{\mathsf{C}}^{\mathfrak{C}}$ by

$$\mathsf{I}\binom{d}{\underline{c}} = \begin{cases} \mathbb{1} & \text{if } \underline{c} = d, \\ \varnothing & \text{otherwise} \end{cases} \tag{8.1.9}$$

for $\binom{d}{\underline{c}} \in \Sigma_{\mathfrak{C}}^{\mathsf{op}} \times \mathfrak{C}$.

EXAMPLE 8.1.10. Suppose $\mathfrak{C}$ is a single-element set $\{*\}$. In this case, we write $\Sigma_{\mathfrak{C}}$ as Σ and its objects as n for $n \geq 0$. For $X, Y \in \mathsf{SSeq}_{\mathsf{C}}^{\mathfrak{C}}$, we have

$$Y^m(k) = \int^{(k_1,\ldots,k_m) \in (\Sigma^{\mathsf{op}})^{\times m}} \Sigma^{\mathsf{op}}(k_1 + \cdots + k_m; k) \cdot \Big[\bigotimes_{j=1}^{m} Y(k_j)\Big],$$

$$(X \circ Y)(k) = \int^{m \in \Sigma} X(m) \otimes Y^m(k)$$

for $m, k \in \Sigma$. The object $\mathsf{I} \in \mathsf{SSeq}_{\mathsf{C}}^{\mathfrak{C}}$ has $\mathsf{I}(1) = \mathbb{1}$ and $\mathsf{I}(k) = \varnothing$ for $k \neq 1$. ◇

PROPOSITION 8.1.11. *For each set $\mathfrak{C}$, $\big(\mathsf{SSeq}_{\mathsf{C}}^{\mathfrak{C}}, \circ, I\big)$ is a monoidal category.*

PROOF. Suppose $X, Y, Z \in \mathsf{SSeq}_{\mathsf{C}}^{\mathfrak{C}}$. To exhibit the associativity isomorphism, first note that for $\underline{c} = (c_1, \ldots, c_m), \underline{b} \in \Sigma_{\mathfrak{C}}$, there exist natural isomorphisms:

$$\begin{aligned}
(Y \circ Z)^{\underline{c}}(\underline{b}) &= \int^{\underline{a}_1,\ldots,\underline{a}_m} \Sigma_{\mathfrak{C}}^{\mathsf{op}}(\underline{a}_1, \ldots, \underline{a}_m; \underline{b}) \cdot \Big[\bigotimes_{j=1}^{m} (Y \circ Z)\tbinom{c_j}{\underline{a}_j}\Big] \\
&= \int^{\underline{a}_1,\ldots,\underline{a}_m} \Sigma_{\mathfrak{C}}^{\mathsf{op}}(\underline{a}; \underline{b}) \cdot \Big[\bigotimes_{j=1}^{m} \int^{\underline{d}_j} Y\tbinom{c_j}{\underline{d}_j} \otimes Z^{\underline{d}_j}(\underline{a}_j)\Big] \\
&\cong \int^{\underline{a}_1,\ldots,\underline{a}_m} \int^{\underline{d}_1,\ldots,\underline{d}_m} \Sigma_{\mathfrak{C}}^{\mathsf{op}}(\underline{a}; \underline{b}) \cdot \bigotimes_{j=1}^{m} \Big[Y\tbinom{c_j}{\underline{d}_j} \otimes Z^{\underline{d}_j}(\underline{a}_j)\Big] \\
&\cong \int^{\underline{d}_1,\ldots,\underline{d}_m} \Big[\bigotimes_{j=1}^{m} Y\tbinom{c_j}{\underline{d}_j}\Big] \otimes \Big[\int^{\underline{a}_1,\ldots,\underline{a}_m} \Sigma_{\mathfrak{C}}^{\mathsf{op}}(\underline{a}; \underline{b}) \cdot \bigotimes_{j=1}^{m} Z^{\underline{d}_j}(\underline{a}_j)\Big] \\
&= \int^{\underline{d}_1,\ldots,\underline{d}_m} \Big[\bigotimes_{j=1}^{m} Y\tbinom{c_j}{\underline{d}_j}\Big] \otimes Z^{\underline{d}}(\underline{b})
\end{aligned}$$

$$\begin{aligned}
&\cong \int^{\underline{d}_1,\ldots,\underline{d}_m} \Big[\bigotimes_{j=1}^m Y\tbinom{c_j}{\underline{d}_j}\Big] \otimes \Big[\int^{\underline{e}} \Sigma^{\mathsf{op}}_{\mathfrak{C}}(\underline{d};\underline{e}) \cdot Z^{\underline{e}}(\underline{b})\Big]\\
&\cong \int^{\underline{e}} \Big[\int^{\underline{d}_1,\ldots,\underline{d}_m} \Sigma^{\mathsf{op}}_{\mathfrak{C}}(\underline{d};\underline{e}) \cdot \bigotimes_{j=1}^m Y\tbinom{c_j}{\underline{d}_j}\Big] \otimes Z^{\underline{e}}(\underline{b})\\
&= \int^{\underline{e}} Y^{\underline{c}}(\underline{e}) \otimes Z^{\underline{e}}(\underline{b})
\end{aligned}$$

Here $\underline{a} = (\underline{a}_1, \ldots, \underline{a}_m)$ and $\underline{d} = (\underline{d}_1, \ldots, \underline{d}_m)$ are the concatenations. The third natural isomorphism follows from the universal property of coends. The other three natural isomorphisms follow from our assumption on C that its monoidal product commutes with colimits on both sides, as well as the commutation of coends.

The above canonical isomorphisms lead to the entries of the associativity isomorphism

$$\begin{aligned}
\big((X \circ Y) \circ Z\big)\tbinom{d}{\underline{a}} &= \int^{\underline{c}} (X \circ Y)\tbinom{d}{\underline{c}} \otimes Z^{\underline{c}}(\underline{a})\\
&= \int^{\underline{c}}\int^{\underline{b}} X\tbinom{d}{\underline{b}} \otimes Y^{\underline{b}}(\underline{c}) \otimes Z^{\underline{c}}(\underline{a})\\
&\cong \int^{\underline{b}} X\tbinom{d}{\underline{b}} \otimes \Big[\int^{\underline{c}} Y^{\underline{b}}(\underline{c}) \otimes Z^{\underline{c}}(\underline{a})\Big]\\
&\cong \int^{\underline{b}} X\tbinom{d}{\underline{b}} \otimes (Y \circ Z)^{\underline{b}}(\underline{a})\\
&= \big(X \circ (Y \circ Z)\big)\tbinom{d}{\underline{a}}
\end{aligned}$$

for $\binom{d}{\underline{a}} \in \Sigma^{\mathsf{op}}_{\mathfrak{C}} \times \mathfrak{C}$. The left and right unit isomorphisms are constructed similarly. The monoidal category axioms boil down to those in C. □

DEFINITION 8.1.12. The category of *$\mathfrak{C}$-colored operads in* C is the category

$$\mathsf{Op}^{\mathfrak{C}}_{\mathsf{C}} = \mathsf{Mon}\big(\mathsf{SSeq}^{\mathfrak{C}}_{\mathsf{C}}, \circ, \mathsf{I}\big)$$

of monoids in the monoidal category $\big(\mathsf{SSeq}^{\mathfrak{C}}_{\mathsf{C}}, \circ, \mathsf{I}\big)$. Furthermore:

- If $\mathfrak{C}$ has $n < \infty$ elements, then we refer to objects in $\mathsf{Op}^{\mathfrak{C}}_{\mathsf{C}}$ as *n-colored operads in* C.
- If $\mathfrak{C} = \{*\}$, then the entries of a one-colored operad O are denoted by O_n for $n \geq 0$.

An operad can be described more explicitly in terms of generating operations. To describe it in details, recall the block permutations and block sums in Definition 6.1.1.

PROPOSITION 8.1.13. *A* $\mathfrak{C}$*-colored operad in* C *is equivalent to a triple* $(\mathsf{O}, \gamma, \mathsf{1})$ *consisting of the following data.*

- $\mathsf{O} \in \mathsf{SSeq}_{\mathsf{C}}^{\mathfrak{C}}$.
- *For* $\binom{d}{\underline{c}} \in \Sigma_{\mathfrak{C}}^{\mathsf{op}} \times \mathfrak{C}$ *with* $|\underline{c}| = n \geq 1$, $\underline{b}_j \in \mathsf{Prof}(\mathfrak{C})$ *for* $1 \leq j \leq n$, *and* $\underline{b} = (\underline{b}_1, \ldots, \underline{b}_n)$ *their concatenation, it is equipped with an* operadic composition

$$\mathsf{O}\tbinom{d}{\underline{c}} \otimes \bigotimes_{j=1}^{n} \mathsf{O}\tbinom{c_j}{\underline{b}_j} \xrightarrow{\ \gamma\ } \mathsf{O}\tbinom{d}{\underline{b}} \in \mathsf{C}. \tag{8.1.14}$$

- *For each* $c \in \mathfrak{C}$, *it is equipped with a* c-colored unit

$$\mathbb{1} \xrightarrow{\ \mathsf{1}_c\ } \mathsf{O}\tbinom{c}{c} \in \mathsf{C}. \tag{8.1.15}$$

These data are required to satisfy the following associativity, unity, and equivariance axioms.

Associativity: *Suppose that:*

- *in* (8.1.14) $\underline{b}_j = \left(b_1^j, \ldots, b_{k_j}^j\right) \in \mathsf{Prof}(\mathfrak{C})$ *with at least one* $k_j > 0$;
- $\underline{a}_i^j \in \mathsf{Prof}(\mathfrak{C})$ *for each* $1 \leq j \leq n$ *and* $1 \leq i \leq k_j$;
- *for each* $1 \leq j \leq n$,

$$\underline{a}_j = \begin{cases} \left(\underline{a}_1^j, \ldots, \underline{a}_{k_j}^j\right) & \text{if } k_j > 0, \\ \varnothing & \text{if } k_j = 0 \end{cases}$$

with $\underline{a} = (\underline{a}_1, \ldots, \underline{a}_n)$ *their concatenation.*

Then the associativity diagram

$$\begin{array}{ccc} \mathsf{O}\tbinom{d}{\underline{c}} \otimes \left[\bigotimes_{j=1}^{n} \mathsf{O}\tbinom{c_j}{\underline{b}_j}\right] \otimes \bigotimes_{j=1}^{n}\bigotimes_{i=1}^{k_j} \mathsf{O}\tbinom{b_i^j}{\underline{a}_i^j} & \xrightarrow{(\gamma,\mathrm{Id})} & \mathsf{O}\tbinom{d}{\underline{b}} \otimes \bigotimes_{j=1}^{n}\bigotimes_{i=1}^{k_j} \mathsf{O}\tbinom{b_i^j}{\underline{a}_i^j} \\ \text{permute}\ \downarrow\ \cong & & \\ \mathsf{O}\tbinom{d}{\underline{c}} \otimes \bigotimes_{j=1}^{n}\left[\mathsf{O}\tbinom{c_j}{\underline{b}_j} \otimes \bigotimes_{i=1}^{k_j} \mathsf{O}\tbinom{b_i^j}{\underline{a}_i^j}\right] & & \downarrow\ \gamma \\ (\mathrm{Id},\otimes_j\gamma)\ \downarrow & & \\ \mathsf{O}\tbinom{d}{\underline{c}} \otimes \bigotimes_{j=1}^{n} \mathsf{O}\tbinom{c_j}{\underline{a}_j} & \xrightarrow{\quad\gamma\quad} & \mathsf{O}\tbinom{d}{\underline{a}} \end{array} \tag{8.1.16}$$

in C *is commutative.*

Unity: *Suppose $d \in \mathfrak{C}$.*

(1) For each $\underline{c} = (c_1, \ldots, c_n) \in \mathsf{Prof}(\mathfrak{C})$ with $n \geq 1$, the right unity diagram

(8.1.17)
$$\begin{array}{ccc}
\mathsf{O}\binom{d}{\underline{c}} \otimes \mathbb{1}^{\otimes n} & \xrightarrow{\cong} & \mathsf{O}\binom{d}{\underline{c}} \\
{\scriptstyle (\mathrm{Id}, \otimes 1_{c_j})} \downarrow & & \Vert \\
\mathsf{O}\binom{d}{\underline{c}} \otimes \bigotimes_{j=1}^{n} \mathsf{O}\binom{c_j}{c_j} & \xrightarrow{\gamma} & \mathsf{O}\binom{d}{\underline{c}}
\end{array}$$

in C is commutative.

(2) For each $\underline{b} \in \mathsf{Prof}(\mathfrak{C})$ the left unity diagram

(8.1.18)
$$\begin{array}{ccc}
\mathbb{1} \otimes \mathsf{O}\binom{d}{\underline{b}} & \xrightarrow{\cong} & \mathsf{O}\binom{d}{\underline{b}} \\
{\scriptstyle (1_d, \mathrm{Id})} \downarrow & & \Vert \\
\mathsf{O}\binom{d}{d} \otimes \mathsf{O}\binom{d}{\underline{b}} & \xrightarrow{\gamma} & \mathsf{O}\binom{d}{\underline{b}}
\end{array}$$

in C is commutative.

Equivariance: *Suppose that in* (8.1.14) $|\underline{b}_j| = k_j \geq 0$.

(1) For each permutation $\sigma \in \Sigma_n$, the top equivariance diagram

(8.1.19)
$$\begin{array}{ccc}
\mathsf{O}\binom{d}{\underline{c}} \otimes \bigotimes_{j=1}^{n} \mathsf{O}\binom{c_j}{\underline{b}_j} & \xrightarrow{(\sigma, \sigma^{-1})} & \mathsf{O}\binom{d}{\underline{c}\sigma} \otimes \bigotimes_{j=1}^{n} \mathsf{O}\binom{c_{\sigma(j)}}{\underline{b}_{\sigma(j)}} \\
\gamma \downarrow & & \downarrow \gamma \\
\mathsf{O}\binom{d}{\underline{b}_1, \ldots, \underline{b}_n} & \xrightarrow{\sigma\langle k_{\sigma(1)}, \ldots, k_{\sigma(n)}\rangle} & \mathsf{O}\binom{d}{\underline{b}_{\sigma(1)}, \ldots, \underline{b}_{\sigma(n)}}
\end{array}$$

in C is commutative. In the top horizontal morphism, σ is the equivariant structure morphism of O corresponding to $\sigma \in \Sigma_n$, and σ^{-1} permutes the n tensor factors from the left. The bottom horizontal morphism is the equivariant structure morphism of O corresponding to the block permutation

$$\sigma\langle k_{\sigma(1)}, \ldots, k_{\sigma(n)}\rangle \in \Sigma_{k_1 + \cdots + k_n}$$

induced by $\sigma \in \Sigma_n$.

(2) Given permutations $\tau_j \in \Sigma_{k_j}$ for $1 \leq j \leq n$, the bottom equivariance diagram

(8.1.20)
$$\begin{array}{ccc}
\mathsf{O}\binom{d}{\underline{c}} \otimes \bigotimes_{j=1}^{n} \mathsf{O}\binom{c_j}{\underline{b}_j} & \xrightarrow{(\mathrm{Id}, \otimes \tau_j)} & \mathsf{O}\binom{d}{\underline{c}} \otimes \bigotimes_{j=1}^{n} \mathsf{O}\binom{c_j}{\underline{b}_j \tau_j} \\
\gamma \downarrow & & \downarrow \gamma \\
\mathsf{O}\binom{d}{\underline{b}_1, \ldots, \underline{b}_n} & \xrightarrow{\tau_1 \times \cdots \times \tau_n} & \mathsf{O}\binom{d}{\underline{b}_1 \tau_1, \ldots, \underline{b}_n \tau_n}
\end{array}$$

in C is commutative. In the top horizontal morphism, each τ_j is the equivariant structure morphism of O corresponding to $\tau_j \in \Sigma_{k_j}$. The bottom horizontal morphism is the equivariant structure morphism of O corresponding to the block sum

$$\tau_1 \times \cdots \times \tau_n \in \Sigma_{k_1+\cdots+k_n}$$

induced by the τ_j's.

Moreover, a morphism of $\mathfrak{C}$-colored operads is a morphism of the underlying $\mathfrak{C}$-colored symmetric sequences that is compatible with the operadic compositions and the colored units in the obvious sense.

PROOF. For $\mathsf{O} \in \mathsf{SSeq}_{\mathsf{C}}^{\mathfrak{C}}$, the entries of $\mathsf{O} \circ \mathsf{O}$ are

(8.1.21)

$$\begin{aligned}(\mathsf{O} \circ \mathsf{O})\binom{d}{\underline{b}} &= \int^{\underline{c} \in \Sigma_{\mathfrak{C}}} \mathsf{O}\binom{d}{\underline{c}} \otimes \mathsf{O}^{\underline{c}}(\underline{b}) \\ &\cong \int^{\underline{c} \in \Sigma_{\mathfrak{C}}} \int^{\underline{a}_1,\ldots,\underline{a}_{|\underline{c}|} \in \Sigma_{\mathfrak{C}}^{\mathsf{op}}} \Sigma_{\mathfrak{C}}^{\mathsf{op}}(\underline{a};\underline{b}) \cdot \mathsf{O}\binom{d}{\underline{c}} \otimes \bigotimes_{j=1}^{|\underline{c}|} \mathsf{O}\binom{c_j}{\underline{a}_j}\end{aligned}$$

for $\binom{d}{\underline{b}} \in \Sigma_{\mathfrak{C}}^{\mathsf{op}} \times \mathfrak{C}$, where $\underline{a} = (\underline{a}_1, \ldots, \underline{a}_m)$ is the concatenation. So the multiplication $\mu : \mathsf{O} \circ \mathsf{O} \longrightarrow \mathsf{O}$ yields the entrywise operadic composition γ in (8.1.14). The monoid associativity of $\mathsf{O} \in \mathsf{Mon}(\mathsf{SSeq}_{\mathsf{C}}^{\mathfrak{C}})$ corresponds to the associativity diagram (8.1.16). For each $c \in \mathfrak{C}$, the c-colored unit $\mathbb{1}_c$ in (8.1.15) corresponds to the $\binom{c}{c}$-entry of the unit morphism $\varepsilon : \mathsf{I} \longrightarrow \mathsf{O}$ of the monoid O. The unity of the monoid O corresponds to the unity diagrams (8.1.17) and (8.1.18).

The top equivariance diagram (8.1.19) corresponds to the $\underline{c}$-variable in the coend (8.1.21) and the fact that the multiplication $\mu : \mathsf{O} \circ \mathsf{O} \longrightarrow \mathsf{O}$ is a morphism of $\mathfrak{C}$-colored symmetric sequences. Similarly, the bottom equivariance diagram (8.1.20) corresponds to the $\underline{a}_j$ variables in the coend (8.1.21). □

EXAMPLE 8.1.22. The symmetric circle product is far from symmetric, even in the one-colored case. With the color set $\mathfrak{C} = \{*\}$, we write $\Sigma_{\mathfrak{C}}$ as Σ. For one-colored symmetric sequences X and Y, the entry $(X \circ Y)(k)$ is

$$\int^{n \in \Sigma} \int^{k_1,\ldots,k_n \in \Sigma^{\mathsf{op}}} \Sigma^{\mathsf{op}}\big(k_1 + \cdots + k_n; k\big) \cdot X(n) \otimes Y(k_1) \otimes \cdots \otimes Y(k_n)$$

for $k \geq 0$. In particular, its 0th entry is

$$(X \circ Y)(0) = \int^{n \in \Sigma} X(n) \otimes Y(0)^{\otimes n}.$$

This expression is not symmetric, even up to isomorphism, in X and Y. For example, if the only non-$\varnothing$ entries in X and Y are $X(0)$ and $Y(1)$, then

$$(X \circ Y)(0) \cong \varnothing \quad \text{and} \quad (Y \circ X)(0) \cong Y(1) \otimes X(0),$$

which are in general not isomorphic. $\diamond$

8.2 Algebras

In this section, we discuss algebras over an operad in a symmetric monoidal category C, for which we need the following notation.

NOTATION 8.2.1. For a $\mathfrak{C}$-colored object $X = \{X_c\}_{c \in \mathfrak{C}}$ in C and a $\mathfrak{C}$-profile $\underline{c} = (c_1, \ldots, c_m)$, we write

$$X_{\underline{c}} = \begin{cases} X_{c_1} \otimes \cdots \otimes X_{c_m} \in \mathsf{C} & \text{if } m > 0, \\ \mathbb{1} & \text{if } \underline{c} = \varnothing. \end{cases}$$

LEMMA 8.2.2. *Suppose O is a $\mathfrak{C}$-colored operad in C. Then it induces a monad whose functor is*

$$\mathsf{O} \circ - : \mathsf{C}^{\mathfrak{C}} \longrightarrow \mathsf{C}^{\mathfrak{C}},$$

and whose multiplication and unit are induced by the operadic composition and the colored units of O, respectively.

PROOF. Suppose $Y = \{Y_c\}_{c \in \mathfrak{C}}$ is a $\mathfrak{C}$-colored object in C, regarded as a $\mathfrak{C}$-colored symmetric sequence with entries

$$Y\binom{d}{\underline{c}} = \begin{cases} Y_d & \text{if } \underline{c} = \varnothing, \\ \varnothing & \text{otherwise.} \end{cases}$$

Then

$$Y^{\underline{c}}(\underline{b}) = \begin{cases} Y_{\underline{c}} & \text{if } \underline{b} = \varnothing, \\ \varnothing & \text{if } \underline{b} \neq \varnothing \end{cases}$$

for $\underline{b}, \underline{c} \in \Sigma_{\mathfrak{C}}$. This implies that

$$(\mathsf{O} \circ Y)\binom{d}{\underline{b}} = \begin{cases} \displaystyle\int^{\underline{c} \in \Sigma_{\mathfrak{C}}} \mathsf{O}\binom{d}{\underline{c}} \otimes Y_{\underline{c}} & \text{if } \underline{b} = \varnothing, \\ \varnothing & \text{if } \underline{b} \neq \varnothing. \end{cases}$$

So the functor

$$\mathsf{O} \circ - : \mathsf{SSeq}_{\mathsf{C}}^{\mathfrak{C}} \longrightarrow \mathsf{SSeq}_{\mathsf{C}}^{\mathfrak{C}}$$

restricts to a functor $\mathsf{C}^{\mathfrak{C}} \longrightarrow \mathsf{C}^{\mathfrak{C}}$. The monad associativity and unity axioms of $\mathsf{O} \circ -$ now follow from the associativity and unity axioms of the operad O. □

DEFINITION 8.2.3. Suppose O is a $\mathfrak{C}$-colored operad in C. The category $\mathsf{Alg}(\mathsf{O})$ of O-*algebras* is defined as the category of $(\mathsf{O} \circ -)$-algebras for the monad $\mathsf{O} \circ -$ in Lemma 8.2.2.

Unpacking the above definition, we obtain the following more explicit description of an algebra over an operad.

PROPOSITION 8.2.4. *Suppose $(\mathsf{O}, \gamma, 1)$ is a $\mathfrak{C}$-colored operad in C. Then an O-algebra is exactly a pair (X, θ) consisting of*

- *a $\mathfrak{C}$-colored object $X = \{X_c\}_{c \in \mathfrak{C}} \in \mathsf{C}^{\mathfrak{C}}$ and*
- *an O-action structure morphism*

$$\mathsf{O}\tbinom{d}{\underline{c}} \otimes X_{\underline{c}} \xrightarrow{\theta} X_d \in \mathsf{C} \quad \text{for} \quad \tbinom{d}{\underline{c}} \in \Sigma_{\mathfrak{C}}^{\mathsf{op}} \times \mathfrak{C} \tag{8.2.5}$$

such that the following associativity, unity, and equivariance axioms hold.

Associativity: *For $\binom{d}{\underline{c}} \in \Sigma_{\mathfrak{C}}^{\mathsf{op}} \times \mathfrak{C}$ with $|\underline{c}| = n \geq 1$, $\underline{b}_j \in \Sigma_{\mathfrak{C}}$ for $1 \leq j \leq n$, and $\underline{b} = (\underline{b}_1, \ldots, \underline{b}_n)$ their concatenation, the associativity diagram*

$$\begin{array}{ccc} \mathsf{O}\tbinom{d}{\underline{c}} \otimes \Big[\bigotimes_{j=1}^{n} \mathsf{O}\tbinom{c_j}{\underline{b}_j}\Big] \otimes X_{\underline{b}} & \xrightarrow{(\gamma,\mathrm{Id})} & \mathsf{O}\tbinom{d}{\underline{b}} \otimes X_{\underline{b}} \\ \text{permute}\ \downarrow \cong & & \\ \mathsf{O}\tbinom{d}{\underline{c}} \otimes \bigotimes_{j=1}^{n} \Big[\mathsf{O}\tbinom{c_j}{\underline{b}_j} \otimes X_{\underline{b}_j}\Big] & & \downarrow \theta \\ (\mathrm{Id},\otimes_j\theta)\ \downarrow & & \\ \mathsf{O}\tbinom{d}{\underline{c}} \otimes X_{\underline{c}} & \xrightarrow{\theta} & X_d \end{array} \tag{8.2.6}$$

in C is commutative.

Unity: *For each $c \in \mathfrak{C}$, the unity diagram*

$$\begin{array}{ccc} \mathbb{1} \otimes X_c & \xrightarrow[\cong]{\lambda} & X_c \\ (1_c,\mathrm{Id})\ \downarrow & & \| \\ \mathsf{O}\tbinom{c}{c} \otimes X_c & \xrightarrow{\theta} & X_c \end{array} \tag{8.2.7}$$

in C is commutative.

Equivariance: *For each* $\binom{d}{\underline{c}} \in \Sigma_{\mathfrak{C}}^{\mathrm{op}} \times \mathfrak{C}$ *and each permutation* $\sigma \in \Sigma_{|\underline{c}|}$, *the equivariance diagram*

(8.2.8)
$$\begin{array}{ccc} \mathsf{O}\binom{d}{\underline{c}} \otimes X_{\underline{c}} & \xrightarrow{(\sigma,\sigma^{-1})} & \mathsf{O}\binom{d}{\underline{c}\sigma} \otimes X_{\underline{c}\sigma} \\ \downarrow{\scriptstyle \theta} & & \downarrow{\scriptstyle \theta} \\ X_d & = & X_d \end{array}$$

in C *is commutative. In the top horizontal morphism,* σ^{-1} *is the right permutation on the tensor factors in* $X_{\underline{c}}$ *induced by* $\sigma \in \Sigma_{|\underline{c}|}$.

Moreover, a morphism of O*-algebras*

$$f : (X, \theta^X) \longrightarrow (Y, \theta^Y)$$

is exactly a morphism $f : X \longrightarrow Y$ *of* $\mathfrak{C}$*-colored objects in* C *such that the diagram*

$$\begin{array}{ccc} \mathsf{O}\binom{d}{\underline{c}} \otimes X_{\underline{c}} & \xrightarrow{(\mathrm{Id},\otimes f)} & \mathsf{O}\binom{d}{\underline{c}} \otimes Y_{\underline{c}} \\ \downarrow{\scriptstyle \theta^X} & & \downarrow{\scriptstyle \theta^Y} \\ X_d & \xrightarrow{f} & Y_d \end{array}$$

in C *is commutative for all* $\binom{d}{\underline{c}} \in \Sigma_{\mathfrak{C}}^{\mathrm{op}} \times \mathfrak{C}$.

PROOF. For a $\mathfrak{C}$-colored object $X = \{X_c\}_{c \in \mathfrak{C}}$ in C, the $\mathfrak{C}$-colored object $\mathsf{O} \circ X$ has entries

$$(\mathsf{O} \circ X)_d = \int^{\underline{c} \in \Sigma_{\mathfrak{C}}} \mathsf{O}\binom{d}{\underline{c}} \otimes X_{\underline{c}} \quad \text{for} \quad d \in \mathfrak{C}.$$

So for the monad $(\mathsf{O} \circ -)$, an $(\mathsf{O} \circ -)$-algebra (X, ϕ) has O-action structure morphisms

$$\begin{array}{ccc} \mathsf{O}\binom{d}{\underline{c}} \otimes X_{\underline{c}} & \xrightarrow{\theta} & X_d \\ \downarrow{\scriptstyle \omega_{\underline{c}}} & & \uparrow{\scriptstyle \phi} \\ \int^{\underline{c} \in \Sigma_{\mathfrak{C}}} \mathsf{O}\binom{d}{\underline{c}} \otimes X_{\underline{c}} & = & (\mathsf{O} \circ X)_d \end{array}$$

as in (8.2.5). The associativity axiom (8.2.6) and the unity axiom (8.2.7) correspond to those of an $(\mathsf{O} \circ -)$-algebra in (1.3.4). The equivariance diagram (8.2.8) comes from the $\underline{c}$-variable in the coend above. The reverse identification is the same. □

Next we provide some basic examples of operads and their algebras.

EXAMPLE 8.2.9 (Endomorphism Operad). Suppose that the underlying symmetric monoidal category C is closed in the sense of Definition 1.2.33, and the right adjoint of

$$- \otimes X : \mathsf{C} \longrightarrow \mathsf{C}$$

for $X \in \mathsf{C}$ is the internal hom $[X, -]$. For each $\mathfrak{C}$-colored object $X = \{X_c\}_{c \in \mathfrak{C}}$ in C, the *endomorphism operad* is the $\mathfrak{C}$-colored operad $\mathsf{End}(X)$ with entries

$$\mathsf{End}(X)\tbinom{d}{\underline{c}} = [X_{\underline{c}}, X_d] \quad \text{for} \quad \tbinom{d}{\underline{c}} \in \Sigma_{\mathfrak{C}}^{\mathsf{op}} \times \mathfrak{C}.$$

- Its equivariant structure is induced by permutations of the tensor factors in $X_{\underline{c}}$.
- Its d-colored unit

  $$\mathbb{1} \longrightarrow [X_d, X_d]$$

  is adjoint to the left unit isomorphism $\lambda : \mathbb{1} \otimes X_d \xrightarrow{\cong} X_d$.
- Its operadic composition γ is induced by the adjunction between $- \otimes X$ and the internal hom.

Via this adjunction, for a $\mathfrak{C}$-colored operad O in C, an O-algebra structure (X, θ) is equivalent to a morphism

$$\theta' : \mathsf{O} \longrightarrow \mathsf{End}(X)$$

of $\mathfrak{C}$-colored operads. ◇

EXAMPLE 8.2.10 (Operad for Monoids). The *associative operad* is the one-colored operad

$$\mathsf{As} = \Big\{ \coprod_{\Sigma_n} \mathbb{1} \Big\}_{n \geq 0}$$

in C in which:

- The operadic composition is induced by

  $$\big(\sigma; \tau_1, \ldots, \tau_n\big) = \underbrace{\sigma\langle k_1, \ldots, k_n\rangle}_{\text{block permutation}} \cdot \underbrace{\big(\tau_1 \times \cdots \times \tau_n\big)}_{\text{block sum}} \in \Sigma_{k_1+\cdots+k_n}$$

  for $\sigma \in \Sigma_n$ with $n \geq 1$, $\tau_i \in \Sigma_{k_i}$ for $1 \leq i \leq n$, and $k_i \geq 0$.
- The operadic unit corresponds to the identity permutation $\mathrm{id}_1 \in \Sigma_1$.
- The symmetric sequence structure corresponds to the group multiplication in Σ_n.

The category of As-algebras is canonically isomorphic to the category of monoids in C. ◇

EXAMPLE 8.2.11 (Operad for Commutative Monoids). The *commutative operad* is the one-colored operad

$$\mathsf{Com} = \{\mathbb{1}\}_{n\geq 0}$$

in C in which:

- The operadic composition is induced by the unit isomorphism $\mathbb{1}\otimes\mathbb{1}\cong\mathbb{1}$.
- The operadic unit is the identity morphism of $\mathbb{1}$.
- The symmetric sequence structure is given by the trivial Σ_n-action on $\mathbb{1}$.

The category of Com-algebras is canonically isomorphic to the category of commutative monoids in C. ◇

EXAMPLE 8.2.12 (Operad for Diagrams of Monoids). Suppose D is a small category with object set $\mathfrak{D}$. For $\binom{d}{\underline{c}} = \binom{d}{c_1,\ldots,c_n} \in \Sigma_{\mathfrak{D}}^{\mathsf{op}} \times \mathfrak{D}$, we use the abbreviation

$$\overline{\mathsf{D}}(\underline{c};d) = \overline{\mathsf{D}}\tbinom{d}{\underline{c}} = \begin{cases} \Sigma_n \times \prod\limits_{j=1}^{n} \mathsf{D}(c_j,d) & \text{if } n\geq 1,\\ \{*\} & \text{if } n=0.\end{cases} \tag{8.2.13}$$

There is a $\mathfrak{D}$-colored operad As^{D} in C with entries

$$\mathsf{As}^{\mathsf{D}}\tbinom{d}{\underline{c}} = \coprod_{\overline{\mathsf{D}}(\underline{c};d)} \mathbb{1} \quad \text{for} \quad \tbinom{d}{\underline{c}} \in \Sigma_{\mathfrak{D}}^{\mathsf{op}} \times \mathfrak{D}.$$

A coproduct summand corresponding to an element $(\sigma,\underline{f}) \in \overline{\mathsf{D}}(\underline{c};d)$ is denoted by $\mathbb{1}_{(\sigma,\underline{f})}$. We describe the operad structure on As^{D} in terms of the subscripts.

- For $\tau \in \Sigma_{|\underline{c}|}$, the right τ-action sends $\mathbb{1}_{(\sigma,\underline{f})}$ to $\mathbb{1}_{(\sigma\tau,\underline{f}\tau)}$.
- Its c-colored unit corresponds to $\mathbb{1}_{(\mathrm{id}_1,\mathrm{Id}_c)}$ for $c \in \mathfrak{D}$.
- Its operadic composition

$$\mathsf{As}^{\mathsf{D}}\tbinom{d}{\underline{c}} \otimes \bigotimes_{j=1}^{n} \mathsf{As}^{\mathsf{D}}\tbinom{c_j}{\underline{b}_j} \xrightarrow{\ \gamma\ } \mathsf{As}^{\mathsf{D}}\tbinom{d}{\underline{b}}$$

corresponds to

$$\Big((\sigma,\underline{f}); \{(\tau_j,\underline{g}_j)\}_{j=1}^{n}\Big) \longmapsto \Big((\sigma;\tau_1,\ldots,\tau_n), (f_1\underline{g}_1,\ldots,f_n\underline{g}_n)\Big).$$

Here

$$f_j \underline{g}_j = \left(f_j g_{j1}, \ldots, f_j g_{jk_j}\right) \in \prod_{i=1}^{k_j} \mathsf{D}(b_{ji}, d) \quad \text{if}$$

$$\underline{g}_j = \left(g_{j1}, \ldots, g_{jk_j}\right) \in \prod_{i=1}^{k_j} \mathsf{D}(b_{ji}, c_j),$$

and

$$\left(\sigma; \tau_1, \ldots, \tau_n\right) = \sigma\langle k_1, \ldots, k_n\rangle \cdot \left(\tau_1 \times \cdots \times \tau_n\right)$$

is as in the operadic composition in the associative operad As in Example 8.2.10.

There is a canonical isomorphism

$$\mathsf{Alg}(\mathsf{As}^{\mathsf{D}}) \xrightarrow{\cong} \mathsf{Mon}(\mathsf{C})^{\mathsf{D}}$$

from the category of As^{D}-algebras to the category of D-diagrams in $\mathsf{Mon}(\mathsf{C})$, the category of monoids in C. Indeed, each As^{D}-algebra (X, θ) has a restricted structure morphism

$$\begin{array}{ccc} X_{\underline{c}} & \xrightarrow{\theta(\sigma,\underline{f})} & X_d \\ {\scriptstyle (\sigma,\underline{f})\ \text{inclusion}}\big\downarrow & & \big\uparrow{\scriptstyle \theta} \\ \coprod_{\overline{\mathsf{D}}(\underline{c};d)} X_{\underline{c}} & \xrightarrow{\cong} & \mathsf{As}^{\mathsf{D}}\binom{d}{\underline{c}} \otimes X_{\underline{c}} \end{array}$$

for each $(\sigma, \underline{f}) \in \overline{\mathsf{D}}(\underline{c}; d)$.

- For a morphism $f : c \longrightarrow d \in \mathsf{D}$, there is a restricted structure morphism

$$X_c \xrightarrow{\theta(\mathrm{id}_1, f)} X_d \quad \in \mathsf{C}.$$

The associativity and unity axioms of (X, θ) imply that this is a D-diagram in C.

- For each $c \in \mathfrak{D}$, the restricted structure morphisms

$$X_c \otimes X_c \xrightarrow{\theta(\mathrm{id}_2, \{\mathrm{Id}_c, \mathrm{Id}_c\})} X_c \quad \text{and} \quad \mathbb{1} \xrightarrow{\theta(\mathrm{id}_0, *)} X_c$$

give X_c the structure of a monoid in C, once again by the associativity and unity axioms of (X, θ). An inspection shows that this gives a D-diagram of monoids in C. In other words, the morphisms $\theta_{(\mathrm{id}_1, f)}$ are compatible with the entrywise monoid structures.

The $\mathfrak{D}$-colored operad As^{D} is generated by the operations described in the previous two items. Therefore, As^{D} is the $\mathfrak{D}$-colored operad whose algebras are D-diagrams of monoids in C. ◇

EXAMPLE 8.2.14 (Operad for Diagrams of Commutative Monoids). Suppose D is a small category with object set $\mathfrak{D}$. There is a $\mathfrak{D}$-colored operad $\mathsf{Com}^{\mathsf{D}}$ in C with entries

$$\mathsf{Com}^{\mathsf{D}}\binom{d}{\underline{c}} = \coprod_{\prod_{j=1}^{n} \mathsf{D}(c_j, d)} \mathbb{1} \quad \text{for} \quad \binom{d}{\underline{c}} = \binom{d}{c_1, \ldots, c_n} \in \Sigma_{\mathfrak{D}}^{\mathsf{op}} \times \mathfrak{D}.$$

Its operad structure is defined as in Example 8.2.12 by ignoring the first component. Just like Example 8.2.12, there is a canonical isomorphism

$$\mathsf{Alg}(\mathsf{Com}^{\mathsf{D}}) \xrightarrow{\cong} \mathsf{CMon}(\mathsf{C})^{\mathsf{D}}$$

from the category of $\mathsf{Com}^{\mathsf{D}}$-algebras to the category of D-diagrams of commutative monoids in C. To check this, a key fact is that the monoid multiplication

$$X_c \otimes X_c \xrightarrow{\mu_c = \theta_{\{\mathrm{Id}_c, \mathrm{Id}_c\}}} X_c \in \mathsf{C}$$

is commutative. This is true because the pair $\{\mathrm{Id}_c, \mathrm{Id}_c\}$ is fixed by the permutation $(1, 2) \in \Sigma_2$. So the equivariance axiom (8.2.8) implies that μ_c is commutative. In summary, $\mathsf{Com}^{\mathsf{D}}$ is the $\mathfrak{D}$-colored operad whose algebras are D-diagrams of commutative monoids in C. ◇

8.3 Involutive Operads as Involutive Monoids

For the rest of this chapter, C is an involutive strict symmetric monoidal category; see the conventions stated just before Section 8.1. The purposes of this section are to define involutive operads and to provide some examples.

For a set $\mathfrak{C}$, recall from Definition 8.1.12 that the category $\mathsf{Op}_{\mathsf{C}}^{\mathfrak{C}}$ of $\mathfrak{C}$-colored operads in C is the category $\mathsf{Mon}(\mathsf{SSeq}_{\mathsf{C}}^{\mathfrak{C}}, \circ, \mathsf{I})$ of monoids in the monoidal category $\mathsf{SSeq}_{\mathsf{C}}^{\mathfrak{C}}$ of $\mathfrak{C}$-colored symmetric sequences in C. Involutive operads will be defined as involutive objects in $\mathsf{Op}_{\mathsf{C}}^{\mathfrak{C}}$ equipped with a suitable involutive structure. First we equip the category $\mathsf{SSeq}_{\mathsf{C}}^{\mathfrak{C}}$ with an involutive structure.

DEFINITION 8.3.1. We equip $\mathsf{C}^{\Sigma_{\mathfrak{C}}^{\mathsf{op}} \times \mathfrak{C}} = \mathsf{SSeq}_{\mathsf{C}}^{\mathfrak{C}}$ with the following structures.

Involution Functor: Extend the involution functor $I : \mathsf{C} \longrightarrow \mathsf{C}$ to a functor

$$I : \mathsf{SSeq}_{\mathsf{C}}^{\mathfrak{C}} \longrightarrow \mathsf{SSeq}_{\mathsf{C}}^{\mathfrak{C}}$$

by setting

$$(IX)\binom{d}{\underline{c}} = I\bigl(X\tbinom{d}{\underline{c}}\bigr)$$

for $X \in \mathsf{SSeq}_{\mathsf{C}}^{\mathfrak{C}}$ and $\binom{d}{\underline{c}} \in \Sigma_{\mathfrak{C}}^{\mathsf{op}} \times \mathfrak{C}$, and similarly for morphisms in $\mathsf{SSeq}_{\mathsf{C}}^{\mathfrak{C}}$.

Unit: Extend the unit $\iota : \mathrm{Id}_{\mathsf{C}} \xrightarrow{\cong} I^2$ componentwise to a natural isomorphism

$$\iota : \mathrm{Id}_{\mathsf{SSeq}_{\mathsf{C}}^{\mathfrak{C}}} \xrightarrow{\cong} I^2$$

by setting

$$(\iota_X)\tbinom{d}{\underline{c}} = \iota_{X\binom{d}{\underline{c}}} : X\tbinom{d}{\underline{c}} \xrightarrow{\cong} I^2X\tbinom{d}{\underline{c}}$$

for $X \in \mathsf{SSeq}_{\mathsf{C}}^{\mathfrak{C}}$ and $\binom{d}{\underline{c}} \in \Sigma_{\mathfrak{C}}^{\mathsf{op}} \times \mathfrak{C}$.

PROPOSITION 8.3.2. *$\bigl(\mathsf{SSeq}_{\mathsf{C}}^{\mathfrak{C}}, \circ, \mathsf{I}, I, \iota\bigr)$ is an involutive monoidal category.*

PROOF. By Proposition 8.1.11, $\mathsf{SSeq}_{\mathsf{C}}^{\mathfrak{C}}$ is a monoidal category. With the structure in Definition 8.3.1, $(\mathsf{SSeq}_{\mathsf{C}}^{\mathfrak{C}}, I, \iota)$ is an involutive category by Proposition 2.1.7. It remains to check the commutativity of the diagrams (4.1.3), (4.1.4), (4.1.5), and (4.1.6) in $\mathsf{SSeq}_{\mathsf{C}}^{\mathfrak{C}}$.

The diagram (4.1.5) in this case is

$$\begin{array}{ccc}
X \circ Y & \xrightarrow{\iota_{X \circ Y}} & I^2(X \circ Y) \\
{\scriptstyle \iota_X \circ \iota_Y}\downarrow & & \uparrow{\scriptstyle I(I_2)} \\
I^2X \circ I^2Y & \xrightarrow{I_2} & I(IX \circ IY)
\end{array}$$

for $X, Y \in \mathsf{SSeq}_{\mathsf{C}}^{\mathfrak{C}}$. At a typical $\binom{d}{\underline{b}}$-entry, the above diagram becomes the following diagram in C.

(8.3.3)

$$\begin{array}{ccc}
\int^{\underline{c}} X\binom{d}{\underline{c}} \otimes Y^{\underline{c}}(\underline{b}) & \xrightarrow{\iota} & I^2\Big[\int^{\underline{c}} X\binom{d}{\underline{c}} \otimes Y^{\underline{c}}(\underline{b})\Big] \\
\Big\downarrow{\scriptstyle ((\iota_X)_*, (\iota_Y)_*)} & & \Big\uparrow{\scriptstyle \cong} \\
\int^{\underline{c}} I^2X\binom{d}{\underline{c}} \otimes (I^2Y)^{\underline{c}}(\underline{b}) & & I\Big[\int^{\underline{c}} I\big(X\binom{d}{\underline{c}} \otimes Y^{\underline{c}}(\underline{b})\big)\Big] \\
\Big\downarrow{\scriptstyle \cong} & & \Big\uparrow{\scriptstyle I(I_2)_*} \\
\int^{\underline{c}} I^2X\binom{d}{\underline{c}} \otimes I\big((IY)^{\underline{c}}(\underline{b})\big) & & I\Big[\int^{\underline{c}} IX\binom{d}{\underline{c}} \otimes I\big(Y^{\underline{c}}(\underline{b})\big)\Big] \\
\Big\downarrow{\scriptstyle (I_2)_*} & & \Big\uparrow{\scriptstyle \cong} \\
\int^{\underline{c}} I\Big(IX\binom{d}{\underline{c}} \otimes (IY)^{\underline{c}}(\underline{b})\Big) & \xrightarrow{\cong} & I\Big[\int^{\underline{c}} IX\binom{d}{\underline{c}} \otimes (IY)^{\underline{c}}(\underline{b})\Big]
\end{array}$$

In the diagram (8.3.3):

- The middle left and the lower right vertical isomorphisms come from the natural isomorphisms

$$\begin{aligned}
(IY)^{\underline{c}}(\underline{b}) &= \int^{\{\underline{a}_j\}} \Sigma^{\mathsf{op}}_{\mathfrak{C}}(\underline{a};\underline{b}) \cdot \bigotimes_{j=1}^{|\underline{c}|} IY\binom{c_j}{\underline{a}_j} \\
&\cong \int^{\{\underline{a}_j\}} \Sigma^{\mathsf{op}}_{\mathfrak{C}}(\underline{a};\underline{b}) \cdot I\Big[\bigotimes_{j=1}^{|\underline{c}|} Y\binom{c_j}{\underline{a}_j}\Big] \\
&\cong I\Big[\int^{\{\underline{a}_j\}} \Sigma^{\mathsf{op}}_{\mathfrak{C}}(\underline{a};\underline{b}) \cdot \bigotimes_{j=1}^{|\underline{c}|} Y\binom{c_j}{\underline{a}_j}\Big] \\
&= I\big(Y^{\underline{c}}(\underline{b})\big)
\end{aligned}$$

  with $\underline{a} = (\underline{a}_1, \ldots, \underline{a}_{|\underline{c}|})$. Here the first isomorphism uses Corollary 4.1.10, which says that I is a strong monoidal functor. The second isomorphism uses Proposition 2.1.4, which states that I preserves all the colimits that exist in C because it is adjoint to itself.
- The bottom horizontal isomorphism and the upper right vertical isomorphism also follow from the fact that I preserves colimits.

To prove the commutativity of (8.3.3), we may restrict to an object $X\binom{d}{\underline{c}} \otimes Y^{\underline{c}}(\underline{b})$, in which case the commutativity follows from that of (4.1.5) in C.

The commutativity of the diagrams (4.1.3), (4.1.4), and (4.1.6) in $\mathsf{SSeq}^{\mathfrak{C}}_{\mathsf{C}}$ follows similarly from that in C. □

COROLLARY 8.3.4. *The category* $\mathsf{Op}^{\mathfrak{C}}_{\mathsf{C}}$ *of* $\mathfrak{C}$*-colored operads in* C *inherits an involutive structure from* C.

PROOF. Since $\mathsf{Op}^{\mathfrak{C}}_{\mathsf{C}}$ is the category $\mathsf{Mon}\big(\mathsf{SSeq}^{\mathfrak{C}}_{\mathsf{C}}\big)$ of monoids in the category $\mathsf{SSeq}^{\mathfrak{C}}_{\mathsf{C}}$ of $\mathfrak{C}$-colored symmetric sequences in C, the assertion follows from Theorem 4.5.1 applied to the involutive monoidal category $\mathsf{SSeq}_{\mathsf{C}}^{\mathfrak{C}}$ in Proposition 8.3.2. □

DEFINITION 8.3.5. The category of $\mathfrak{C}$-*colored involutive operads* in C is defined as the category

$$\mathsf{IOp}_{\mathsf{C}}^{\mathfrak{C}} = \mathsf{Inv}(\mathsf{Op}_{\mathsf{C}}^{\mathfrak{C}})$$

of involutive objects in the involutive category $\mathsf{Op}_{\mathsf{C}}^{\mathfrak{C}}$.

In other words, the category of $\mathfrak{C}$-colored involutive operads in C,

$$\mathsf{IOp}_{\mathsf{C}}^{\mathfrak{C}} = \mathsf{Inv}\Big(\mathsf{Mon}(\mathsf{SSeq}_{\mathsf{C}}^{\mathfrak{C}})\Big) = \mathsf{IMon}(\mathsf{SSeq}_{\mathsf{C}}^{\mathfrak{C}}),$$

is the category of involutive monoids in $\mathsf{SSeq}_{\mathsf{C}}^{\mathfrak{C}}$. Let us unpack this definition.

PROPOSITION 8.3.6. *A $\mathfrak{C}$-colored involutive operad in C is exactly a tuple*

$$(\mathsf{O}, \gamma, \mathsf{1}, j)$$

with

- $(\mathsf{O}, \gamma, \mathsf{1})$ *a $\mathfrak{C}$-colored operad in C, and*
- $\big(\mathsf{O}, j : \mathsf{O} \longrightarrow I\mathsf{O}\big)$ *an involutive object in* $\mathsf{SSeq}_{\mathsf{C}}^{\mathfrak{C}}$

such that the diagrams

(8.3.7)
$$\begin{array}{ccc}
\mathbb{1} & \xrightarrow{\ \mathsf{1}_c\ } & \mathsf{O}\binom{c}{c} \\
{\scriptstyle I_0}\big\downarrow & & \big\downarrow{\scriptstyle j} \\
I\mathbb{1} & \xrightarrow{\ I\mathsf{1}_c\ } & I\mathsf{O}\binom{c}{c}
\end{array}$$

for $c \in \mathfrak{C}$ and

(8.3.8)
$$\begin{array}{ccccc}
\mathsf{O}\binom{d}{\underline{c}} \otimes \bigotimes_{j=1}^{n} \mathsf{O}\binom{c_j}{\underline{b}_j} & & \xrightarrow{\qquad\qquad\gamma\qquad\qquad} & & \mathsf{O}\binom{d}{\underline{b}} \\
{\scriptstyle (j,\otimes j)}\big\downarrow & & & & \big\downarrow{\scriptstyle j} \\
I\mathsf{O}\binom{d}{\underline{c}} \otimes \bigotimes_{j=1}^{n} I\mathsf{O}\binom{c_j}{\underline{b}_j} & \xrightarrow{\ I_2\ } & I\Big[\mathsf{O}\binom{d}{\underline{c}} \otimes \bigotimes_{j=1}^{n} \mathsf{O}\binom{c_j}{\underline{b}_j}\Big] & \xrightarrow{\ I\gamma\ } & I\mathsf{O}\binom{d}{\underline{b}}
\end{array}$$

with γ as in (8.1.14)*, are commutative.*

Moreover, a morphism of $\mathfrak{C}$-colored involutive operads in C is exactly a morphism of the underlying $\mathfrak{C}$-colored operads in C that is also a morphism of involutive objects in $\mathsf{SSeq}_{\mathsf{C}}^{\mathfrak{C}}$.

PROOF. By definition the category of (involutive) $\mathfrak{C}$-colored operads in C is the category of (involutive) monoids in $\mathsf{SSeq}_{\mathsf{C}}^{\mathfrak{C}}$. When we write $(\mathsf{O}, \gamma, \mathsf{1})$, we are using the description of a $\mathfrak{C}$-colored operad in C in Proposition 8.1.13. We apply Proposition 4.5.5 to the involutive monoidal category $\mathsf{SSeq}_{\mathsf{C}}^{\mathfrak{C}}$. The diagrams (8.3.7) and (8.3.8) correspond to the diagrams (4.5.6) and (4.5.7), respectively. □

The rest of this section contains examples of involutive operads.

EXAMPLE 8.3.9 (Trivial Involution). Suppose that the involutive structure on C is trivial, i.e., $I = \mathrm{Id}_{\mathsf{C}}$ and $\iota = \mathrm{Id}_{\mathrm{Id}_{\mathsf{C}}}$. Then a $\mathfrak{C}$-colored involutive operad in C is a pair (O, j) such that:

- O is a $\mathfrak{C}$-colored operad in C.
- $j : \mathsf{O} \longrightarrow \mathsf{O}$ is a morphism of $\mathfrak{C}$-colored operads in C.
- $j \circ j = \mathrm{Id}_{\mathsf{O}}$.

Indeed, the involutive structure on $\mathsf{SSeq}_{\mathsf{C}}^{\mathfrak{C}}$ is also trivial. By Proposition 8.3.6, $j : \mathsf{O} \longrightarrow \mathsf{O}$ is a morphism of $\mathfrak{C}$-colored symmetric sequences in C such that $j \circ j = \mathrm{Id}_{\mathsf{O}}$. The diagrams (8.3.7) and (8.3.8) mean that j preserves the colored units and the operadic composition, so j is actually a morphism of $\mathfrak{C}$-colored operads. ◇

EXAMPLE 8.3.10 (Involutive Associative Operad). Recall from Example 8.2.10 the associative operad $\mathsf{As} = \bigl\{\coprod_{\Sigma_n} \mathbb{1}\bigr\}_{n \geq 0}$, which is a one-colored operad in C whose algebras are exactly the monoids in C. A copy of $\mathbb{1}$ in As_n corresponding to a permutation $\sigma \in \Sigma_n$ is denoted by $\mathbb{1}_\sigma$, and similarly for $I\mathsf{As}$. We equip As with the involutive structure

$$j^{\mathrm{rev}} : \mathsf{As} \longrightarrow I\mathsf{As}$$

defined entrywise by the commutative diagram

(8.3.11)
$$\begin{array}{ccccc}
\mathbb{1}_\sigma & \xrightarrow{I_0} & I\mathbb{1}_{\pi_n\sigma} & \xrightarrow{\text{inclusion}} & \coprod_{\Sigma_n} I\mathbb{1} \\
{\scriptstyle\text{inclusion}}\downarrow & & & & \downarrow\cong \\
\mathsf{As}_n & \xrightarrow{j^{\mathrm{rev}}} & I(\mathsf{As}_n) & = & I\bigl[\coprod_{\Sigma_n} \mathbb{1}\bigr]
\end{array}$$

in C for $\sigma \in \Sigma_n$. Here $\pi_n \in \Sigma_n$ is the *interval-reversing permutation* given by

(8.3.12)
$$\pi_n(i) = n + 1 - i \quad \text{for} \quad 1 \leq i \leq n.$$

The right vertical natural isomorphism is from Proposition 2.1.4.

We now use Proposition 8.3.6 to check that $(\mathsf{As}, j^{\mathrm{rev}})$ is a one-colored involutive operad in C.

- That

$$j^{\mathrm{rev}} : \mathsf{As} \longrightarrow I\mathsf{As}$$

is a morphism in $\mathsf{SSeq}_{\mathsf{C}}^{\mathfrak{C}}$ follows from the fact that in both As and $I\mathsf{As}$, the right Σ_n-action is given by the multiplication in Σ_n in the subscripts.

- The involutive object axiom

$$Ij^{\mathsf{rev}} \circ j^{\mathsf{rev}} = \iota$$

follows from

  - the permutation identity $\pi_n^2 = \mathrm{id}_n \in \Sigma_n$ and
  - the equality $I(I_0) \circ I_0 = \iota_{\mathbb{1}}$ in (4.1.6).

- The unit diagram (8.3.7) is commutative because the operadic unit

$$1 : \mathbb{1} \longrightarrow \mathsf{As}_1$$

is the identity morphism of $\mathbb{1}$.

- Using the notations in Definition 6.1.1, the diagram (8.3.8) for $(\mathsf{As}, j^{\mathsf{rev}})$ boils down to the permutation identity

$$\begin{aligned} &\pi_k \cdot \sigma\langle k_1, \ldots, k_n\rangle \cdot \big(\tau_1 \times \cdots \times \tau_n\big) \\ &= \underbrace{(\pi_n\sigma)\langle k_1, \ldots, k_n\rangle}_{\text{block permutation}} \cdot \underbrace{\big(\pi_{k_1}\tau_1 \times \cdots \times \pi_{k_n}\tau_n\big)}_{\text{block sum}} \in \Sigma_k \end{aligned} \tag{8.3.13}$$

for $\sigma \in \Sigma_n$, $\tau_i \in \Sigma_{k_i}$ for $1 \leq i \leq n$, and $k = k_1 + \cdots + k_n$. This equality follows from a direct inspection.

Therefore, $(\mathsf{As}, j^{\mathsf{rev}})$ is a one-colored involutive operad in C, called the *involutive associative operad*. We will return to this involutive operad in Proposition 8.5.4 and Example 8.6.8. ◇

EXAMPLE 8.3.14 (Non-Reversing Involutive Associative Operad). There is another one-colored involutive operad structure

$$j : \mathsf{As} \longrightarrow I\mathsf{As}$$

on the associative operad defined as in (8.3.11) by removing the interval-reversing permutation π_n. In other words, it is determined by the morphism

$$I_0 : \mathbb{1}_\sigma \longrightarrow I\mathbb{1}_\sigma \quad \text{for} \quad \sigma \in \Sigma_n.$$

We call (As, j) the *non-reversing involutive associative operad.* ◇

EXAMPLE 8.3.15 (Involutive Operad for Diagrams). Suppose D is a small category with object set $\mathfrak{D}$, and As^{D} is the $\mathfrak{D}$-colored operad in C in Example 8.2.12 whose algebras are D-diagrams of monoids in C. We equip As^{D} with the involutive structure

$$j^{\mathsf{rev}} : \mathsf{As}^{\mathsf{D}} \longrightarrow I\mathsf{As}^{\mathsf{D}}$$

determined by the commutative diagram

$$\begin{array}{ccccc}
\mathbb{1}_{(\sigma,\underline{f})} & \xrightarrow{I_0} & I\mathbb{1}_{(\pi_n\sigma,\underline{f})} & \xrightarrow{\text{inclusion}} & \coprod_{\overline{\mathsf{D}}(\underline{c};d)} I\mathbb{1} \\
\downarrow{\scriptstyle \text{inclusion}} & & & & \downarrow{\scriptstyle \cong} \\
\mathsf{As}^{\mathsf{D}}\binom{d}{\underline{c}} & \xrightarrow{j^{\mathsf{rev}}} & I\mathsf{As}^{\mathsf{D}}\binom{d}{\underline{c}} & = & I\Big[\coprod_{\overline{\mathsf{D}}(\underline{c};d)} \mathbb{1}\Big]
\end{array}$$

in C for $\binom{d}{\underline{c}} \in \Sigma_{\mathfrak{D}}^{\mathsf{op}} \times \mathfrak{D}$ with $|\underline{c}| = n$, $(\sigma, \underline{f}) \in \overline{\mathsf{D}}(\underline{c}; d)$, and $\pi_n \in \Sigma_n$ the interval-reversing permutation in (8.3.12). Using Proposition 8.3.6, essentially the same reasoning as in Example 8.3.10 shows that $(\mathsf{As}^{\mathsf{D}}, j^{\mathsf{rev}})$ is a $\mathfrak{C}$-colored involutive operad in C. We will return to this involutive operad in Theorem 8.5.6, Example 8.6.8, and Definition 8.7.4.

As in Example 8.3.14, there is also a non-reversing version $(\mathsf{As}^{\mathsf{D}}, j)$ whose involutive structure j is defined as above, but without the interval-reversing permutation π_n. ◇

EXAMPLE 8.3.16 (Involutive Commutative Operad). Similar to the previous two examples, the commutative operad Com in Example 8.2.11 and the $\mathfrak{D}$-colored operad $\mathsf{Com}^{\mathsf{D}}$ in Example 8.2.14 have canonical involutive operad structures. In each case, the involution structure j is simply given by the morphism $I_0 : \mathbb{1} \longrightarrow I\mathbb{1}$, without changing the indexing subscript in the $\mathsf{Com}^{\mathsf{D}}$ case. We call the one-colored involutive operad (Com, j) in C the *involutive commutative operad.* ◇

More examples of involutive operads will be given in the coming sections.

8.4 Involutive Monads from Involutive Operads

In Lemma 8.2.2 we saw that each $\mathfrak{C}$-colored operad O in C induces a monad

$$\mathsf{O} \circ - : \mathsf{C}^{\mathfrak{C}} \longrightarrow \mathsf{C}^{\mathfrak{C}},$$

whose algebras are defined as O-algebras. The purpose of this section is to show that an involutive operad has an associated involutive monad as in Definition 4.8.1. The involutive algebras of this involutive monad will be defined as the involutive algebras of the involutive operad.

The involutive structure (I, ι) on C is extended to the product category $\mathsf{C}^{\mathfrak{C}}$ entrywise. To show that an involutive operad has an associated involutive monad, first we equip it with a distributive law.

DEFINITION 8.4.1. Fix a $\mathfrak{C}$-colored involutive operad $(\mathsf{O}, \gamma, \mathsf{1}, j)$ in C. Define a natural transformation

$$\mathsf{O} \circ I \xrightarrow{\ \nu\ } I(\mathsf{O} \circ -)$$

on $\mathsf{C}^{\mathfrak{C}}$ by the commutative diagram

(8.4.2)
$$\begin{array}{ccccc}
[\mathsf{O} \circ (IX)]_d & \xrightarrow{\ \nu_d^X\ } & I(\mathsf{O} \circ X)_d & = & I\Big[\int^{\underline{c}} \mathsf{O}\binom{d}{\underline{c}} \otimes X_{\underline{c}}\Big] \\
\Big\| & & & & \uparrow \cong \\
\int^{\underline{c}} \mathsf{O}\binom{d}{\underline{c}} \otimes (IX)_{\underline{c}} & & & & \int^{\underline{c}} I\Big[\mathsf{O}\binom{d}{\underline{c}} \otimes X_{\underline{c}}\Big] \\
\uparrow \omega_{\underline{c}} & & & & \uparrow \omega_{\underline{c}} \\
\mathsf{O}\binom{d}{\underline{c}} \otimes (IX)_{\underline{c}} & \xrightarrow[\cong]{(j, I_2)} & I\mathsf{O}\binom{d}{\underline{c}} \otimes I(X_{\underline{c}}) & \xrightarrow[\cong]{I_2} & I\Big[\mathsf{O}\binom{d}{\underline{c}} \otimes X_{\underline{c}}\Big]
\end{array}$$

in C for $X \in \mathsf{C}^{\mathfrak{C}}$ and $d \in \mathfrak{C}$.

LEMMA 8.4.3. *ν in Definition 8.4.1 is a well-defined natural transformation.*

PROOF. For each color $d \in \mathfrak{C}$, the morphism ν_d^X in C in (8.4.2) is well-defined by

- the commutation of I with colimits,
- the naturality of $I_2 : I(-) \otimes I(-) \longrightarrow I(- \otimes -)$, and
- that $j : \mathsf{O} \longrightarrow I\mathsf{O}$ is a morphism of $\mathfrak{C}$-colored symmetric sequences.

The naturality of ν with respect to X follows from the naturality of I_2. □

LEMMA 8.4.4. *$(\mathsf{O} \circ -, \nu) : \mathsf{C}^{\mathfrak{C}} \longrightarrow \mathsf{C}^{\mathfrak{C}}$ is an involutive functor.*

PROOF. It suffices to check the involutive functor axiom (2.2.2) for $X \in \mathsf{C}^{\mathfrak{C}}$ and $d \in \mathfrak{C}$. The diagram (2.2.2) boils down to the outer-most diagram below for $\underline{c} \in \Sigma_{\mathfrak{C}}$.

$$\begin{array}{ccccc}
\mathsf{O}\binom{d}{\underline{c}} \otimes X_{\underline{c}} & & \xrightarrow{\quad\iota\quad} & & I^2[\mathsf{O}\binom{d}{\underline{c}} \otimes X_{\underline{c}}] \\
\Big\downarrow{\scriptstyle (\mathrm{Id},\iota_*)} & \searrow{\scriptstyle (\iota,\iota)} & & & \Big\uparrow{\scriptstyle I(I_2)} \\
\mathsf{O}\binom{d}{\underline{c}} \otimes (I^2X)_{\underline{c}} & & I^2\mathsf{O}\binom{d}{\underline{c}} \otimes I^2(X_{\underline{c}}) & \xrightarrow{I_2} & I[I\mathsf{O}\binom{d}{\underline{c}} \otimes I(X_{\underline{c}})] \\
\Big\downarrow{\scriptstyle (j,I_2)} & \nearrow{\scriptstyle (Ij,I(I_2))} & & & \Big\uparrow{\scriptstyle I(j,I_2)} \\
I\mathsf{O}\binom{d}{\underline{c}} \otimes I(IX)_{\underline{c}} & & \xrightarrow{\quad I_2\quad} & & I[\mathsf{O}\binom{d}{\underline{c}} \otimes (IX)_{\underline{c}}]
\end{array}$$

The left triangle is commutative by the functoriality of $\otimes$, (4.1.5), and the involutive object axiom $Ij \circ j = \iota$ in (2.5.2). The top right and the bottom right trapezoids are commutative by (4.1.5) and the naturality of I_2, respectively. □

THEOREM 8.4.5. *For each $\mathfrak{C}$-colored involutive operad $(\mathsf{O}, \gamma, \mathbb{1}, j)$ in C, when equipped with the involutive structure ν in Definition 8.4.1, $(\mathsf{O} \circ -, \nu)$ is an involutive monad on $\mathsf{C}^{\mathfrak{C}}$.*

PROOF. $\mathsf{O} \circ -$ is a monad on $\mathsf{C}^{\mathfrak{C}}$ by Lemma 8.2.2, and $(\mathsf{O} \circ -, \nu)$ is an involutive functor by Lemma 8.4.4. It remains to check that the multiplication and the unit of the monad $\mathsf{O} \circ -$ are both involutive natural transformations in the sense of Proposition 4.8.2.

For $X \in \mathsf{C}^{\mathfrak{C}}$ and $d \in \mathfrak{C}$, the diagram (4.8.4) for $(\mathsf{O} \circ -, \nu)$ boils down to the outer-most diagram below.

$$\begin{array}{ccccc}
IX_d & \xrightarrow{I\lambda^{-1}} & I(\mathbb{1} \otimes X_d) & \xrightarrow{I(\mathbb{1}_d,\mathrm{Id})} & I(\mathsf{O}\binom{d}{d} \otimes X_d) \\
\Big\downarrow{\scriptstyle \lambda^{-1}} & & \Big\uparrow{\scriptstyle I_2} & & \Big\uparrow{\scriptstyle I_2} \\
\mathbb{1} \otimes IX_d & \xrightarrow{(I_0,\mathrm{Id})} & I\mathbb{1} \otimes IX_d & \xrightarrow{(I\mathbb{1}_d,\mathrm{Id})} & I\mathsf{O}\binom{d}{d} \otimes IX_d \\
\Big\downarrow{\scriptstyle (\mathbb{1}_d,\mathrm{Id})} & & & & \Big\uparrow{\scriptstyle (j,\mathrm{Id})} \\
\mathsf{O}\binom{d}{d} \otimes IX_d & & = = = = = & & \mathsf{O}\binom{d}{d} \otimes IX_d
\end{array}$$

The top left square, top right square, and the bottom rectangle are commutative by the left unity axiom in (4.1.4), the naturality of I_2, and the compatibility (8.3.7) of j with the colored units, respectively.

Using the notation in (8.2.6), the diagram (4.8.3) for $(\mathsf{O} \circ -, \nu)$ boils down to the outer-most diagram below.

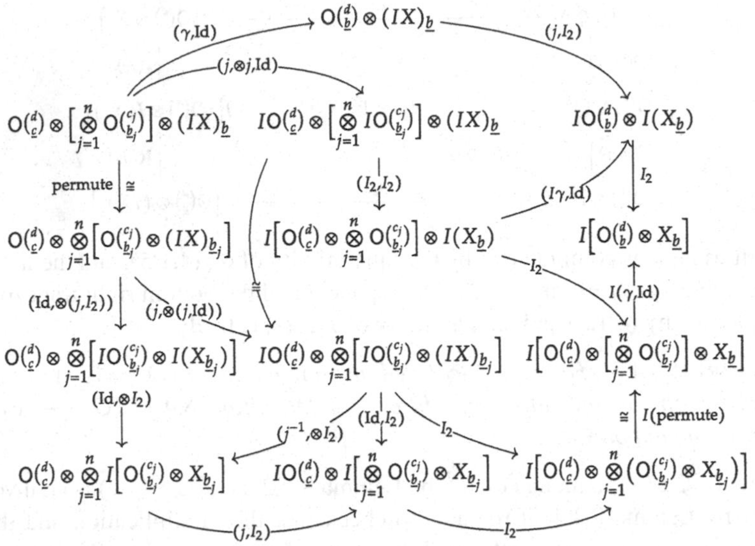

The top sub-diagram is commutative by (8.3.8). All other sub-diagrams are commutative because (i) C is a symmetric monoidal category, and (ii) (I, I_2, I_0) is a symmetric monoidal functor on C. □

8.5 Involutive Algebras of Involutive Operads

The first purpose of this section is to define involutive algebras of an involutive operad. Then we observe that reversing involutive monoids can be identified with the involutive algebras over the involutive associative operad, and similarly for the diagram case and the commutative case.

DEFINITION 8.5.1. For a $\mathfrak{C}$-colored involutive operad $(\mathsf{O}, \gamma, \mathsf{1}, j)$ in C, the category

$$\mathsf{IAlg}(\mathsf{O}, \gamma, \mathsf{1}, j)$$

of *involutive* $(\mathsf{O}, \gamma, \mathsf{1}, j)$*-algebras* is defined as the category of involutive $(\mathsf{O}\circ -, \nu)$-algebras, in the sense of Definition 4.8.7, for the involutive monad $(\mathsf{O} \circ -, \nu)$ in Theorem 8.4.5. If there is no danger of confusion, we will use the shorter notation $\mathsf{IAlg}(\mathsf{O})$, and call its objects *involutive* O*-algebras.*

Using Propositions 4.8.8 and 8.2.4, let us unwrap the previous definition.

PROPOSITION 8.5.2. *For a $\mathfrak{C}$-colored involutive operad $(\mathsf{O}, \gamma, \mathsf{1}, j)$ in C, an involutive* O*-algebra is exactly a tuple (X, θ, j^X) with*

- *(X, θ) an* O*-algebra and*

- (X, j^X) *an involutive object in* $\mathsf{C}^{\mathfrak{C}}$

such that the diagram

(8.5.3)
$$\begin{array}{ccccc}
\mathsf{O}\binom{d}{\underline{c}} \otimes X_{\underline{c}} & \xrightarrow{(\mathrm{Id}, j^X_*)} & \mathsf{O}\binom{d}{\underline{c}} \otimes (IX)_{\underline{c}} & \xrightarrow{(j, I_2)} & I\mathsf{O}\binom{d}{\underline{c}} \otimes I(X_{\underline{c}}) \\
\downarrow{\scriptstyle \theta} & & & & \downarrow{\scriptstyle I_2} \\
X_d & \xrightarrow{\quad j^X \quad} & IX_d & \xleftarrow{\quad I\theta \quad} & I\big[\mathsf{O}\binom{d}{\underline{c}} \otimes X_{\underline{c}}\big]
\end{array}$$

is commutative for $\binom{d}{\underline{c}} \in \Sigma_{\mathfrak{C}}^{\mathsf{op}} \times \mathfrak{C}$.

Moreover, a morphism of involutive O*-algebras is exactly a morphism of the underlying* O*-algebras that is also a morphism of involutive objects in* $\mathsf{C}^{\mathfrak{C}}$.

Another description of an involutive O-algebra in terms of the involutive endomorphism operad is given in Exercise 2 in Section 8.8. The rest of this section contains examples of involutive algebras of involutive operads.

Recall from Definition 4.7.2 that $\mathsf{RIMon(C)}$ is the category of reversing involutive monoids in C. Next we observe that reversing involutive monoids in C are exactly the involutive algebras of the involutive associative operad $(\mathsf{As}, j^{\mathsf{rev}})$ in Example 8.3.10.

THEOREM 8.5.4. *There is a canonical isomorphism*

$$\mathsf{RIMon}(C) \cong \mathsf{IAlg}(\mathsf{As}, j^{\mathsf{rev}})$$

between

- *the category of reversing involutive monoids in* C *and*
- *the category of involutive* $(\mathsf{As}, j^{\mathsf{rev}})$*-algebras.*

PROOF. By Proposition 8.5.2 for the involutive associative operad $(\mathsf{As}, j^{\mathsf{rev}})$, an involutive As-algebra is a tuple $(X, \mu, \mathsf{1}, j^X)$ with

- $(X, \mu, \mathsf{1})$ an As-algebra, i.e., a monoid in C, and
- $(X, j^X : X \longrightarrow IX)$ an involutive object in C

such that the diagram

(8.5.5)
$$\begin{array}{ccccc}
\mathsf{As}_n \otimes X^{\otimes n} & \xrightarrow{(\mathrm{Id}, (j^X)^{\otimes n})} & \mathsf{As}_n \otimes (IX)^{\otimes n} & \xrightarrow{(j^{\mathsf{rev}}, I_2)} & I\mathsf{As}_n \otimes I(X^{\otimes n}) \\
\downarrow{\scriptstyle \theta} & & & & \downarrow{\scriptstyle I_2} \\
X & \xrightarrow{\quad j^X \quad} & IX & \xleftarrow{\quad I\theta \quad} & I\big[\mathsf{As}_n \otimes X^{\otimes n}\big]
\end{array}$$

is commutative for $n \geq 0$. Using the description of a reversing involutive monoid in Proposition 4.7.3, it remains to show that the diagrams (8.5.5) for $n \geq 0$ are equivalent to the diagrams (4.5.6) and (4.7.4).

As the first reduction, observe that for each n, the diagram (8.5.5) is determined by its restriction to the coproduct summand

$$\mathbb{1}_{\mathrm{id}_n} \otimes X^{\otimes n} \longrightarrow \coprod_{\sigma \in \Sigma_n} \left(\mathbb{1}_\sigma \otimes X^{\otimes n} \right) \cong \mathsf{As}_n \otimes X^{\otimes n}$$

because of

- the equivariance axiom (8.2.8) of the O-algebra X,
- the naturality of I_2, and
- the symmetry isomorphism in C.

In what follows, when we refer to the diagram (8.5.5), we mean its restriction to the coproduct summand above.

Moreover:

- Since $\mathsf{As}_0 = \mathbb{1} = X^{\otimes 0}$, the diagram (8.5.5) for $n = 0$ is the unit diagram (4.5.6).
- The diagram (8.5.5) for $n = 1$ factors as follows.

$$
\begin{array}{ccccc}
1 \otimes X & \xrightarrow{(I_0, j^X)} & I1 \otimes IX & \xrightarrow{I_2} & I(1 \otimes X) \\
\Big\| & & \uparrow{\scriptstyle (I_0, \mathrm{Id})} & & \Big\downarrow{\scriptstyle I\lambda} \\
1 \otimes X & \xrightarrow{(\mathrm{Id}, j^X)} & 1 \otimes IX & \xrightarrow{\lambda} & IX \\
\Big\downarrow{\scriptstyle \lambda} & & & & \Big\| \\
X & \xrightarrow{\hspace{3cm} j^X \hspace{3cm}} & & & IX
\end{array}
$$

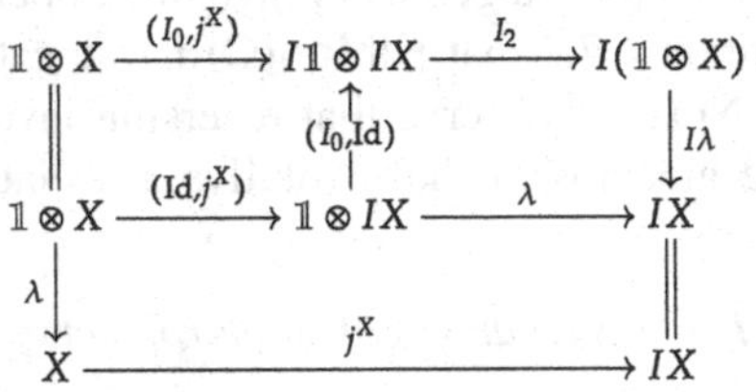

The three sub-diagrams are commutative by the functoriality of $\otimes$, the naturality of λ, and that (I, I_2, I_0) is a monoidal functor (1.2.13). So this imposes no extra conditions on $(X, \mu, 1, j^X)$.

- The diagram (8.5.5) for $n = 2$ reduces to the diagram (4.7.4), in which the symmetry isomorphism ξ arises because the involutive structure on As_2 is the composite

$$
\begin{array}{ccc}
\mathsf{As}_2 & \xrightarrow{\hspace{2cm} j^{\mathrm{rev}} \hspace{2cm}} & I\mathsf{As}_2 \\
\Big\| & & \uparrow{\scriptstyle \cong} \\
\mathbb{1} \sqcup \mathbb{1} & \xrightarrow{\pi_2} \mathbb{1} \sqcup \mathbb{1} \xrightarrow{(I_0, I_0)} & I\mathbb{1} \sqcup I\mathbb{1}.
\end{array}
$$

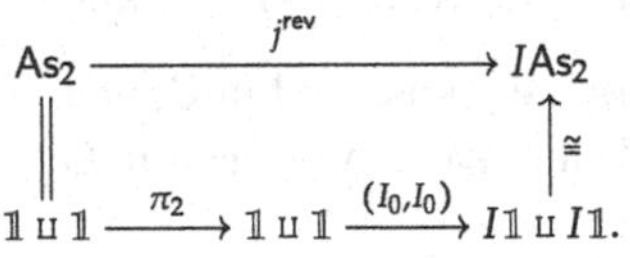

By Proposition 4.7.3, we have shown that $(X, \mu, 1, j^X)$ has the structure of a reversing involutive monoid in C.

It remains to check that the diagrams (8.5.5) for $n \geq 3$ impose no extra conditions on the reversing involutive monoid $(X, \mu, 1, j^X)$. Indeed, for $n \geq 3$, the diagram (8.5.5) reduces to the outer-most diagram below, where j^X is abbreviated to j.

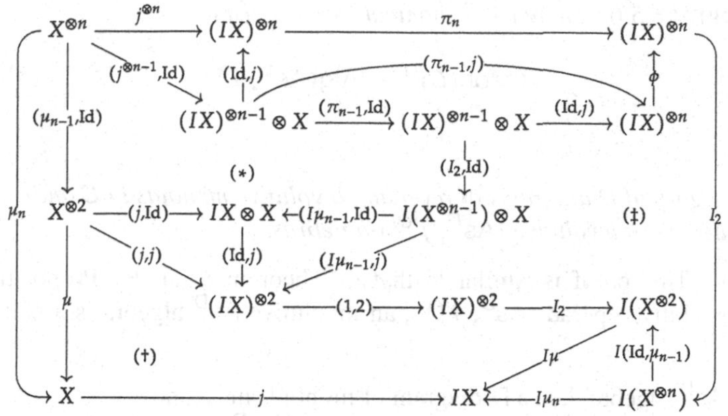

In the above diagram:

- $\mu_n : X^{\otimes n} \longrightarrow X$ is the iterated multiplication on X.
- The top right vertical isomorphism is the permutation of tensor factors induced by the block permutation

$$\phi = (1, 2)\langle n - 1, 1\rangle \in \Sigma_n$$

  that swaps the initial interval of length $n - 1$ with the last entry. See Definition 6.1.1 for the notation. The sub-diagram involving π_n is commutative by the permutation identity

$$\pi_n = \phi \cdot \big(\pi_{n-1} \times \mathrm{id}_1\big) \in \Sigma_n.$$

- The sub-diagram $(*)$ is the diagram (8.5.5) for $n - 1$ tensored with X.
- The sub-diagram $(\dagger)$ is the diagram (8.5.5) for $n = 2$, which is equivalent to the diagram (4.7.4).
- The sub-diagram $(\ddagger)$ is commutative by the assumption that (I, I_2, I_0) is a symmetric monoidal functor.
- All other sub-diagrams are commutative by the functoriality of $\otimes$ and the associativity of μ.

Therefore, by an induction, the diagrams (8.5.5) for $n \geq 3$ are all determined by the case $n = 2$.

We have shown that an involutive $(\mathsf{As}, j^{\mathrm{rev}})$-algebra determines exactly a reversing involutive monoid in C. □

Suppose D is a small category with object set $\mathfrak{D}$, and As^{D} is the $\mathfrak{D}$-colored operad in C in Example 8.2.12 whose algebras are D-diagrams of monoids in C. It is equipped with the involutive operad structure j^{rev} in Example 8.3.15. Next is the D-diagram version of Theorem 8.5.4.

THEOREM 8.5.6. *There is a canonical isomorphism*

$$\mathsf{RIMon}(C)^D \cong \mathsf{IAlg}(\mathsf{As}^D, j^{\mathsf{rev}})$$

between

- *the category of* D*-diagrams of reversing involutive monoids in* C *and*
- *the category of involutive* $(\mathsf{As}^D, j^{\mathsf{rev}})$*-algebras.*

PROOF. This proof is similar to that of Theorem 8.5.4. By Proposition 8.5.2 for the involutive operad $(\mathsf{As}^{\mathsf{D}}, j^{\mathsf{rev}})$, an involutive As^{D}-algebra is a pair (X, j^X) with

- X an As^{D}-algebra, i.e., a D-diagram of monoids in C, and
- $(X, j^X : X \longrightarrow IX)$ an involutive object in $\mathsf{C}^{\mathfrak{D}}$

such that the diagram (8.5.3) is commutative for $\mathsf{O} = \mathsf{As}^{\mathsf{D}}$ and $\binom{d}{\underline{c}} \in \Sigma_{\mathfrak{D}}^{\mathsf{op}} \times \mathfrak{D}$. As in the proof of the previous Theorem, these diagrams are determined by:

(i) the $\binom{d}{\varnothing}$ case;
(ii) the coproduct summand $\mathbb{1}_{(\mathrm{id}_1, f)}$ of As^{D} for f a morphism in D;
(iii) the coproduct summand $\mathbb{1}_{(\mathrm{id}_2, (\mathrm{Id}_d, \mathrm{Id}_d))}$ of As^{D} for $d \in \mathfrak{D}$.

Cases (i) and (iii) are equivalent to (4.5.6) and (4.7.4) for the monoid X_d. So each X_d is a reversing involutive monoid by Proposition 4.7.3. Case (ii) means that, for each morphism $f : c \longrightarrow d$ in D, the morphism

$$X_f : X_c \longrightarrow X_d$$

of monoids in C is also a morphism of involutive objects in C. Therefore, X is actually a D-diagram of reversing involutive monoids in D. □

Recall from Example 8.3.15 the $\mathfrak{D}$-colored involutive operad $(\mathsf{As}^{\mathsf{D}}, j)$, whose involutive structure j does not involve the interval-reversing permutation π_n.

PROPOSITION 8.5.7. *There is a canonical isomorphism*

$$\mathsf{IMon}(C)^D \cong \mathsf{IAlg}(\mathsf{As}^D, j)$$

between

- *the category of* D*-diagrams of involutive monoids in* C *and*
- *the category of involutive* (As^D, j)*-algebras.*

PROOF. We reuse the proof of Theorem 8.5.6 and replace Proposition 4.7.3 by the description of involutive monoids in Proposition 4.5.5. □

Recall from Example 8.3.16 the involutive commutative operad (Com, j) and more generally its D-diagram version $(\mathsf{Com}^{\mathsf{D}}, j)$. Also recall from Definition 4.6.2

the category $\mathsf{ICMon}(\mathsf{C})$ of involutive commutative monoids in C. The following commutative analogue of Theorem 8.5.6 is proved by essentially the same argument.

PROPOSITION 8.5.8. *For each small category D, there is a canonical isomorphism*

$$\mathsf{ICMon}(C)^D \cong \mathsf{IAlg}(\mathsf{Com}^D, j)$$

between

- *the category of D-diagrams of involutive commutative monoids in C and*
- *the category of involutive* (Com^D, j)*-algebras.*

8.6 Change of Involutive Symmetric Monoidal Categories

For an involutive symmetric monoidal category C, the category $\mathsf{SSeq}_{\mathsf{C}}^{\mathfrak{C}}$ of $\mathfrak{C}$-colored symmetric sequences in C is an involutive monoidal category by Proposition 8.3.2. Its involutive monoids are defined as $\mathfrak{C}$-colored involutive operads in C. The purpose of this section is to observe that an involutive symmetric monoidal functor induces an involutive functor between the categories of $\mathfrak{C}$-colored involutive operads. As we will see later, this procedure produces many interesting examples of involutive operads. First we make the following preliminary observation.

PROPOSITION 8.6.1. *Suppose* $(F, F_2, F_0, \nu) : C \longrightarrow D$ *is an involutive symmetric monoidal functor between involutive symmetric monoidal categories. Then it induces an involutive monoidal functor*

$$\mathsf{SSeq}_C^{\mathfrak{C}} \xrightarrow{\;F'\;} \mathsf{SSeq}_D^{\mathfrak{C}}$$

in which $\mathsf{SSeq}_C^{\mathfrak{C}}$ *and* $\mathsf{SSeq}_D^{\mathfrak{C}}$ *have the involutive monoidal structures in Proposition 8.3.2.*

PROOF. The induced functor F' is defined entrywise as

$$(F'X)\tbinom{d}{\underline{c}} = F\big(X\tbinom{d}{\underline{c}}\big) \in \mathsf{D}$$

for $X \in \mathsf{SSeq}_{\mathsf{C}}^{\mathfrak{C}}$ and $\binom{d}{\underline{c}} \in \Sigma_{\mathfrak{C}}^{\mathsf{op}} \times \mathfrak{C}$, and similarly for morphisms in $\mathsf{SSeq}_{\mathsf{C}}^{\mathfrak{C}}$. Since

$$\mathsf{SSeq}_{\mathsf{C}}^{\mathfrak{C}} = \mathsf{C}^{\Sigma_{\mathfrak{C}}^{\mathsf{op}} \times \mathfrak{C}} \quad \text{and} \quad \mathsf{SSeq}_{\mathsf{D}}^{\mathfrak{C}} = \mathsf{D}^{\Sigma_{\mathfrak{C}}^{\mathsf{op}} \times \mathfrak{C}}$$

are diagram categories, F' is an involutive functor by Proposition 2.2.10, whose distributive law $\nu^{F'}$ is inherited from that of F.

Next we show that F' inherits a monoidal functor structure from F. The structure morphism

$$F'_0 : \mathsf{I}_{\mathsf{D}} \longrightarrow F'\mathsf{I}_{\mathsf{C}}$$

is the morphism $F_0 : \mathbb{1}^{\mathsf{D}} \longrightarrow F\mathbb{1}^{\mathsf{C}}$ at the $\binom{c}{c}$-entries for $c \in \mathfrak{C}$. Here I_{C} and I_{D} are the monoidal units in $\mathsf{SSeq}_{\mathsf{C}}^{\mathfrak{C}}$ and $\mathsf{SSeq}_{\mathsf{D}}^{\mathfrak{C}}$ as defined in (8.1.9), and $\mathbb{1}^{\mathsf{C}}$ and $\mathbb{1}^{\mathsf{D}}$ are the monoidal units in C and D, respectively. At all other entries, F'_0 is the unique morphism from an initial object in C.

To define the structure morphism F'_2, suppose $X, Y \in \mathsf{SSeq}_{\mathsf{C}}^{\mathfrak{C}}$. First observe that, for $\underline{b} \in \Sigma_{\mathfrak{C}}^{\mathsf{op}}$ and $\underline{c} = (c_1, \ldots, c_m) \in \Sigma_{\mathfrak{C}}$, there is a natural morphism F_* defined as the composite

$$\begin{array}{ccc}
(F'Y)^{\underline{c}}(\underline{b}) & \xrightarrow{\quad F_* \quad} & F\big(Y^{\underline{c}}(\underline{b})\big) \\
\| & & \| \\
\displaystyle\int^{\{\underline{a}_j\}_{j=1}^m} \Sigma_{\mathfrak{C}}^{\mathsf{op}}(\underline{a};\underline{b}) \cdot \bigotimes_{j=1}^m FY\tbinom{c_j}{\underline{a}_j} & & F\Big[\displaystyle\int^{\{\underline{a}_j\}_{j=1}^m} \Sigma_{\mathfrak{C}}^{\mathsf{op}}(\underline{a};\underline{b}) \cdot \bigotimes_{j=1}^m Y\tbinom{c_j}{\underline{a}_j}\Big] \\
{\scriptstyle (F_2)_*}\big\downarrow & & \big\uparrow \\
& \displaystyle\int^{\{\underline{a}_j\}_{j=1}^m} \Sigma_{\mathfrak{C}}^{\mathsf{op}}(\underline{a};\underline{b}) \cdot F\Big[\bigotimes_{j=1}^m Y\tbinom{c_j}{\underline{a}_j}\Big] &
\end{array}$$

in D with $\underline{a} = (\underline{a}_1, \ldots, \underline{a}_m) \in \Sigma_{\mathfrak{C}}^{\mathsf{op}}$ the concatenation. Here $(F_2)_*$ is induced by the natural transformation $F_2 : F(-) \otimes F(-) \longrightarrow F(- \otimes -)$. The lower right natural morphism exists by the functoriality of F and the universal properties of coproducts and coends. Note that F_* is natural with respect to $\underline{c} \in \Sigma_{\mathfrak{C}}$ because F is a symmetric monoidal functor.

Using the natural morphism F_*, we define F'_2 entrywise as the composite

$$\begin{array}{ccc}
(F'X \circ F'Y)\tbinom{d}{\underline{b}} & \xrightarrow{\quad F'_2 \quad} & [F'(X \circ Y)]\tbinom{d}{\underline{b}} \\
\| & & \| \\
\displaystyle\int^{\underline{c}} FX\tbinom{d}{\underline{c}} \otimes (F'Y)^{\underline{c}}(\underline{b}) & & F\Big[\displaystyle\int^{\underline{c}} X\tbinom{d}{\underline{c}} \otimes Y^{\underline{c}}(\underline{b})\Big] \\
{\scriptstyle \int^{\underline{c}} \mathrm{Id} \otimes F_*}\big\downarrow & & \big\uparrow \\
\displaystyle\int^{\underline{c}} FX\tbinom{d}{\underline{c}} \otimes F\big[Y^{\underline{c}}(\underline{b})\big] & \xrightarrow{\quad (F_2)_* \quad} & \displaystyle\int^{\underline{c}} F\big[X\tbinom{d}{\underline{c}} \otimes Y^{\underline{c}}(\underline{b})\big]
\end{array}$$

in D for $\binom{d}{\underline{b}} \in \Sigma_{\mathfrak{C}}^{\mathsf{op}} \times \mathfrak{C}$. Since F'_2 is entrywise defined by natural morphisms and the natural transformation F_2, it defines a natural transformation

$$F'_2 : F'(-) \circ F'(-) \longrightarrow F'(- \circ -).$$

By naturality of the constructions, the monoidal functor axioms for (F', F'_2, F'_0) boil down to the symmetric monoidal functor axioms for (F, F_2, F_0).

Using Lemma 4.3.2, it remains to check the commutativity of the diagrams (4.3.3) and (4.3.4) for $(F', F'_2, F'_0, \nu^{F'})$. The diagram (4.3.3) in this case is

$$\begin{array}{ccccc}
F'IX \circ F'IY & \xrightarrow{F'_2} & F'(IX \circ IY) & \xrightarrow{F'(I_2)} & F'I(X \circ Y) \\
\Big\downarrow{\scriptstyle (\nu^{F'}_X, \nu^{F'}_Y)} & & & & \Big\downarrow{\scriptstyle \nu^{F'}_{X \circ Y}} \\
IF'X \circ IF'Y & \xrightarrow{I_2} & I(F'X \circ F'Y) & \xrightarrow{I(F'_2)} & IF'(X \circ Y)
\end{array}$$

for $X, Y \in \mathsf{SSeq}^{\mathfrak{C}}_{\mathsf{C}}$. By the naturality of the symmetric monoidal functors F and I in both C and D, the above diagram boils down to the diagram

$$\begin{array}{ccccc}
FIX\binom{d}{\underline{c}} \otimes (F'IY)^{\underline{c}}(\underline{b}) & \xrightarrow{(\mathrm{Id},F_*)} & FIX\binom{d}{\underline{c}} \otimes F(IY)^{\underline{c}}(\underline{b}) & \xrightarrow{F_2} & F\big[IX\binom{d}{\underline{c}} \otimes (IY)^{\underline{c}}(\underline{b})\big] \\
\Big\downarrow{\scriptstyle (\nu,\nu_*)} & & \Big\downarrow{\scriptstyle (\mathrm{Id},F(I_*))} & & \Big\downarrow{\scriptstyle F(\mathrm{Id},I_*)} \\
IFX\binom{d}{\underline{c}} \otimes (IF'Y)^{\underline{c}}(\underline{b}) & & FIX\binom{d}{\underline{c}} \otimes FI\big[Y^{\underline{c}}(\underline{b})\big] & \xrightarrow{F_2} & F\big[IX\binom{d}{\underline{c}} \otimes I\big[Y^{\underline{c}}(\underline{b})\big]\big] \\
\Big\downarrow{\scriptstyle (\mathrm{Id},I_*)} & & \Big\downarrow{\scriptstyle (\nu,\nu)} & & \Big\downarrow{\scriptstyle F(I_2)} \\
IFX\binom{d}{\underline{c}} \otimes I(F'Y)^{\underline{c}}(\underline{b}) & \xrightarrow{(\mathrm{Id},I(F_*))} & IFX\binom{d}{\underline{c}} \otimes IF\big[Y^{\underline{c}}(\underline{b})\big] & & FI\big[X\binom{d}{\underline{c}} \otimes Y^{\underline{c}}(\underline{b})\big] \\
\Big\downarrow{\scriptstyle I_2} & & \Big\downarrow{\scriptstyle I_2} & & \Big\downarrow{\scriptstyle \nu} \\
I\big[FX\binom{d}{\underline{c}} \otimes (F'Y)^{\underline{c}}(\underline{b})\big] & \xrightarrow{I(\mathrm{Id},F_*)} & I\big[FX\binom{d}{\underline{c}} \otimes F\big[Y^{\underline{c}}(\underline{b})\big]\big] & \xrightarrow{I(F_2)} & IF\big[X\binom{d}{\underline{c}} \otimes Y^{\underline{c}}(\underline{b})\big]
\end{array}$$

in D for $d \in \mathfrak{C}$ and $\underline{b}, \underline{c} \in \Sigma_{\mathfrak{C}}$. The upper left and the lower right rectangles are commutative by (4.3.3) for (F, F_2, F_0, ν). The lower left and the upper right rectangles are commutative by the naturality of I_2 and F_2, respectively. A similar argument shows that the diagram (4.3.4) for $(F', F'_2, F'_0, \nu^{F'})$ is commutative. □

THEOREM 8.6.2. *Every involutive symmetric monoidal functor*

$$(F, F_2, F_0, \nu) : \mathsf{C} \longrightarrow \mathsf{D}$$

between involutive symmetric monoidal categories induces an involutive functor

$$\mathsf{IOp}^{\mathfrak{C}}_{\mathsf{C}} \xrightarrow{\;\mathsf{IOp}(F)\;} \mathsf{IOp}^{\mathfrak{C}}_{\mathsf{D}} .$$

PROOF. We apply Proposition 4.5.10 to the involutive monoidal functor

$$\mathsf{SSeq}_{\mathsf{C}}^{\mathfrak{C}} \xrightarrow{F'} \mathsf{SSeq}_{\mathsf{D}}^{\mathfrak{C}}$$

in Proposition 8.6.1 to obtain an involutive functor

$$\mathsf{IOp}_{\mathsf{C}}^{\mathfrak{C}} = \mathsf{IMon}(\mathsf{SSeq}_{\mathsf{C}}^{\mathfrak{C}}) \xrightarrow{\mathsf{IOp}(F)} \mathsf{IMon}(\mathsf{SSeq}_{\mathsf{D}}^{\mathfrak{C}}) = \mathsf{IOp}_{\mathsf{D}}^{\mathfrak{C}} ,$$

in which the equalities are the definitions of $\mathsf{IOp}_{\mathsf{C}}^{\mathfrak{C}}$ and $\mathsf{IOp}_{\mathsf{D}}^{\mathfrak{C}}$. □

EXPLANATION 8.6.3. In the context of Theorem 8.6.2, for a $\mathfrak{C}$-colored involutive operad $(\mathsf{O}, \gamma, \mathsf{1}, j)$ in C, its image under $\mathsf{IOp}(F)$ is the $\mathfrak{C}$-colored involutive operad in D with entries

$$(F\mathsf{O})\binom{d}{\underline{c}} = F\left(\mathsf{O}\binom{d}{\underline{c}}\right) \quad \text{for} \quad \binom{d}{\underline{c}} \in \Sigma_{\mathfrak{C}}^{\mathsf{op}} \times \mathfrak{C}.$$

For $\sigma \in \Sigma_{|\underline{c}|}$, its σ-action is

$$F\mathsf{O}\binom{d}{\underline{c}} \xrightarrow{F\sigma} F\mathsf{O}\binom{d}{\underline{c}\sigma} .$$

For $c \in \mathfrak{C}$, its c-colored unit is the composite

$$\mathbb{1}^{\mathsf{D}} \xrightarrow{F_0} F\mathbb{1}^{\mathsf{C}} \xrightarrow{F\mathsf{1}_c} F\mathsf{O}\binom{c}{c} .$$

Its operadic composition is the composite

$$F\mathsf{O}\binom{d}{\underline{c}} \otimes \bigotimes_{j=1}^{n} F\mathsf{O}\binom{c_j}{\underline{b}_j} \xrightarrow{F_2} F\left[\mathsf{O}\binom{d}{\underline{c}} \otimes \bigotimes_{j=1}^{n} \mathsf{O}\binom{c_j}{\underline{b}_j}\right] \xrightarrow{F\gamma} F\mathsf{O}\binom{d}{\underline{b}}$$

for $\binom{d}{\underline{c}} \in \Sigma_{\mathfrak{C}}^{\mathsf{op}} \times \mathfrak{C}$ with $|\underline{c}| = n \geq 1$, $\underline{b}_j \in \mathsf{Prof}(\mathfrak{C})$, and $\underline{b} = (\underline{b}_1, \ldots, \underline{b}_n)$. Its involutive structure is the composite

$$F\mathsf{O}\binom{d}{\underline{c}} \xrightarrow{Fj} FI_{\mathsf{C}}\mathsf{O}\binom{d}{\underline{c}} \xrightarrow{\upsilon} I_{\mathsf{D}}F\mathsf{O}\binom{d}{\underline{c}}$$

for each $\binom{d}{\underline{c}} \in \Sigma_{\mathfrak{C}}^{\mathsf{op}} \times \mathfrak{C}$. ◇

As an illustration of Theorem 8.6.2, we consider the following functor.

DEFINITION 8.6.4. Suppose $(\mathsf{C}, \otimes, \mathbb{1})$ is a monoidal category. Define the functor

$$\mathsf{Set} \xrightarrow{\Phi} \mathsf{C} \quad \text{by} \quad \Phi(X) = \coprod_X \mathbb{1} \tag{8.6.5}$$

for $X \in \mathsf{Set}$.

We regard $(\mathsf{Set}, \times, *)$ as an involutive monoidal category with the trivial involutive structure $(\mathrm{Id}_{\mathsf{Set}}, \mathrm{Id}_{\mathrm{Id}_{\mathsf{Set}}})$.

LEMMA 8.6.6. *For each involutive symmetric monoidal category C, the functor* $\Phi : \mathsf{Set} \longrightarrow \mathsf{C}$ *is canonically an involutive strong symmetric monoidal functor.*

PROOF. The monoidal functor structure morphism

$$\Phi_0 : \mathbb{1} \longrightarrow \Phi(*) = \mathbb{1}$$

is $\mathrm{Id}_{\mathbb{1}}$. The other structure morphism is given by the composite

$$\begin{array}{ccc} \Phi(X) \otimes \Phi(Y) & \xrightarrow{\Phi_2} & \Phi(X \times Y) \\ \Vert & & \uparrow{\scriptstyle \amalg \lambda_1} \\ \left(\coprod_X \mathbb{1}\right) \otimes \left(\coprod_Y \mathbb{1}\right) & \xrightarrow{\cong} & \coprod_{X \times Y} \mathbb{1} \otimes \mathbb{1} \end{array}$$

in C for $X, Y \in \mathsf{Set}$. The bottom horizontal isomorphism follows from the assumption that the monoidal product in C commutes with colimits, in particular coproducts, on both sides. The symmetric monoidal category axioms in C imply that (Φ, Φ_2, Φ_0) is a strong symmetric monoidal functor.

We equip Φ with the distributive law ν^{Φ} defined as the composite

$$\begin{array}{ccccc} \Phi \mathrm{Id}_{\mathsf{Set}}(X) & & \xrightarrow{\nu^{\Phi}_X} & & I_{\mathsf{C}}\Phi(X) \\ \Vert & & & & \Vert \\ \coprod_X \mathbb{1} & \xrightarrow{\amalg I_0} & \coprod_X I\mathbb{1} & \xrightarrow{\cong} & I\left[\coprod_X \mathbb{1}\right] \end{array}$$

for $X \in \mathsf{Set}$. The bottom right horizontal isomorphism uses Proposition 2.1.4(2). To check that this makes Φ into an involutive strong symmetric monoidal functor, we use Lemma 4.3.2. The diagram (4.3.4) for Φ is commutative because both composites are equal to I_0.

The diagram (4.3.3) for Φ boils down to the outer-most diagram below.

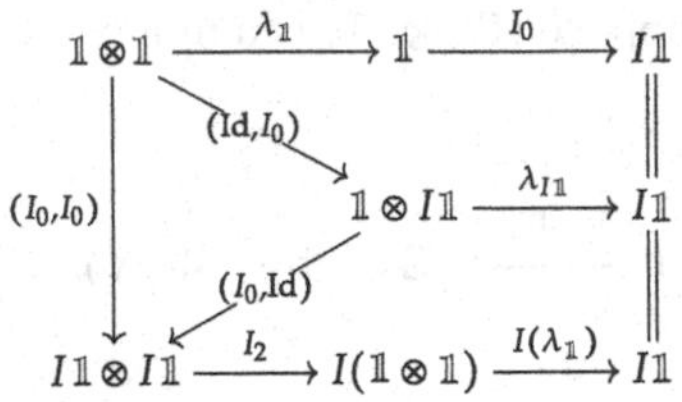

The lower right trapezoid is commutative by the left unity diagram (4.1.4) for I. The upper right trapezoid is commutative by the naturality of λ. The left triangle is commutative by definition. □

COROLLARY 8.6.7. *For each involutive symmetric monoidal category* C*, the involutive strong symmetric monoidal functor* $\Phi : \mathsf{Set} \longrightarrow \mathsf{C}$ *induces an involutive functor*

$$\mathsf{IOp}^{\mathfrak{C}}_{\mathsf{Set}} \xrightarrow{\ \mathsf{IOp}(\Phi)\ } \mathsf{IOp}^{\mathfrak{C}}_{\mathsf{C}} .$$

PROOF. We apply Theorem 8.6.2 to Φ in Lemma 8.6.6. □

EXAMPLE 8.6.8 (Involutive Associative and Commutative Operads). By Explanation 8.6.3, the following involutive operads are all in the image of the involutive functor $\mathsf{IOp}(\Phi)$ in Corollary 8.6.7:

- The involutive associative operad $(\mathsf{As}, j^{\mathsf{rev}})$ in Example 8.3.10.
- The non-reversing involutive associative operad (As, j) in Example 8.3.14.
- The D-diagram versions $(\mathsf{As}^{\mathsf{D}}, j^{\mathsf{rev}})$ and $(\mathsf{As}^{\mathsf{D}}, j)$ in Example 8.3.15.
- The commutative versions (Com, j) and $(\mathsf{Com}^{\mathsf{D}}, j)$ in Example 8.3.16.

The color set $\mathfrak{C}$ is

- $\{*\}$ for $(\mathsf{As}, j^{\mathsf{rev}})$, (As, j), and (Com, j);
- $\mathsf{Ob}(\mathsf{D})$ for $(\mathsf{As}^{\mathsf{D}}, j^{\mathsf{rev}})$, $(\mathsf{As}^{\mathsf{D}}, j)$, and $(\mathsf{Com}^{\mathsf{D}}, j)$.

In other words, they all come from their corresponding involutive operads in Set under the involutive strong symmetric monoidal functor $\Phi : \mathsf{Set} \longrightarrow \mathsf{C}$. ◇

The next section contains more examples of Corollary 8.6.7.

8.7 AQFT Involutive Operad

The purpose of this section is to discuss an involutive operadic approach to algebraic quantum field theory (AQFT) from [BSW∞, BSW19a, BSW19b]. We observe that the AQFT involutive operads arise from the involutive functor in Corollary 8.6.7. An algebraic quantum field theory is a functor that goes from a category of geometric objects to a category of algebraic objects. The following definition is a way to formalize the idea of disjoint subsets in geometric objects.

DEFINITION 8.7.1. Suppose D is a small category with object set $\mathfrak{D}$. An *orthogonality relation* on D is a set T of pairs (f_1, f_2), called an *orthogonal pair*, of morphisms in D that satisfy the following conditions:

Common Codomain: f_1 and f_2 have the same codomain.
Symmetry: If $(f_1, f_2) \in T$, then $(f_2, f_1) \in T$.
Closure: If $(f_1, f_2) \in T$ and if h_1, h_2, g are morphisms in C, then $(gf_1, gf_2) \in T$ and $(f_1h_1, f_2h_2) \in T$ as long as the compositions are defined.

If T is an orthogonality relation on D, then (D, T) is called an *orthogonal category*.

EXAMPLE 8.7.2 (Lorentzian Manifolds). Suppose Loc is a small category that is equivalent to the category of oriented, time-oriented, and globally hyperbolic Lorentzian manifolds of a fixed dimension $d \geq 2$. We always assume that our manifolds are Hausdorff and second-countable. The reader is referred to [BGP07] for an introduction to Lorentzian manifolds.

Suppose C is the subset of Cauchy morphisms in Loc, and

$$\mathsf{Loc}' = \mathsf{Loc}[C^{-1}]$$

is the category of fractions, in which the morphisms in C are formally inverted. Since Loc is a small category, so is Loc'. Part of the construction of the category of fractions is a functor

$$L : \mathsf{Loc} \longrightarrow \mathsf{Loc}'$$

that satisfies the following universal property. For each functor $F : \mathsf{Loc} \longrightarrow \mathsf{D}$ such that $F(g)$ is an invertible morphism in D for each $g \in C$, there exists a unique functor $F' : \mathsf{Loc}' \longrightarrow \mathsf{D}$ such that the diagram

$$\begin{array}{ccc} \mathsf{Loc} & \xrightarrow{L} & \mathsf{Loc}' \\ {\scriptstyle F} \downarrow & \swarrow {\scriptstyle F'} & \\ \mathsf{D} & & \end{array}$$

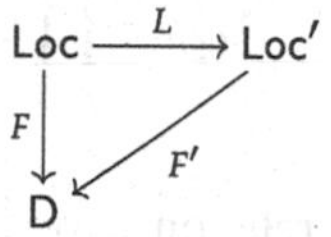

is commutative. For an object or a morphism in Loc, its image under the functor L is denoted by the same symbol. For discussion of category of fractions, see [Bor94a] (Section 5.2) or the original source [GZ67] (Chapter 1).

Define T as the set of pairs of morphisms in Loc' of the form

$$(gf_1h_1, gf_2h_2)$$

in which:

- $f_1 : X \longrightarrow Z$ and $f_2 : Y \longrightarrow Z$ are morphisms in Loc such that the images $f_1(X)$ and $f_2(Y)$ are causally disjoint subsets in Z.
- g, h_1, h_2 are morphisms in Loc' such that the compositions gf_1h_1 and gf_2h_2 are defined.

This defines an orthogonality relation T on Loc'. ◇

EXAMPLE 8.7.3 (Oriented and Riemannian Manifolds). For a fixed integer $d \geq 1$, suppose Man is a small category that is equivalent to the category of d-dimensional oriented manifolds, with orientation-preserving open embeddings as morphisms. See [ONe83] for discussion of manifolds. Define T as the set of pairs

$$(f_1 : X \longrightarrow Z, f_2 : Y \longrightarrow Z)$$

of morphisms in Man such that the images of f_1 and f_2 are disjoint in Z. This defines an orthogonality relation on Man.

Similarly, suppose Riem is a small category that is equivalent to the category of d-dimensional oriented, Riemannian manifolds, with orientation-preserving isometric open embeddings as morphisms. Define T as the set of pairs (f_1, f_2) of morphisms with a common codomain and disjoint images. This defines an orthogonality relation on Riem. ◇

We now associate an involutive operad in Set to each orthogonal category.

DEFINITION 8.7.4. Suppose (D, T) is an orthogonal category with $\mathsf{Ob}(\mathsf{D}) = \mathfrak{D}$. Define the following structures.

Entries: For each $\binom{d}{\underline{c}} \in \mathsf{Prof}(\mathfrak{D}) \times \mathfrak{D}$, define the quotient set

$$\mathsf{O}_{(\mathsf{D},T)}\tbinom{d}{\underline{c}} = \overline{\mathsf{D}}\tbinom{d}{\underline{c}} \Big/ \sim$$

with

$$\overline{\mathsf{D}}\tbinom{d}{\underline{c}} = \Sigma_{|\underline{c}|} \times \prod_{j=1}^{|\underline{c}|} \mathsf{D}(c_j, d)$$

as in (8.2.13). The equivalence relation $\sim$ on $\overline{\mathsf{D}}\binom{d}{\underline{c}}$ is defined as

$$(\sigma, \underline{f}) \sim (\sigma', \underline{f}')$$

if and only if the following two conditions hold.

- $\underline{f} = \underline{f}'$.
- The permutation

$$\sigma\underline{f} \xrightarrow{\sigma'\sigma^{-1}} \sigma'\underline{f}$$

is generated by transpositions of adjacent orthogonal pairs. This means that the permutation $\sigma'\sigma^{-1} \in \Sigma_{|\underline{c}|}$ factors into a product $\tau_k \cdots \tau_1$ of transpositions such that, for each $1 \le i \le k$, the permutation

$$\tau_{i-1} \cdots \tau_1 \sigma \underline{f} \xrightarrow{\tau_i} \tau_i \tau_{i-1} \cdots \tau_1 \sigma \underline{f}$$

is a transposition of an adjacent orthogonal pair in the sequence $\tau_{i-1} \cdots \tau_1 \sigma \underline{f}$.

Other Structure: The right $\Sigma_{|\underline{c}|}$-action, colored units, operadic composition, and involutive structure on $\mathsf{O}_{(\mathsf{D},T)}$ are inherited from those in $(\mathsf{As}^{\mathsf{D}}, j^{\mathsf{rev}})$ in Set, as in Examples 8.2.12 and 8.3.15.

This finishes the definition of $\mathsf{O}_{(\mathsf{D},T)}$.

LEMMA 8.7.5. *$\mathsf{O}_{(\mathsf{D},T)}$ in Definition 8.7.4 is a $\mathfrak{D}$-colored involutive operad in* Set.

PROOF. Since $(\mathsf{As}^{\mathsf{D}}, j^{\mathsf{rev}})$ in Set is a $\mathfrak{D}$-colored involutive operad, it suffices to show that its structures respect the equivalence relation $\sim$ that defines $\mathsf{O}_{(\mathsf{D},T)}$. We will use the notations in Example 8.2.12.

To show that the operadic composition in As^{D} respects $\sim$, suppose $(\sigma, \underline{f}) \sim (\sigma', \underline{f})$ with

- $\underline{f} = (f_1, \ldots, f_n) \in \prod_{j=1}^{n} \mathsf{D}(c_j, d)$, $n = |\underline{c}|$,
- $\underline{g}_j = (g_{j1}, \ldots, g_{jk_j}) \in \prod_{i=1}^{k_j} \mathsf{D}(b_{ji}, c_j)$, and
- $\tau_j \in \Sigma_{k_j}$ for $1 \le j \le n$.

By assumption there is a factorization of the permutation $\sigma'\sigma^{-1}$ into a product $\tau_k \cdots \tau_1$ of transpositions that connects the two sequences

$$\sigma\underline{f} = \big(f_{\sigma^{-1}(1)}, \ldots, f_{\sigma^{-1}(n)}\big) \quad \text{and} \quad \sigma'\underline{f} = \big(f_{\sigma'^{-1}(1)}, \ldots, f_{\sigma'^{-1}(n)}\big)$$

by successively permuting adjacent orthogonal pairs. We will write $\tau_j \underline{g}_j$ as $\underline{g}_j^{\tau_j}$. We must show that the two sequences

$$\big(f_{\sigma^{-1}(1)}\underline{g}_1^{\tau_1}, \ldots, f_{\sigma^{-1}(n)}\underline{g}_n^{\tau_n}\big) \quad \text{and} \quad \big(f_{\sigma'^{-1}(1)}\underline{g}_1^{\tau_1}, \ldots, f_{\sigma'^{-1}(n)}\underline{g}_n^{\tau_n}\big)$$

are connected by finitely many steps, in which each step permutes an adjacent orthogonal pair. This is achieved using the block permutation

$$(\sigma'\sigma^{-1})\langle k_1, \ldots, k_n\rangle = (\tau_k \cdots \tau_1)\langle k_1, \ldots, k_n\rangle$$

together with the following facts.

- A permutation of two adjacent blocks, say the jth and the $(j+1)$st blocks, can be factored into a finite product of adjacent transpositions with each permuting an element originally from the jth block and an element originally from the $(j+1)$st block.
- If (f_p, f_q) is an orthogonal pair for some $1 \leq p < q \leq n$, then $(f_p g_{pr}, f_q g_{qs})$ is also an orthogonal pair for $1 \leq r \leq k_p$ and $1 \leq s \leq k_q$, by one of the closure properties of an orthogonality relation.

This shows that

$$\Big((\sigma; \tau_1, \ldots, \tau_n), (f_1 \underline{g}_1, \ldots, f_n \underline{g}_n)\Big) \sim \Big((\sigma'; \tau_1, \ldots, \tau_n), (f_1 \underline{g}_1, \ldots, f_n \underline{g}_n)\Big).$$

A similar argument shows that the operadic composition in As^{D} respects the $\sim$-equivalence classes for $(\tau_j, \underline{g}_j) \sim (\tau'_j, \underline{g}_j)$ for $1 \leq j \leq n$.

From Example 8.3.15, over Set the involutive structure $j^{\mathsf{rev}} : \mathsf{As}^{\mathsf{D}} \longrightarrow I\mathsf{As}^{\mathsf{D}}$ is induced by the assignment

$$(\sigma, \underline{f}) \longmapsto (\pi_n \sigma, \underline{f}) \quad \text{for} \quad (\sigma, \underline{f}) \in \overline{\mathsf{D}}\tbinom{d}{\underline{c}},$$

with π_n the interval-reversing permutation in (8.3.12). That j^{rev} descends to the quotient $\mathsf{O}_{(\mathsf{D},T)}$ follows from the permutation identities

$$(j, j+1)\pi_n^{-1} = (j, j+1)\pi_n = \pi_n(n-j, n-j+1)$$

for all adjacent transpositions $(j, j+1)$ with $1 \leq j \leq n-1$. □

COROLLARY 8.7.6. *Suppose C is an involutive symmetric monoidal category, and (D, T) is an orthogonal category with $\mathsf{Ob}(\mathsf{D}) = \mathfrak{D}$. Then*

$$\mathsf{IOp}(\Phi)(O_{(\mathsf{D},T)})$$

is a $\mathfrak{D}$-colored involutive operad in C.

PROOF. We apply the involutive functor $\mathsf{IOp}(\Phi)$ in Corollary 8.6.7 to the $\mathfrak{D}$-colored involutive operad $\mathsf{O}_{(\mathsf{D},T)}$ in Lemma 8.7.5. □

DEFINITION 8.7.7. In the setting of Corollary 8.7.6, we define

$$\mathsf{O}^{\mathsf{C}}_{(\mathsf{D},T)} = \mathsf{IOp}(\Phi)(\mathsf{O}_{(\mathsf{D},T)}) \in \mathsf{IOp}^{\mathfrak{D}}_{\mathsf{C}},$$

called the *AQFT involutive operad* in C of type (D, T).

EXPLANATION 8.7.8. Following Explanation 8.6.3, the AQFT involutive operad $\mathsf{O}^{\mathsf{C}}_{(\mathsf{D},T)}$ has entries

$$\mathsf{O}^{\mathsf{C}}_{(\mathsf{D},T)}\binom{d}{\underline{c}} = \coprod_{\mathsf{O}_{(\mathsf{D},T)}\binom{d}{\underline{c}}} \mathbb{1}$$

for $\binom{d}{\underline{c}} \in \Sigma^{\mathsf{op}}_{\mathfrak{D}} \times \mathfrak{D}$, with operadic and involutive structures induced by those on $\mathsf{O}_{(\mathsf{D},T)}$. ◇

COROLLARY 8.7.9. *In the setting of Definition 8.7.7, an involutive $\mathsf{O}^{\mathsf{C}}_{(\mathsf{D},T)}$-algebra X is exactly a D-diagram of reversing involutive monoids in C such that for each orthogonal pair $(f : a \longrightarrow c, g : b \longrightarrow c)$ in T, the diagram*

(8.7.10)
$$\begin{array}{ccccc} X_a \otimes X_b & \xrightarrow{(X_f, X_g)} & X_c \otimes X_c & \xrightarrow{\mu} & X_c \\ {\scriptstyle (X_f,X_g)}\downarrow & & & & \parallel \\ X_c \otimes X_c & \xrightarrow{\xi} & X_c \otimes X_c & \xrightarrow{\mu} & X_c \end{array}$$

in C is commutative.

PROOF. This follows from Theorem 8.5.6 and the definition of $\mathsf{O}_{(\mathsf{D},T)}$ as an entrywise quotient of As^{D} in Set. □

EXAMPLE 8.7.11 (Locally Covariant QFT with Time-Slice). Suppose (Loc', T) is the orthogonal category in Example 8.7.2. An involutive algebra over the AQFT involutive operad $\mathsf{O}^{\mathsf{C}}_{(\mathsf{Loc}',T)}$ is called a *locally covariant quantum field theory* with the time-slice axiom [BFV03, FV15]. Such an involutive algebra X is a Loc'-diagram of reversing involutive monoids in C that satisfies (8.7.10), which is called the *Einstein causality axiom.* The time-slice axiom refers to the fact that Cauchy morphisms in Loc act as isomorphisms in X because the subset of Cauchy morphisms is inverted in the category Loc' of fractions. In other words, for each Cauchy morphism $f : A \longrightarrow B$ in Loc, the morphism

$$X_f : X_A \longrightarrow X_B \in \mathsf{C}$$

is invertible. For example, if $\mathsf{C} = \mathsf{Vect}(\mathbb{C})$ or $\mathsf{Chain}(\mathbb{C})$, then X is a Loc'-diagram of complex unital (differential graded) $*$-algebras that satisfies the Einstein causality axiom. ◇

EXAMPLE 8.7.12 (Chiral Conformal and Euclidean QFT). For the orthogonal category (Man, T) in Example 8.7.3, involutive algebras over the AQFT involutive operad $\mathsf{O}^{\mathsf{C}}_{(\mathsf{Man},T)}$ are called *chiral conformal quantum field theories*. The case with $d = 1$ and $\mathsf{C} = \mathsf{Vect}(\mathbb{K})$ is studied in [BDH15].

Similarly, for the orthogonal category (Riem, T) in Example 8.7.3, involutive algebras over the AQFT involutive operad $\mathsf{O}^{\mathsf{C}}_{(\mathsf{Riem},T)}$ are called *Euclidean quantum field theories*. The case with $\mathsf{C} = \mathsf{Vect}(\mathbb{K})$ is studied in [Sch99]. ◇

8.8 Exercises and Notes

(1) Check the monoidal category axioms in the proof of Proposition 8.1.11.
(2) In the setting of Example 8.2.9, suppose (X, j) is an involutive object in $\mathsf{C}^{\mathfrak{C}}$.

- Prove that $\mathsf{End}(X)$ inherits from X the structure of a $\mathfrak{C}$-colored involutive operad in C. Hint: Use the fact that I is adjoint to itself (Proposition 2.1.4) and the natural morphism

$$I[A, B] \longrightarrow [IA, IB] \quad \text{for} \quad A, B \in \mathsf{C},$$

which is itself adjoint to the composite

$$I[A, B] \otimes IA \xrightarrow{I_2} I\big([A, B] \otimes A\big) \longrightarrow IB.$$

- For a $\mathfrak{C}$-colored involutive operad O in C, prove that an involutive O-algebra structure on (X, j) is equivalent to a morphism $\mathsf{O} \longrightarrow \mathsf{End}(X)$ of $\mathfrak{C}$-colored involutive operads.

(3) In the proof of Proposition 8.3.2, prove the commutativity of the diagrams (4.1.3), (4.1.4), and (4.1.6) in $\mathsf{SSeq}^{\mathfrak{C}}_{\mathsf{C}}$.
(4) Check the permutation identity (8.3.13).
(5) Suppose $\varphi : \mathfrak{C} \longrightarrow \mathfrak{D}$ is a function between sets, and C is a symmetric monoidal category. Show that there is an induced adjunction

$$\mathsf{Op}^{\mathfrak{C}}_{\mathsf{C}} \underset{\varphi^*}{\overset{\varphi_!}{\rightleftarrows}} \mathsf{Op}^{\mathfrak{D}}_{\mathsf{C}}$$

whose right adjoint sends a $\mathfrak{D}$-colored operad O to the $\mathfrak{C}$-colored operad $\varphi^*\mathsf{O}$ with entries

$$(\varphi^*\mathsf{O})\binom{c_0}{c_1, \ldots, c_n} = \mathsf{O}\binom{\varphi c_0}{\varphi c_1, \ldots, \varphi c_n}$$

for $c_0, \ldots, c_n \in \mathfrak{C}$, and with operad structure induced by that of O. Hint: Use the Adjoint Lifting Theorem [Bor94b] (4.5.6).

(6) Suppose $f : \mathsf{O} \longrightarrow \mathsf{P}$ is a morphism of $\mathfrak{C}$-colored operads in a symmetric monoidal category C. Show that there is an induced adjunction

$$\mathsf{Alg}(\mathsf{O}) \underset{f^*}{\overset{f_!}{\rightleftarrows}} \mathsf{Alg}(\mathsf{P})$$

whose right adjoint sends a P-algebra (X, θ) to the O-algebra $(X, f^*\theta)$ with structure morphism

$$\mathsf{O}\binom{d}{\underline{c}} \otimes X_{\underline{c}} \xrightarrow{(f,\mathrm{Id})} \mathsf{P}\binom{d}{\underline{c}} \otimes X_{\underline{c}} \xrightarrow{\theta} X_d, \qquad f^*\theta = \theta \circ (f,\mathrm{Id})$$

for $\binom{d}{\underline{c}} \in \Sigma_{\mathfrak{C}}^{\mathsf{op}} \times \mathfrak{C}$.

(7) For an involutive symmetric monoidal category C, consider the category $\mathsf{Inv}(\mathsf{C})$ of involutive objects in C with the involutive symmetric monoidal structure in Theorem 4.4.1.

- Prove that there is a canonical isomorphism

$$\mathsf{Inv}(\mathsf{SSeq}_{\mathsf{C}}^{\mathfrak{C}}) \cong \mathsf{SSeq}_{\mathsf{Inv}(\mathsf{C})}^{\mathfrak{C}}$$

of involutive monoidal categories.

- Prove that there is a canonical isomorphism

$$\mathsf{IOp}_{\mathsf{C}}^{\mathfrak{C}} \cong \mathsf{Op}_{\mathsf{Inv}(\mathsf{C})}^{\mathfrak{C}}$$

of categories. In other words, involutive operads are operads in involutive objects.

- Conclude that:
  - Exercise 5 has an analogue for involutive operads.
  - Exercise 6 has an analogue for involutive algebras and a morphism f of $\mathfrak{C}$-colored involutive operads.

(8) In the proof of Theorem 8.5.6, check that the diagram (8.5.3) for As^{D} is actually determined by the cases (i), (ii), and (iii) as stated.

(9) Give detailed proofs of Propositions 8.5.7, 8.5.8, and Corollary 8.7.9.

Notes. The concept of an involutive operad is due to [BSW19b], and most of this chapter is based on that paper, the main exceptions being Theorem 8.4.5 and Section 8.6.

The book [Yau16] is a gentle introduction to colored operads. A compact list of definitions regarding one-colored operads and their algebras is in [May97a,

May97b]. For broad surveys of operads and their applications, the reader may consult the article [Mar08] and the book [MSS02]. For variations of operads, such as props and wheeled props, the reader is referred to [YJ15]. The reader may also be interested in applications of operads to homotopy theory [WY18], wiring diagrams [Yau18], homotopical aspects of quantum field theory [Yau20], and ∞-operads and monoidal categories with group equivariance [Yau∞].

What we call an operad in Definition 8.1.12 is also called a *symmetric operad* and a *symmetric multicategory* in the literature. A $\mathfrak{C}$-colored operad as in Proposition 8.1.13, but without the symmetric group actions and the equivariance axioms, is called a *multicategory*, which is originally due to Lambek [Lam69]. The term *operad* is due to May [May72], who defined one-colored operad as in Proposition 8.1.13 in topological spaces. The $\mathfrak{C}$-colored circle product in (8.1.8) is the colored version of the circle product defined by Kelly [Kel05a].

Symbol Definition List

$\mathbb{Z}$		The set of integers
$\mathbb{Z}_{\geq 0}$		The set of non-negative integers
$\mathbb{K}$		A field
$\mathbb{C}$		The field of complex numbers

Chapter 1	**Page**	**Description**
$\mathsf{Ob}(\mathsf{C})$	1	Class of objects in a category C
$\mathsf{C}(X, Y)$, $\mathsf{C}(X; Y)$	1	Set of morphisms from X to Y in C
$\mathsf{dom}(f)$	1	Domain of a morphism f
$\mathsf{cod}(f)$	1	Codomain of a morphism f
Id_X	2	Identity morphism of X
$\mathsf{Mor}(\mathsf{C})$	2	Collection of morphisms in C
gf or $g \circ f$	2	Composition of morphisms
C^{op}	2	Opposite category of C
Cat	3	Category of small categories
$\theta * \theta'$	4	Horizontal composition of natural transformations
C^{D}	4	Diagram category of functors $\mathsf{D} \longrightarrow \mathsf{C}$
$L \dashv R$	4	An adjunction
$\mathsf{colim}\, F$, $\mathsf{colim}_{x \in \mathsf{D}}\, Fx$	6	Colimit of $F : \mathsf{D} \longrightarrow C$
$\coprod$, $\amalg$	6	Coproducts
$\otimes$	7	Monoidal product
$\mathbb{1}$	7	Monoidal unit
α	7	Associativity isomorphism
λ, ρ	7	Left and right unit isomorphisms
$(X, \mu, 1)$	9	A monoid

© The Author(s), under exclusive license to Springer Nature Switzerland AG 2020

D. Yau, *Involutive Category Theory*, Lecture Notes in Mathematics 2279,
https://doi.org/10.1007/978-3-030-61203-0

$\mathsf{Mon}(\mathsf{C})$	9	Category of monoids in C
(F, F_2, F_0)	9	A monoidal functor
ξ	12	Symmetry isomorphism
$\mathsf{CMon}(\mathsf{C})$	14	Category of commutative monoids in C
Set	15	Category of sets
Poset	15	Category of partially ordered sets
Rel	15	Category of relations
$\mathsf{Chain}(\mathbb{K})$	15	Category of chain complexes over $\mathbb{K}$
$\mathsf{Vect}(\mathbb{K})$	15	Category of $\mathbb{K}$-vector spaces
Hilb	15	Category of complex Hilbert spaces
$[-,-]$	16	Internal hom
$\mathsf{End}(\mathsf{C})$	16	Category of endofunctors on C
(T, μ, ϵ)	16	A monad
$\overset{\mathsf{ob}}{\sim}$	19	Equivalence relation on $\mathsf{Ob}(\mathsf{D})$
$\overline{\mathsf{Mor}}(\mathsf{D})$	19	Non-empty finite sequences of morphisms
$\widetilde{\mathsf{Mor}}(\mathsf{D})$	19	$\overset{\mathsf{ob}}{\sim}$-composable sequences
$\simeq$	19	Equivalence relation on $\widetilde{\mathsf{Mor}}(\mathsf{D})$
$[\underline{f}]$	20	$\simeq$-equivalence class
$\mathbf{1}$	24	Discrete category with one object
$\mathbf{2}$	24	Category with one non-identity morphism

Chapter 2

(C, I, ι)	27	An involutive category
I, I_{C}	27	Involution functor
$\iota, \iota^{\mathsf{C}}$	27	Unit of the involution functor
$\mathrm{Id}_{\mathrm{Id}_{\mathsf{C}}}$	29	Identity natural transformation of Id_{C}
$\overline{V}$	30	Conjugate vector space of V
(F, ν)	31	An involutive functor
ν, ν^F	31	Distributive law
ICat	34	Category of small involutive categories
$\mathsf{IEnd}(\mathsf{C})$	35	Category of involutive functors on C
(X, j^X)	40	An involutive object
$\mathsf{Inv}(\mathsf{C})$	40	Category of involutive objects in C
Cat_i	41	A subcategory of Cat

Chapter 3

ICat_0	52	Subcategory of ICat of strict involutive functors
$\mathsf{ICat}^{\mathsf{st}}$	52	Category of small strict involutive categories
$\mathcal{F}^{\mathsf{si}}$	53	Left adjoint of $U : \mathsf{ICat}^{\mathsf{st}} \to \mathsf{Cat}$
$\mathcal{F}^{\mathsf{cosi}}$	53	Right adjoint of $U : \mathsf{ICat}^{\mathsf{st}} \to \mathsf{Cat}$
$\mathcal{F}^{\mathsf{icat}}(\mathsf{C})$	54	Free involutive category of a category C
I_{icat}	55	Involution functor of $\mathcal{F}^{\mathsf{icat}}(\mathsf{C})$
ι^{icat}	55	Unit of I_{icat}

Chapter 4

IMCat	81	Category of small involutive monoidal categories
IMCat_0	81	Subcategory of IMCat of strict involutive strict monoidal functors
$\mathsf{ISymCat}$	82	Symmetric version of IMCat
$\mathsf{ISymCat}_0$	82	Symmetric version of IMCat_0
$(I_{\mathsf{mon}}, \iota^{\mathsf{mon}})$	86	Involutive structure on $\mathsf{Mon}(\mathsf{C})$
$\mathsf{IMon}(\mathsf{C})$	87	Category of involutive monoids in C
$\mathsf{ICMon}(\mathsf{C})$	92	Category of involutive commutative monoids in C
$(I_{\mathsf{rmon}}, \iota^{\mathsf{rmon}})$	94	Another involutive structure on $\mathsf{Mon}(\mathsf{C})$
$\mathsf{RIMon}(\mathsf{C})$	95	Category of reversing involutive monoids in C
(T, μ, ϵ, ν)	97	An involutive monad
(I_T, ι^T)	98	Involutive structure on $\mathsf{Alg}(T)$
$\mathsf{IAlg}(T)$	99	Category of involutive T-algebras
$(T_{\mathsf{inv}}, \mu^{\mathsf{inv}}, \epsilon^{\mathsf{inv}})$	100	Induced monad in $\mathsf{Inv}(\mathsf{C})$
$(T_{\mathsf{mon}}, \mu, \epsilon)$	102	Free monoid monad

Chapter 5

$\mathsf{W}_{\mathsf{im}}(S)$	107	Involutive monoidal words generated by S
$\mathsf{word}(X)$	108	Underlying word of X
$\mathsf{word}_1(X)$	108	Underlying integer sequence of X
$\mathcal{F}_{\mathsf{im}}(\mathsf{C})$	109	Free involutive monoidal category of C
$\mathsf{W}_{\mathsf{istm}}(S)$	121	Involutive strict monoidal words generated by S
$\mathcal{F}_{\mathsf{istm}}(\mathsf{C})$	121	Free involutive strict monoidal category of C
$\mathsf{Nat}(F; G)$	124	Collection of natural transformations $F \longrightarrow G$
$\mathsf{It}(\mathsf{C})$	124	Category of functors on C
C_{st}	129	Strictification of C
$\underline{x}^{\otimes}$	129	Left-normalized monoidal product
$\otimes^{\mathsf{st}}$	130	Monoidal product in C_{st}
I_{st}	131	Involution functor on C_{st}
ι^{st}	132	Unit of I_{st}
$\mathsf{W}^{\mathsf{i}}_{\mathsf{im}}(S)$	134	Involutive monoidal words of $I_S : S \to S$
$\mathcal{F}^{\mathsf{i}}_{\mathsf{im}}(\mathsf{C})$	135	Free involutive monoidal category of an involutive category
$\mathsf{W}^{\mathsf{i}}_{\mathsf{istm}}(S)$	141	Involutive strict monoidal words of $I_S : S \to S$

Chapter 6

$(y_1, \ldots, y_k)\sigma$	146	Right symmetric group action
$\sigma\langle k_1, \ldots, k_n\rangle$	146	Block permutation
$\tau_1 \times \cdots \times \tau_n$	146	Block sum
$\mathcal{F}_{\mathsf{isym}}(\mathsf{C})$	146	Free involutive symmetric monoidal category
$\mathcal{F}_{\mathsf{iss}}(\mathsf{C})$	153	Free involutive strict symmetric monoidal category
$\mathcal{F}^{\mathsf{i}}_{\mathsf{isym}}(\mathsf{C})$	157	Free involutive symmetric monoidal category of an involutive category
$\mathcal{F}^{\mathsf{i}}_{\mathsf{iss}}(\mathsf{C})$	160	Free involutive strict symmetric monoidal category of an involutive category

Chapter 7

$\mathsf{IP}(\mathsf{C})$	166	Category of inner product objects in C
$\mathsf{MIP}(\mathsf{C})$	168	Category of monoid inner product objects in C
$\mathsf{QP}(\mathsf{C})$	172	Category of quantum probability objects in C

Chapter 8

$\mathsf{Prof}(\mathfrak{C})$	187	Set of $\mathfrak{C}$-profiles
$\underline{a}$	187	A profile
$\sigma\underline{a}, \underline{a}\sigma$	187	Left and right permutations of $\underline{a}$
$\mathsf{SSeq}^{\mathfrak{C}}_{\mathsf{C}} = \mathsf{C}^{\Sigma^{\mathsf{op}}_{\mathfrak{C}} \times \mathfrak{C}}$	187	Category of $\mathfrak{C}$-colored symmetric sequences in C
$\binom{d}{\underline{c}}$	187	An object in $\Sigma^{\mathsf{op}}_{\mathfrak{C}} \times \mathfrak{C}$
$X\binom{d}{\underline{c}}$	187	An entry of a $\mathfrak{C}$-colored symmetric sequence X
$\mathsf{C}^S = \prod_S \mathsf{C}$	188	Category of S-colored objects
$\int^{d \in \mathsf{D}} F(d, d)$	188	Coend of $F : \mathsf{D}^{\mathsf{op}} \times \mathsf{D} \longrightarrow \mathsf{B}$
$S \cdot X$	189	S-indexed coproduct
$X \circ Y$	189	Symmetric circle product
I	190	Monoidal unit in $\mathsf{SSeq}^{\mathfrak{C}}_{\mathsf{C}}$
$\mathsf{Op}^{\mathfrak{C}}_{\mathsf{C}}$	191	Category of $\mathfrak{C}$-colored operads in C
γ	192	Operadic composition
$\mathsf{1}_c$	192	c-colored unit
$X_{\underline{c}}$	195	$X_{c_1} \otimes \cdots \otimes X_{c_m}$ or $\mathbb{1}$
$\mathsf{O} \circ -$	195	Monad of an operad O
$\mathsf{Alg}(\mathsf{O})$	196	Category of O-algebras
$\mathsf{End}(X)$	198	Endomorphism operad of X
As	198	Associative operad
$(\sigma; \tau_1, \ldots, \tau_n)$	198	Operadic composition in As
Com	199	Commutative operad
$\overline{\mathsf{D}}(\underline{c}; d)$	199	$\Sigma_n \times \prod_{j=1}^n \mathsf{D}(c_j, d)$

As^{D}	199	Operad for D-diagrams of monoids
$f_j \underline{g}_j$	200	$(f_j g_{j1}, \ldots, f_j g_{jk_j})$
$\mathsf{Com}^{\mathsf{D}}$	201	Operad for D-diagrams of commutative monoids
$\mathsf{IOp}_{\mathsf{C}}^{\mathfrak{C}}$	204	Category of $\mathfrak{C}$-colored involutive operads in C
$j^{\mathsf{rev}} : \mathsf{As} \to I\mathsf{As}$	205	Involutive structure on As
$j^{\mathsf{rev}} : \mathsf{As}^{\mathsf{D}} \to I\mathsf{As}^{\mathsf{D}}$	207	Involutive structure on As^{D}
$\mathsf{IOp}(F)$	217	Change-of-category functor on involutive operads
Loc	221	Category of Lorentzian manifolds
Loc'	221	Category of fractions $\mathsf{Loc}[C^{-1}]$
Man	222	Category of oriented manifolds
Riem	222	Category of oriented Riemannian manifolds
$\mathsf{O}_{(\mathsf{D},T)}$	222	An entrywise quotient of As^{D} in Set
$\mathsf{O}_{(\mathsf{D},T)}^{\mathsf{C}}$	224	AQFT involutive operad

Bibliography

[AC04] S. Abramsky, B. Coecke, A categorical semantics of quantum protocols, in *Proceding of the 19th IEEE Conferencde on Logic in Computer Science* (IEEE Computer Science Press, New York, 2004)

[AC08] S. Abramsky, B. Coecke, Categorical quantum mechanics, in *Handbook of Quantum Logic and Quantum Structures*, vol. II (Elsevier, Berlin, 2008), pp. 261–323

[AR94] J. Adámek, J. Rosický, in *Locally Presentable and Accessible Categories*. Lecture Note Series, vol. 189 (London Mathematical Society, Cambridge, 1994)

[Arv76] W. Arveson, *An Invitation to C*-Algebras*, Graduate Texts in Mathematics, vol. 39 (Springer, Berlin, 1976)

[Awo10] S. Awodey, *Category Theory* (Oxford University, Oxford, 2010)

[Bae06] J.C. Baez, Quantum quandaries: a category-theoretic perspective, in *Structural Foundations of Quantum Gravity*, ed. by S. French, D. Rickles, J. Saatsi (Oxford University, Oxford, 2006), pp. 240–265

[BGP07] C. Bär, N. Ginoux, F. Pfäffle, *Wave Equations on Lorentzian Manifolds and Quantization*. ESI Lectures in Mathematical Physics (European Mathematical Society, Zürich, 2007)

[Bar08] D.J. Barnes, *Rational Equivariant Spectra*. PhD dissertation (Sheffield University, Sheffield, 2008). https://arxiv.org/abs/0802.0954v1.

[BW85] M. Barr, C. Wells, Toposes, triples and theories, *Grundlehren der Mathematischen Wissenschaften*, vol. 278 (Springer, New York, 1985). Reprints in Theory Appl. Categ. (2005), pp. 1–289

[BDH15] A. Bartels, C.L. Douglas, A. Henriques, Conformal nets I: coordinate-free nets. Int. Math. Res. Not. **13**, 4975–5052 (2015)

[BBP99] M.A. Bednarczyk, A.M. Borzyszkowski, W. Pawlowski, Generalized congruences–epimorphisms in *Cat*. Theory Appl. Categ. **5**, 266–280 (1999)

[BM09] E.J. Beggs, S. Majid, Bar categories and star operations. Algebr. Represent. Theory **12**, 103–152 (2009)

[Ben67] J. Bénabou, Introduction to bicategories, in *Reports of the Midwest Category Seminar*. Lecture Notes in Mathematics, vol. 47 (Springer, Berlin, 1967), pp. 1–77

[BBS19] M. Benini, S. Bruinsma, A. Schenkel, Linear Yang-Mills theory as a homotopy AQFT. Commun. Math. Phys. (2019). https://doi.org/10.1007/s00220-019-03640-z

[BSW∞] M. Benini, A. Schenkel, L. Woike, Operads for algebraic quantum field theory. preprint, arXiv:1709.08657

[BSW19a] M. Benini, A. Schenkel, L. Woike, Homotopy theory of algebraic quantum field theories. Lett. Math. Phys. https://doi.org/10.1007/s11005-018-01151-x

© The Author(s), under exclusive license to Springer Nature Switzerland AG 2020

D. Yau, *Involutive Category Theory*, Lecture Notes in Mathematics 2279,
https://doi.org/10.1007/978-3-030-61203-0

[BSW19b] M. Benini, A. Schenkel, L. Woike, Involutive categories, colored $*$-operads and quantum field theory. Theory Appl. Categ. **34**, 13–57 (2019)

[BNY18] J. Bichon, S. Neshveyev, M. Yamashita, Graded twisting of comodule algebras and module categories. J. Noncommutative Geometry **12**, 331–368 (2018). https://doi.org/10.4171/JNCG/278

[Bor94a] F. Borceux, *Handbook of Categorical Algebra 1: Basic Category Theory* (University of Cambridge, Cambridge, 1994)

[Bor94b] F. Borceux, *Handbook of Categorical Algebra 2: Categories and Structures* (University of Cambridge, Cambridge, 1994)

[BW98] M. Bordemann, S. Waldmann, Formal GNS construction and states in deformation quantization. Comm. Math. Phys. **195**, 549–583 (1998)

[BFV03] R. Brunetti, K. Fredenhagen, R. Verch, The generally covariant locality principle: a new paradigm for local quantum field theory. Comm. Math. Phys. **237**, 31–68 (2003)

[CG17] K. Costello, O. Gwilliam, *Factorization Algebras in Quantum Field Theory*, vol. 1. New Mathematical Monographs, vol. 31 (University of Cambridge, Cambridge, 2017)

[DGNO10] V. Drinfeld, S. Gelaki, D. Nikshych, V. Ostrik, On braided fusion categories I. Selecta Math. **16**, 1–119 (2010). https://doi.org/10.1007/s00029-010-0017-z

[Egg11] J.M. Egger, On involutive monoidal categories. Theory Appl. Categ. **25**, 368–393 (2011)

[Eps66] D.B.A. Epstein, Functors between tensored categories. Invent. Math. **1**, 221–228 (1966)

[FV15] C.J. Fewster, R. Verch, Algebraic quantum field theory in curved spacetimes, in *Advances in Algebraic Quantum Field Theory*, ed. by R. Brunetti et al. (Springer, Heidelberg, 2015), pp.125–189

[GZ67] P. Gabriel, M. Zisman, *Calculus of Fractions and Homotopy Theory* (Springer, Berlin, 1967)

[Gal17] C. Galindo, Coherence for monoidal G-categories and braided G-crossed categories. J. Algebra **487**, 118–137 (2017)

[GN43] I. Gelfand, M. Neumark, On the imbedding of normed rings into the ring of operators in Hilbert space. Matematicheskii Sbornik **12**, 197–217 (1943)

[Gra18] M. Grandis, *Category Theory and Applications: A Textbook for Beginners* (World Scientific, Singapore, 2018). https://doi.org/10.1142/10737.

[Haa96] R. Haag, *Local Quantum Physics: Fields, Particles, Algebras*, Texts and Monographs in Physics (Springer, Berlin, 1996)

[Hov99] M. Hovey, *Model categories*, Mathematical Surveys and Monographs, vol. 63 (American Mathematical Society, Providence, 1999)

[Jac12] B. Jacobs, Involutive categories and monoids, with a GNS-correspondence. Found. Phys. **42**, 874–895 (2012)

[JY21] N. Johnson, D. Yau, *2-dimensional categories* (Oxford University Press, New York, 2021)

[JS93] A. Joyal, R. Street, Braided tensor categories. Adv. Math. **102**, 20–78 (1993)

[Kel64] G.M. Kelly, On Mac Lane's conditions for coherence of natural associativities, commutativities, etc. J. Algebra **1**, 397–402 (1964)

[Kel05a] G.M. Kelly, On the operads of J.P. May. Repr. Theory Appl. Categ. **13**, 1–13 (2005)

[KM15] I. Khavkine, V. Moretti, Algebraic QFT in Curved Spacetime and Quasifree Hadamard States: An Introduction, in *Advances in Algebraic Quantum Field Theory*, ed. by R. Brunetti et al. Mathematical Physics Studies (Springer, Berlin, 2015), pp. 191–251

[Lac10] S. Lack, A 2-Categories Companion, in *Towards Higher Categories*, ed. by J.C. Baez, J.P. May (Springer, New York, 2010)

[Lam69] J. Lambek, Deductive systems and categories. II. Standard constructions and closed categories, in *1969 Category Theory, Homology Theory and their Applications, I (Battelle Instutite Conference, Seattle, Wash.)* (Springer, Berlin, 1969), pp. 76–122

[Lei14] T. Leinster, *Basic Category Theory* (University of Cambridge, Cambridge, 2014)
[Lor∞] F. Loregian, This is the (co)end, my only (co)friend. arXiv:1501.02503
[Mac63] S. Mac Lane, Natural associativity and commutativity. Rice Univ. Studies **49**, 28–46 (1963)
[Mac98] S. Mac Lane, *Categories for the working mathematician*, Graduate Texts in Mathematics, vol. 5, 2nd ed. (Springer, New York, 1998)
[Mar08] M. Markl, Operads and PROPs, in *Handbook of Algebra*, vol. 5 (Elsevier, Amsterdam, 2008), pp.87–140
[MSS02] M. Markl, S. Shnider, J. Stasheff, *Operads in Algebra, Topology and Physics*, Mathematical Surveys and Monographs, vol. 96 (American Mathematical Society, Providence, 2002)
[May72] J.P. May, *The geometry of iterated loop spaces*, Lecture Notes in Mathematics, vol. 271 (Springer, New York, 1972)
[May97a] J.P. May, Definitions: operads, algebras and modules. Contemp. Math. **202**, 1–7 (1997)
[May97b] J.P. May, Operads, algebras, and modules. Contemp. Math. **202**, 15–31 (1997)
[Mor18] V. Moretti, *Spectral Theory and Quantum Mechanics: Mathematical Structure of Quantum Theories, Symmetries and Introduction to the Algebraic Formulation*, 2nd ed. (Springer, Berlin, 2018)
[ONe83] B. O'Neill, *Semi-Riemannian Geometry: With Applications to Relativity* (Academic Press, San Diego, 1983)
[Par18] A.J. Parzygnat, From Observables and States to Hilbert Space and Back: A 2-Categorical Adjunction. Appl. Categ. Struct. **26**, 1123–1157 (2018)
[Pin14] C.C. Pinter, *A Book of Set Theory* (Dover, New York, 2014)
[Rie16] E. Riehl, *Category Theory in Context* (Courier Dover, New York, 2016)
[Rom17] S. Roman, *An Introduction to the Language of Category Theory*, Compact Textbooks in Mathematics (Birkhäuser, Basel, 2017)
[Sch99] D. Schlingemann, From Euclidean field theory to quantum field theory. Rev. Math. Phys. **11**, 1151–1178 (1999)
[Sch90] K. Schmüdgen, *Unbounded operator algebras and representation theory*, Operator Theory: Advances and Applications, vol. 37 (Birkhäuser, Basel, 1990)
[Sch72] H. Schubert, *Categories* (Springer, Berlin, 1972)
[Seg47] I.E. Segal, Irreducible representations of operator algebras. Bull. Amer. Math. Soc. **53**, 73–88 (1947)
[Sel07] P. Selinger, Dagger compact closed categories and completely positive maps. Elect. Notes Theoret. Comp. Sci. **170**, 139–163 (2007)
[Sel10] P. Selinger, Autonomous categories in which $A \cong A^*$ (extended abstract), in *Proceedings of the 7th International Workshop on Quantum Physics and Logic (QPL 2010), Oxford* (2010), pp. 151–160
[SV02] A. Shen, N.K. Vereshchagin, *Basic Set Theory*, Student Mathematical Library, vol. 17 (American Mathematical Society, Rhode Island, 2002)
[Sim11] H. Simmons, *An Introduction to Category Theory* (University of Cambridge, Cambridge, 2011)
[Tam01] D. Tambara, Invariants and semi-direct products for finite group actions on tensor categories. J. Math. Soc. Japan **53**, 429–456 (2001)
[Wei97] C.A. Weibel, *An Introduction to Homological Algebra*. Cambridge Studies in Advanced Mathematics, vol. 38 (University of Cambridge, Cambridge, 1997)
[Wei80] J. Weidmann, *Linear operators in Hilbert spaces*, Graduate Texts in Mathematics, vol. 68 (Springer, Berlin, 1980)
[WY18] D. White, D. Yau, Bousfield localization and algebras over colored operads. Appl. Categ. Struct. **26**, 153–203 (2018)

[Yau16] D. Yau, *Colored Operads*, AMS Graduate Studies in Mathematics, vol. 170 (American Mathematical Society, Providence, 2016)

[Yau18] D. Yau, *Operads of Wiring Diagrams*. Lecture Notes in Mathematics, vol. 2192 (Springer, Switzerland, 2018)

[Yau20] D. Yau, *Homotopical Quantum Field Theories* (World Scientific, Singapore, 2020)

[Yau∞] D. Yau, *Infinity Operads and Monoidal Categories with Group Equivariance*. arXiv:1903.03839

[YJ15] D. Yau, M.W. Johnson, *A Foundation for PROPs, Algebras, and Modules*, Mathematical Surveys and Monographs, vol. 203 (American Mathematical Society, Providence, 2015)

Index

© The Author(s), under exclusive license to Springer Nature Switzerland AG 2020

D. Yau, *Involutive Category Theory*, Lecture Notes in Mathematics 2279,
https://doi.org/10.1007/978-3-030-61203-0

Printed in the United States
By Bookmasters